Microbiology Monographs

Volume 32

Series editor
Alexander Steinbüchel
Münster, Germany

More information about this series at http://www.springer.com/series/7171

Eun Yeol Lee
Editor

Methanotrophs

Microbiology Fundamentals and Biotechnological Applications

Springer

Editor
Eun Yeol Lee
Department of Chemical Engineering
Kyung Hee University
Yongin-si, Gyeonggi-do
Republic of Korea

ISSN 1862-5576 ISSN 1862-5584 (electronic)
Microbiology Monographs
ISBN 978-3-030-23260-3 ISBN 978-3-030-23261-0 (eBook)
https://doi.org/10.1007/978-3-030-23261-0

This Springer imprint is published by the registered company Springer Nature Switzerland AG.
The registered company address is: Gewerbestrasse 11, 6330 Cham, Switzerland

Preface

Methanotrophs are aerobic proteobacteria that can utilize methane as sole carbon and energy source. In ecology, methanotrophs have an essential role in the global carbon cycle by removing methane generated geothermally and by methanogens. Methanotrophs have been employed as the biocatalyst for mitigating methane as greenhouse gas and remediating halogenated hydrocarbons in soil and underground water. Recently, methane is considered the next-generation carbon feedstock for industrial biotechnology because of its high abundance and low price. Methanotrophs can be used as the biocatalyst for the production of chemicals, fuels, and biomaterials from methane under environmentally benign production conditions.

Despite the growing importance of basic and applied researches on methanotrophs, there was no comprehensive textbook for senior undergraduate and postgraduate levels. *Methanotroph: Microbiological Fundamentals and Biotechnological Applications* was written in an attempt to give the readers a systematic and comprehensive overview for both basic and applied aspects of methanotrophs. Thus, this book can be used as a reference for microbiologists and biochemists dealing with ecology, environmental functioning, and understanding of physiology and metabolism of methanotrophs as well as the fundamentals and applications of methane monooxygenases. This book is also valuable to biotechnologists and biochemical engineers who research on the omics-based understanding of methane metabolism, metabolic engineering for strain improvement, methanobactin biosynthesis, and environmental applications of methanotrophs.

I would like to thank the people who were instrumental in the writing of this book. First of all, I would like to thank the contributors for taking their valuable time in writing the chapters. I appreciate very much Professor Alexander Steinbüchel of the Institute of Microbiology at Münster University, the *Microbiology Monographs* Series Editor, for giving me the opportunity to publish this book. I would like to acknowledge C1 gas refinery R&D program of the National Research Foundation of Korea for supporting the research fund for development of methanotrophic platform strains for bioconversion of methane and methanol. Lastly, I am grateful to my beloved family and lab members for their consideration and support so that I could concentrate on writing books comfortably.

Yongin-si, Gyeonggi-do, Republic of Korea — Eun Yeol Lee
April 2019

Contents

List of Contributors

Samuel Imisi Awala Department of Microbiology, Chungbuk National University, Cheongju, Republic of Korea

Paul L. E. Bodelier Department of Microbial Ecology, Netherlands Institute of Ecology (NIOO-KNAW), Wageningen, The Netherlands

Sunney I. Chan Noyes Laboratory, California Institute of Technology, Pasadena, CA, USA
Institute of Chemistry, Academia Sinica, Nankang, Taipei, Taiwan

Ludmila Chistoserdova Department of Chemical Engineering, University of Washington, Seattle, WA, USA

Alan A. DiSpirito Roy J. Carver Department of Biochemistry, Biophysics and Molecular Biology, Iowa State University, Ames, IA, USA

Adrian Ho Institute for Microbiology, Leibniz Universität Hannover, Hannover, Germany

Marcus A. Horn Institute for Microbiology, Leibniz Universität Hannover, Hannover, Germany

Marina G. Kalyuzhnaya Department of Biology, San Diego State University, San Diego, CA, USA
Viral Information Institute, San Diego State University, San Diego, CA, USA

Sascha M. B. Krause Johann Heinrich von Thünen Institute, Federal Research Institute for Rural Areas, Forestry and Fisheries, Braunschweig, Germany

Miye Kwon Department of Civil and Environmental Engineering, KAIST, Daejeon, South Korea

Eun Yeol Lee Department of Chemical Engineering, Kyung Hee University, Yongin-si, South Korea

Ok Kyung Lee Department of Chemical Engineering, Kyung Hee University, Yongin-si, South Korea

Seung Jae Lee Department of Chemistry and Institute for Molecular Biology and Genetics, Chonbuk National University, Jeonju, Republic of Korea

Snehal Nariya Department of Biology, San Diego State University, San Diego, CA, USA

Diep T. N. Nguyen Department of Chemical Engineering, Kyung Hee University, Yongin-si, South Korea

Ngoc-Loi Nguyen Department of Microbiology, Chungbuk National University, Cheongju, Republic of Korea

German Pérez Department of Microbial Ecology, Netherlands Institute of Ecology (NIOO-KNAW), Wageningen, The Netherlands

Sung-Keun Rhee Department of Microbiology, Chungbuk National University, Cheongju, Republic of Korea

Jeremy D. Semrau Department of Civil and Environmental Engineering, University of Michigan, Ann Arbor, MI, USA

Annelies J. Veraart Department of Aquatic Ecology and Environmental Biology, Institute for Water and Wetland Research, Radboud University, Nijmegen, The Netherlands

Sukhwan Yoon Department of Civil and Environmental Engineering, KAIST, Daejeon, South Korea

Yue Zheng Department of Chemical Engineering, University of Washington, Seattle, WA, USA
CAS Key Laboratory of Urban Pollutant Conversion, Institute of Urban Environment, Chinese Academy of Sciences, Xiamen, China
University of Chinese Academy of Sciences, Beijing, China

Methanotroph Ecology, Environmental Distribution and Functioning

Paul L. E. Bodelier, German Pérez, Annelies J. Veraart, and Sascha M. B. Krause

Contents

Abstract The dynamics of methane concentrations in the atmosphere in recent decades has demonstrated many anomalies which are poorly understood. The only biological way of degrading this potent greenhouse gas is by microbial oxidation. Aerobic methanotrophic bacteria (MB) play an important role in many ecosystems worldwide degrading methane before it can escape to the atmosphere. This group of bacteria has intensively been studied as a model microbial functional guild because there is a strong link between the consumption of methane and the composition of MB communities, facilitating the study of microbial "behavior" in the environment. These studies have revealed a strong biogeography of MB which is displayed in their phylogeny not only on the basis of single functional marker genes but also on

P. L. E. Bodelier (✉) · G. Pérez
Department of Microbial Ecology, Netherlands Institute of Ecology (NIOO-KNAW), Wageningen, The Netherlands
e-mail: P.Bodelier@nioo.knaw.nl

A. J. Veraart
Department of Aquatic Ecology and Environmental Biology, Institute for Water and Wetland Research, Radboud University, Nijmegen, The Netherlands

S. M. B. Krause
Johann Heinrich von Thünen Institute, Federal Research Institute for Rural Areas, Forestry and Fisheries, Braunschweig, Germany

E. Y. Lee (ed.), *Methanotrophs*, Microbiology Monographs 32,
https://doi.org/10.1007/978-3-030-23261-0_1

genome sequence basis. Novel environmental controlling factors have been revealed (e.g. rare earth metals) as well as novel organisms with as yet unknown traits for MB. The resistance and resilience of methane consumption and methane consuming communities have been shown to depend on specific community members. The current knowledge on environmental distribution and of MB has led to propose a life-history scheme, classifying MB communities on their collective traits rather than singly on their capacity the oxidise methane alone.

1 Introduction

Working towards a sustainable future for our planet requires a drastic change in the ways society addresses its primary needs: energy, food, resources and climate. Finite stocks of fossil fuels and corresponding related climate issues leaves only low-carbon economies with high contribution of renewables as alternatives, as recently acknowledged and ratified in the "Paris Agreement" during the United Nations Framework Convention on Climate Change (http://unfccc.int/resource/docs/2015/cop21/eng/l09r01.pdf) . To reach the Paris targets drastic reductions in emissions of greenhouse gases (GHG) are necessary (80% below 1990 levels), including obtaining 20% of our energy from renewable sources. In this context, methane (CH_4) plays a key role, being a 24 times more potent GHG than carbon dioxide, and contributing 40% to global warming on a 20-year timescale (IPCC 2014). Huge amounts of methane are released into the atmosphere associated to gas and oil production and by biological production in natural and agricultural wetlands, landfills, waste water treatment and agricultural production systems, lakes, oceans and termites (Dean et al. 2018). However, a large portion of this methane has already been mitigated before escaping to the atmosphere by the activity of methane-oxidizing bacteria (MB) which utilise methane as carbon and energy source (Bodelier and Steenbergh 2014b). This reaction, turning methane into biomass and metabolites is of high value for production of bio-products (Fig. 1), is unique to these bacteria, and requires a set of enzymes with a range of special features (see Chaps. "Diversity, Physiology, and Biotechnological Potential of Halo(alkali)philic Methane-Consuming Bacteria" and "Metabolic Engineering of Methanotrophs for the Production of Chemicals and Fuels"). Central are the methane monooxygenase (MMO) variants (particulate (pMMO) and soluble methane monooxygenase (sMMO)) which have a very broad substrate range, making them highly suitable for applications in synthetic organic chemistry and for degrading environmental contaminants such as trichloroethylene (TCE) (Pieja et al. 2017). Their importance for climate and industry has spurred intensive research into the environmental distribution, diversity, ecology, physiology and genomics of MB. The long-standing notion that these microbes only belong to the Proteobacteria, and are exclusively obligatory methanotrophic and aerobic, has been proven wrong in the last two decades. Novel Verrucomicrobial (Dunfield et al. 2007; van Teeseling et al. 2014), anaerobic (Ettwig et al. 2010; Haroon et al. 2013; in 't Zandt et al. 2018),

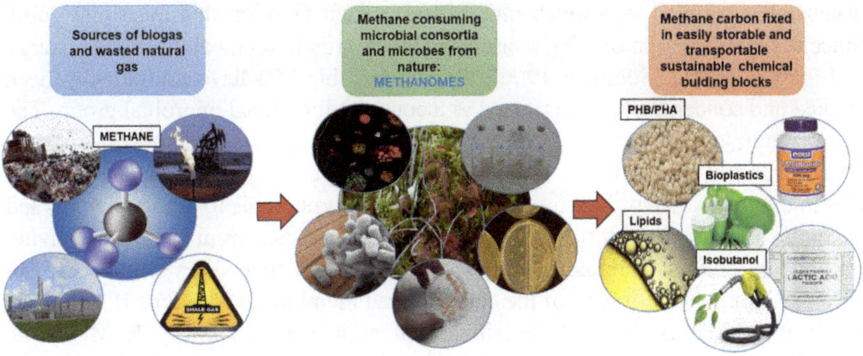

Fig. 1 Methane containing biogas originating from e.g. waste or being flared of during energy production can be turned into sustainable storage of carbon and energy. By using natural consortia of methane consuming microbes or engineered strains, methane can be turned into industrially relevant chemicals which can be stored and transported easily and used for the production of valuable bio-based chemicals

facultative (Crombie and Murrell 2014), and/or denitrifying MB (Dam et al. 2013; Kits et al. 2015) have been discovered. Also, numerous environmental studies facilitated by the development of cultivation independent assessment methods have revealed a vast as yet untapped diversity of MB with distinct biogeography, indicating a link between MB traits and habitat (Knief 2015) of interest for future biotechnological applications.

Crucially, MB have also recently been shown to perform better in natural as well as synthetic multispecies communities (Ho et al. 2014; Schnyder et al. 2018) depending on interactions and exchanges of nutrients with other community members (Zheng et al. 2014; Iguchi et al. 2011) or by exchange of volatile organic compounds (Veraart et al. 2018). However, the cultivation and industrial handling of MB remain difficult, resulting in only a handful of strains currently exploited for industrial purposes. Next to this, global climate models still give poor predictions of future atmospheric methane concentrations (Dean et al. 2018; Nisbet et al. 2014), which display a range of distinct dynamics and anomalies in the past decades. Lack of a deeper understanding of the ecology and functioning of MB in natural and man-made habitats underlies many of the issues we have in using MB in bio-industry and to explain atmospheric methane dynamics. The present chapter will highlight features of the ecology and environmental distribution of MB with a focus on the aerobic MB.

2 Taxonomy and Phylogeny

Currently described species and their phylogeny have extensively been described recently (Dedysh and Knief 2018) and we will not go into detailed taxonomy here. However, insights from comparative genomics and many discoveries in the last two decades has resulted in a much more differentiated view on this microbial guild. Since the first isolation of a bacterium capable of growth on methane as sole energy and carbon source (Söhngen 1906), obligate aerobic MB have traditionally been studied and conceptually been treated as a coherent functional microbial group. The first taxonomic framework for MB emerged after two decades of pioneering work in the last century (Whittenbury et al. 1970; Bowman et al. 1993) basically dividing MB in two main groups based on physiological, morphological, ultrastructural, and chemotaxonomic characteristics. Type I MB distinguished from type II by having intra cytoplasmic membranes (ICM) structured as vesicular discs while these are arranged along the periphery of the cell as paired membranes with type II MB. Next to this, type I MB fix carbon via the ribulose monophosphate pathway (RuMP) while type II use the serine pathway. Furthermore, both groups clearly differed in their fatty acid composition within the polar lipid fraction of the cell membranes (PLFA), with C16 dominated PLFA predominantly in type I and C18 in type II (Bowman et al. 1993; Bodelier et al. 2009) with even PLFA which to date have only been detected in MB (C16:1ω8c and C18:1ω8c) enabling the separate detection of MB from other bacteria in environments (Bodelier et al. 2013). Besides the ability fix molecular nitrogen, which at that time separated type I from type II, the production of resting stages (cysts or spores) have also been considered as taxonomic feature to distinguish more between genera rather than main types. The separation in two main groups at that time was also reflected in the phylogeny based on 16S rRNA, placing type I within the *Gammaproteobacteria* and type II in the *Alphaproteobacteria* (Bowman et al. 1993). However, the successful cultivation of many new species (Knief 2015; Dedysh and Knief 2018) generated information that rendered this dichotomous classification system not representative for the phylogenomic and physiological view we have now. Internal membrane systems do not occur in all MB (Dunfield and Dedysh 2014; Vorobev et al. 2011); PLFA specific for type I MB occur also in type II (Dedysh et al. 2007); nitrogen fixation also occurs in type I MB (*Methylobacter tundripaludum*) and MB have been discovered which have parts or even complete gene sets for both types of carbon fixation pathways (Nguyen et al. 2018), although it has not been demonstrated that both pathways can be operational in either type I or type II MB.

Despite the inadequacy of the "type I/II" classification system as an integrated reflection of physiology, biochemistry and morphology, the system is still in use since current phylogeny based on 16S and MMO (particulate (pMMO) and soluble (sMMO) methane monooxygenase) (Knief 2015; Dedysh and Knief 2018) reflects this classification. Current taxonomic radiation of aerobic MB covers taxonomically classified pure cultures, MB with a candidate status due to the fact that these strains are only available as highly enriched cultures and uncultivated lineages represented by environmental *pmoA* sequences (see Fig. 2). Most MB type strains and classified genera fall with the *Gammaproteobacteria* in the families *Methylococcacea* (type Ia/Ib) and *Methylotermaceae* (type Ic), while the *Alphaproteobacteria* harbor the families of *Methylocystaceae* (type IIa) and *Beijerinckiaceae* (type IIb) (Fig. 2). With the discovery of MB in the phylum of the *Verrucomicrobia* (Dunfield et al. 2007; van Teeseling et al. 2014; Pol et al. 2007) belonging to the family of the

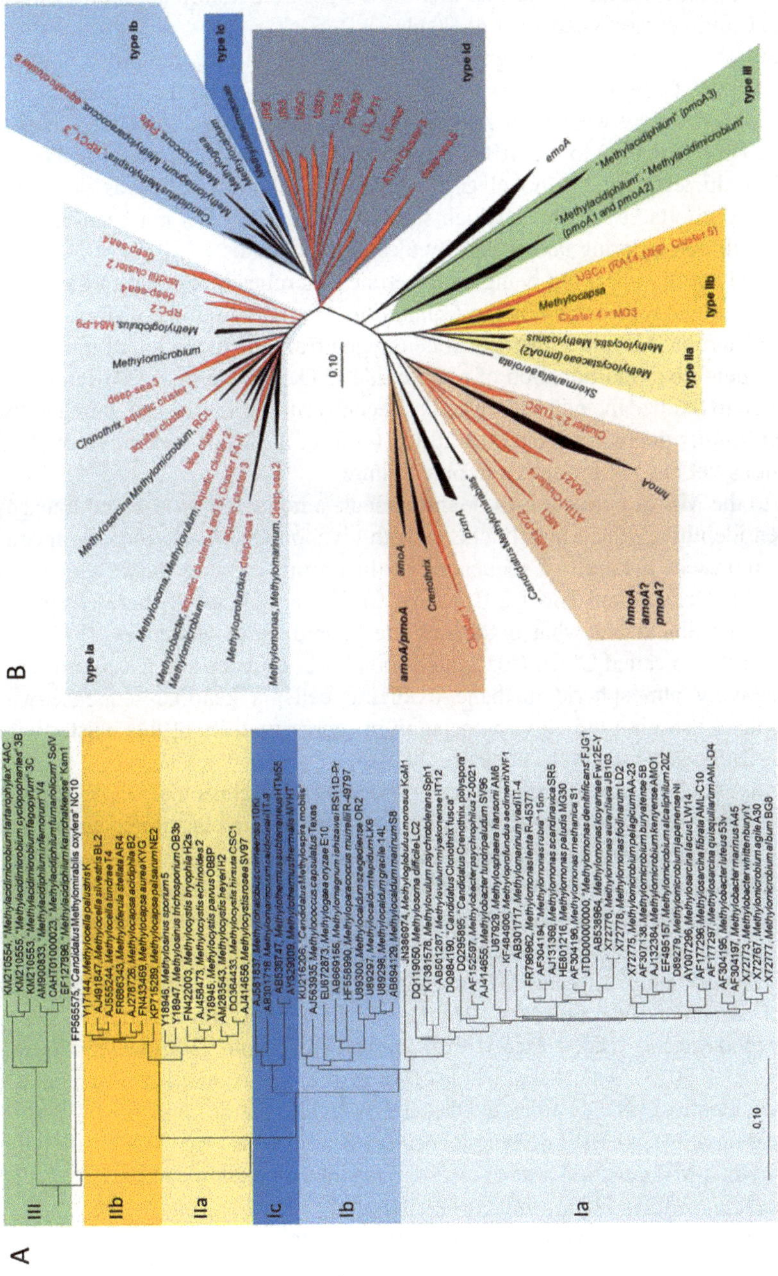

Fig. 2 Phylogeny of described aerobic MB based on 16S rRNA (**a**) and the relatedness of uncultivated lineages in relation described representatives on the basis of the *pmoA* gene (**b**). From Dedysh and Knief (2018) with permission

Methylacidophyliceae a type III classifier was necessary to be in line with the "phylogeny-based" typing. With the advent of this MB phylum another biochemical variation was introduced, by the fact that these MB are autotrophic, fixing carbon using the Calvin cycle (Khadem et al. 2011).

The discovery of the ability to grow on methane of an already described filamentous bacterium *Crenothrix polyspora* (Stoecker et al. 2006) belonging to the *Gammaproteobacteria* was another surprise within the MB realm, adding another family (*Crenotrichideae*) to the MB tree with a deviating *pmoA*, similar to *amoA* of ammonia oxidizers. Later a novel genus within this family with was described *Clonotrix* (Vigliotta et al. 2007), which similar to *Crenotrix* has not been brought into pure culture rendering taxonomically a candidatus status.

Although strictly speaking being an anaerobe, microbes belonging to the novel NC10 phylum and represented by the *Candidatus* Methylomirabilis oxyfera, can be considered aerobic MB since they generate oxygen from the reduction of nitrite and the subsequent disproportionation of nitric oxide (NO) (Ettwig et al. 2010) and use that oxygen to oxidize methane aerobically. Recently, novel species belonging to the *Methylomirabilis* genus have been identified (Graf et al. 2018; Versantvoort et al. 2018) but as yet has not brought into pure culture.

Next to the MB in pure cultures or enrichments a range of *pmoA*-based lineages have been identified which mostly cluster with taxonomically described organisms but also represent a range of sequences distinct from existing *pmoA* sequences (Knief 2015; Dedysh and Knief 2018) (Fig. 2). For almost all of these clusters it remains to be elucidated what organisms are behind these sequences. However, recently for the so called USCα (for upland soil cluster α) using single cell sequencing of actively atmospheric methane oxidizing cells, a genome sequence was obtained demonstrating the close association of representatives of this cluster with the genus *Methylocapsa* (Pratscher et al. 2018).

Recently, microbial taxonomy has been revised by the inclusion of whole genome sequences of taxonomically described microbes as well as metagenome assembled genomes (MAG) from environmental samples (Parks et al. 2018). This extensive effort compared 120 single copy ubiquitous protein coding genes of almost 95,000 genomes. This analyses led to a substantial revision of taxonomy removing many of the inconsistencies and polyphyletic cases of incongruency of phenotypic and single gene based phylogeny. Also for MB, changes were suggested, of which the proposition of dividing the order of *Methylococcales* into three families being *Methylomonadaceae* (Ia), *Methylococcaceae* (Ib) and *Methylothermaceae* (Ic) (Parks et al. 2018). An in detail whole genome analyses, comparing ANI (average nucleotide identity), dDDH (digital DNA-DNA hybridization) and AAI (average amino acid identity) of existing Methylococcales genomes also led to reclassification of polyphyletic MB genera (Orata et al. 2018) leading for example to the new genus *Methylotuvimicrobium*. For a complete revision of the MB taxonomy and phylogeny, the amount of sequenced type strains is still not complete enough. It is obvious though from the GTDB (http://gtdb.ecogenomic.org/) database, where non-MMO carrying bacteria are classified into MB orders (e.g. *Cyclocystaceae* into the order of *Methylococcales*) as well as recent evolutionary studies of copper containing monooxygenases (Osborne and Haritos 2018; Khadka et al. 2018) that methane oxidation is a trait and property that has been passed on by several lateral gene transfer events. However, still remnants and related functions, necessary for performing methanotrophy are still present in ancestors. The latter may lead to more reshuffling as genomes become available and lead more and more to the realization that we are not looking simply at a functional guild but at group of microbes with many more evolutionary selected traits than methane oxidation alone.

3 Environmental Distribution

MB play a crucial role in methane mitigation from a number of important methane emitting sources like wetlands. Hence, in habitats where you have high methane and oxygen in close vicinity MB are expected to flourish. However, there are many habitats where MB would not be expected. In this section, the environmental distribution of MB as well as their biogeography will be discussed. Recent findings in a few key-habitats will be highlighted.

3.1 Biogeography

Principally, MB can be found in all habitats were methane is available. This can be methane producing habitats with close proximity of anoxic and oxic habitats (e.g. wetlands, peat, lake sediments, marine sediments, landfills, etc.), but also in all other habitats in contact with a normal atmosphere which contains approximately 1.86 ppm of methane. The direct connection between MB and climate change has spurred thousands of studies focusing on environmental distribution of MB, their activity and the factors controlling the latter. Many of these aspects have been described in a range of excellent review papers (e.g. Knief 2015; Hanson and Hanson 1996; Conrad 1996, 2007; Bodelier and Laanbroek 2004; Semrau et al. 2010; Kolb 2009; Bodelier and Steenbergh 2014a, b; Ho et al. 2013a, b). The wealth of studies is also due to the fact that the presence and activity of MB can be screened very straightforward by detecting the presence of the *pmoA*/*sMMO* gene or by measuring the methane oxidation potential of incubated samples (e.g. Bodelier et al. 2013). Next to this, a range of studies has shown that community composition of MB is linked to the consumption of methane (Schnyder et al. 2018; Bodelier et al. 2013; Nazaries et al. 2011; Levine et al. 2011) which is enforced by the possibility to trace ^{13}C-labelled methane in taxonomically relevant compounds (e.g. PLFA, DNA, proteins) thereby assessing the active species in complex environments (Bodelier et al. 2013; Boschker et al. 1998; Daebeler et al. 2014). Hence, the direct link between ecosystem function and diversity of the responsible species has made MB a model group within the field of environmental microbiology, being maybe the best described environmental microbial functional group or guild. Using *pmoA* based analysis of all cultivated and uncultivated MB diversity, (Knief 2015) Knief executed a comprehensive biogeography analyses of all MB diversity known at that moment. It showed not only that more than half of all environmental sequences were related to known cultivated species but also that MB display distinct biogeography and habitat preference (Knief 2015). Typically, aquatic habitats (fresh and marine) tend to harbor lineages of *pmoA* sequences which do not occur elsewhere, while terrestrial environments like soil contain MB species with a much broader environmental distribution, maybe reflecting the larger environmental heterogeneity and influencing biotic and abiotic factors affecting MB functioning and survival in terrestrial ecosystems. Typical exceptions, are the *pmoA* sequence lineages (type Id e.g. USCα, USCγ) associated with the uptake of atmospheric methane in soils which were found almost exclusively in upland soils as well as the Verrucomicrobial MB which only occur in extreme (low pH, high temperature) terrestrial soils. The study by Knief was based on 371 studies based on the *pmoA* gene. We utilized the Earth Microbiome Project (EMP) (Thompson et al. 2017) database which is the most comprehensive data collection on microbial 16S rRNA sequencing reads and associated metadata on various habitats, to characterize patterns of methanotrophs across different biomes and habitats on an even bigger scale (Table 1).

When comparing distribution of MB genera in terrestrial and aquatic biomes (Fig. 4; Table 2) it is obvious that terrestrial habitats are populated both by type I and II MB. The most striking difference between terrestrial and aquatic biomes is the far less even relative abundances of detected genera in the aquatic habitats which is due to a strong representation of the genera *Methylobacter*, *Methylocaldum*, and *Methylomonas* (Fig. 4).

In contrast, proportions of absolute read counts were more evenly distributed among major methanotrophic genera in terrestrial biomes (Fig. 4; Table 2), possibly reflecting the differences in selective factors in terrestrial and aquatic habitats. An exception to the broad distribution of the genera belonging to the family *Methylococcaceae* is the genus *Methylosoma*, which exclusively is found in aquatic systems. In contrast to the generally disturbed representatives of the genera *Methylobacter*, *Methylocaldum*, and *Methylomonas*, members of the family *Methylocystaceae*, more in particular of the genus *Methylocystis* are more strongly associated with terrestrial than with aquatic biomes (Fig. 4) although many isolates have been obtained from aquatic environments (Heyer et al. 2002). Members of the genus *Methylosinus* can be found predominantly in aquatic habitats. Within the family of *Beijerinckiaceae*, the genus *Methylocella* is only prevalent in terrestrial biomes. Representatives of this genus do not contain the *pmoA* gene and therefore are generally underrepresented in studies of methanotrophs (Rahman et al. 2010). *Methylocella* is a facultative methanotrophs, which can use other hydrocarbons such as acetate, ethane and propane (Farhan et al. 2018), which may be related to its environmental distribution. It has to be noted that the analyses of the Earth Microbiome dataset are based on relatively short 16S rRNA sequencing reads (90–151 bp), but even when looking only at sequences related cultivated representatives the analyses of Knief (2015) is confirmed, demonstrating clear biogeography of MB genera which is the results of evolution and selection based on many more traits than the capacity to oxidize methane (Ho et al. 2013a, b, 2017a). A very recent study

Table 1 Environments in research studies of the Earth Microbiome Project (EMP) (Thompson et al. 2017) that included methanotrophic genera

	Number of studies
Fresh water	2757
Soil	1581
Rhizosphere	389
Biofilm	387
Freshwater sediment	328
Marine sediment	321
Sand (biofilter)	99
Organic material (both)	71
Mucus (human and animal)	46
Feces (human and animal)	35
Dust and air	34
Animal nest	28
Sebum (human and animal)	22
Underground water	17
Sea and hypersaline water	15
Lake stromatolite mat	13
Bodily fluid (animal)	4
Brackish water	3
Saliva (human and animal)	3
Animal excreta	3

Environmental categories were simplified but largely followed the terminology used in the EMP metadata. Please refer to Fig. 3 for detailed methodology

investigating the biogeography of MB in 697 soil samples distributed over Scotland provides strong evidence that even on a regional scale climo-edaphic factors (Temperature, moisture, soil physico-chemistry) influences distribution of type II MB and USCα related sequence types (Nazaries et al. 2018). Land-use, moisture/rainfall, nutrients and metals were the most important factors influencing distribution. The authors of this study used the climo-edaphic relationships to map MB distribution over Scotland, delivering potential important information to inform models predicting greenhouse gas emission and uptake in soil under future climate or land-use change scenarios (Nazaries et al. 2018). Similar geostatistical approaches were used on a plot scale to map activity, abundance and diversity of MB communities in a wetland soil (Bodelier et al. 2013; Krause et al. 2013; Wang et al. 2012) representing a hydrological gradient. On a plot scale (10 m × 10 m), covering the whole flooding and subsequent moisture gradient, methane oxidation activity and distribution of the active types of MB were very well explained by moisture and the subsequent availability of methane (Bodelier et al. 2013; Wang et al. 2012). However, on a smaller scale (1 m × 1 m and 20 cm × 20 cm) spatial distribution was affected by other, more local parameters not directly influenced by the larger scale hydrological gradients (Krause et al. 2013). Hence, MB display clear biogeography subjected to large scale differences in abiotic factors, but on a small scale local factors may play a more prominent role in niche differentiation and speciation depending on the respective environmental properties.

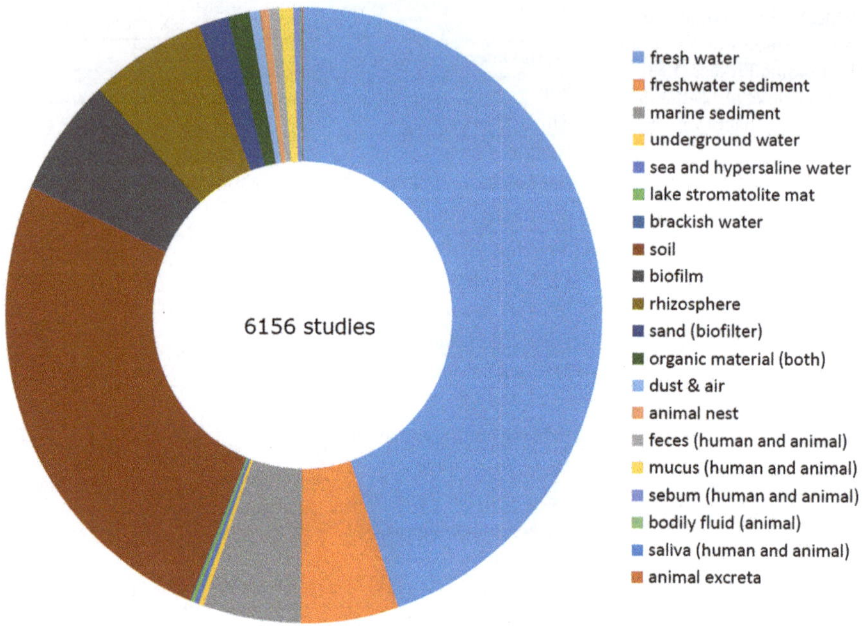

Fig. 3 Environments in research studies of the Earth Microbiome Project (EMP) (Thompson et al. 2017) that included methanotrophic genera. Environmental characterization was simplified and based on available metadata of the EMP project. For data generation, we first used the overview table "Choose Your Own EMP BIOM Table" from the EMP FTP site to select a closed reference OTU based quality filtered biome dataset (file name: emp_cr_silva_16S_123.qc_filtered. biome). Subsequently, a customized R (R Development Core Team 2008) and QIIME scripts was used to extract all methanotrophic genera, read counts, and metadata. Scripts and raw data can be provided upon request and other files are provided at ftp://ftp.microbio.me/emp/release1/. Sample processing, sequencing, and core amplicon data analysis were performed by the Earth Microbiome Project (www.earthmicrobiome.org), and all amplicon sequence data and metadata have been made public through the EMP data portal (qiita.microbio.me/emp)

3.2 *Selected Habitats*

Considering the wide distribution described above, it would be a challenge to describe MB in all habitats were they are detected. Hence, in this chapter we will focus on a few climate relevant habitats with an emphasis on novel recent developments.

3.2.1 Methane Sink Habitats

MB are the only biological way of mitigating methane produced in anoxic habitats but also to take up and oxidize methane from the atmosphere. Uptake of methane in oxic upland soils (forest, grassland, deserts) represents 6% of the sink part in total

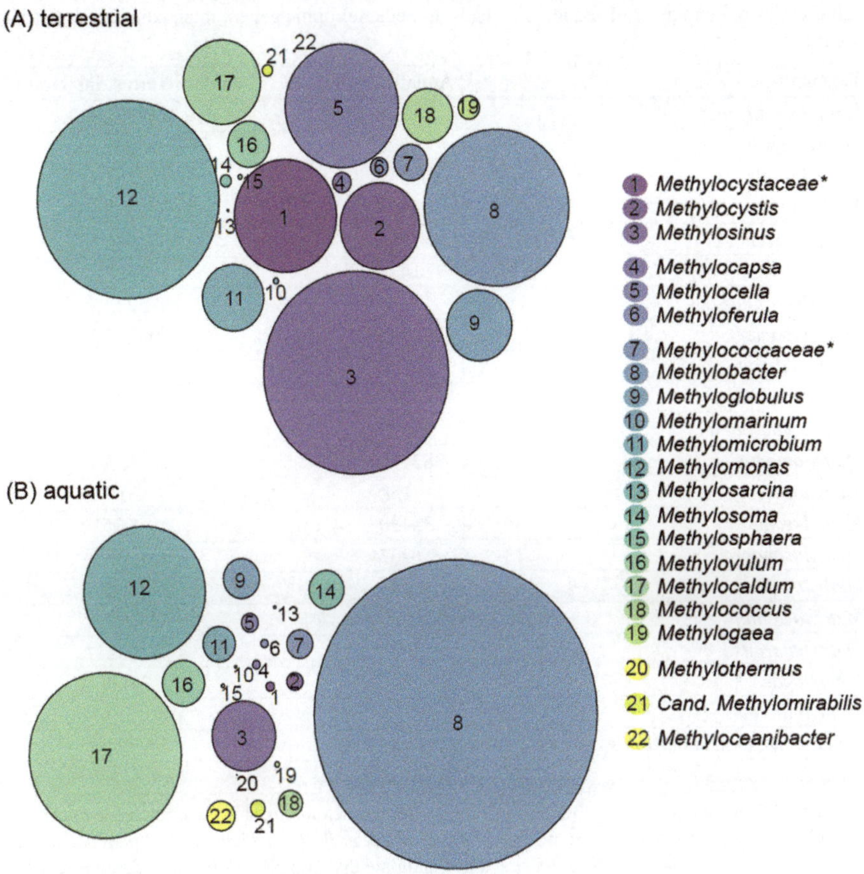

Fig. 4 Distribution of methanotrophic genera in terrestrial (**a**) and aquatic biomes (**b**) according to metadata of the EMP project (Thompson et al. 2017). * These OTUs were only classified to family level in the EMP project (Thompson et al. 2017). Different methanotrophic genera were grouped by their phylogenetic groups of the families *Methylocystaceae*, *Beijerinckiaceae*, *Methylococcaceae*, *Methylothermaceae*, *Methylacidiphilaceae*, and Candidate division NC 10. Circles were arranged using the R function circle Progressive Layout in the R package pack circles version 0.3.3 (R Development Core Team 2008). The size of each circle is proportional to the absolute read counts. Circles were detached for better visibility. Data is based on 4.067,993 16S rRNA gene sequencing reads (Table 2) from 6156 studies published in the Earth Microbiome Project (Thompson et al. 2017). The original data is shown in Table 2. Please refer to legend of Fig. 3 for detailed methodology

global methane balance (Dean et al. 2018) and is a process proposedly carried out by as yet uncultivated MB belonging to type Id and IIb sequence clusters (Dedysh and Knief 2018), better known as USCγ and USCα and related clusters. The process carried out by these so called "high affinity" MB (degrading methane to concentrations below 1.8 ppm) is affected by a range of environmental parameters (CH_4, O_2, moisture, ammonium, agricultural practices, land-use change, plants, organic acids

Table 2 Absolute number of sequencing reads of methanotrophic genera in aquatic and terrestrial biomes

Taxonomy	Aquatic_biome	Terrestrial_biome
Methylocystaceae[a]	1826	51272
Methylocystis	6828	30242
Methylosinus	106567	165801
Methylocapsa	1524	1638
Methylocella	7560	61966
Methyloferula	1425	1457
Methylococcaceae[a]	17302	5272
Methylobacter	2091286	99226
Methyloglobulus	34702	20175
Methylomarinum	161	136
Methylomicrobium	26589	18095
Methylomonas	382196	158153
Methylosarcina	158	15
Methylosoma	32430	564
Methylosphaera	267	85
Methylovulum	46019	8630
Methylocaldum	604837	28607
Methylococcus	14602	12032
Methylogaea	602	2148
Methylothermus	6	0
Candidatus.Methylomirabilis	5485	514
Methyloceanibacter	19585	8

[a]These OTU's were classified to the family level. Different methanotrophic genera were grouped by their phylogenetic groups of the families *Methylocystaceae*, *Beijerinckiaceae*, *Methylococcaceae*, *Methylothermaceae*, *Methylacidiphilaceae*, and Candidate division NC 10. Please refer to Fig. 4 for detailed methodology

and monoterpenes etc.) as has been described in a range of studies (Kolb 2009; Nazaries et al. 2011; Dunfield 2007; Kolb and Horn 2012; Maurer et al. 2008; Menyailo et al. 2010; Wieczorek et al. 2011). Recovery after disturbance can take decades and may depend on the diversity of MB present (Levine et al. 2011). Due to the fact that atmospheric methane alone is not sufficient to grow, all attempts to grow and isolate atmospheric methane MB has so far failed and hence, novel information on the distribution and functioning of these microbes has been very sparse. Novel habitats potentially contributing to atmospheric methane degradation, have been reported recently. Glacier fore field soils (Nauer et al. 2012), high artic cryosols (Lau et al. 2015), thawing permafrost (Singleton et al. 2018), limestone (Waring et al. 2017), Lava (Pratscher et al. 2018) and karst caves (Zhao et al. 2018). Genomes related to USCγ (Edwards et al. 2017) and USCα (Pratscher et al. 2018; Singleton et al. 2018) have been recovered from artic soil and from forest soils. These genomes have confirmed the relatedness of USCα with genus *Methylocapsa* and for USCγ with *Methylocaldum*. The novel USCα genomes recovered from a mire in Northern

Sweden revealed the potential for utilization of acetate, confirming what was already detected by stable isotope probing earlier (Pratscher et al. 2011) as well as the presence of genes necessary for CO consumption (Singleton et al. 2018). Pratscher et al. (2018) was able to visualize the first USCα cell from forest soils using a combination of fluorescence cell sorting using a combination of a suicide substrate and fluorescent in situ hybridization. In this way sufficient cells, actively oxidizing atmospheric methane could be captured to perform a genome analyses which revealed that these microbes only have XoxF (lanthanide dependent) methanol dehydrogenase only, genes involved in exopolymer production and excretion as well as the capacity to form trehalose. The latter may point towards adaptation to dry or cold conditions and growth in biofilms. The latter fitted with the high abundance of these microbes in subterranean cave wall biofilms in caves globally (Pratscher et al. 2018), a survey which was possible with 16S rRNA primers developed using the genome info. All in all, these new developments point into the direction that atmospheric MB thrive in oligotrophic habitats, where they make part of pioneering microflora.

However, atmospheric methane oxidation is not restricted to these exotic, enigmatic microbes but potentially can be carried out by some well-known low affinity (i.e. thrive at high methane concentrations) MB which have been demonstrated to oxidize atmospheric methane in pure culture incubations (Knief and Dunfield 2005) although these MB cannot grow nor survive on atmospheric methane alone. Recently, it was demonstrated that agricultural soils can be turned into a strong sink for atmospheric methane upon addition of organic amendments (e.g. compost, sewage sludge; aquatic plant material) (Ho et al. 2015, 2017b). Although, the mechanisms responsible for this phenomenon are not yet clear, the use of stable isotope probing in combination with PLFA (phospholipid fatty acid) demonstrated that conventional MB (i.e. *Methylocystis*; *Methylosinus*; *Methyloferula*) were responsible for the uptake of methane in these agricultural soils (Ho et al. 2019). Since cell specific activities were lower than what is needed for maintenance of cells, leading to the conclusion that these MB need alternative sources of energy. Conventional MB in rice field soil were able to turn the soil into a persistent and strong sink for atmospheric methane after being spiked with high concentrations of methane (Cai et al. 2016).

Consumption of more methane (i.e. repeated incubation with 10,000 ppmv) led to a soil community that was able to take up methane for more than a 100 days, demonstrating that even a wetland soil can turn into a sink for methane without involvement of high affinity MB. Utilization of storage compounds like PHB (poly hydroxyl butyric acid) that can accumulate during periods of high methane availability were hypothesized to be the source of energy to continue *pmoA* activity under atmospheric methane supply. This may partially also explain the effect of addition of organic amendments in agricultural soil, which may promote internal methane production in soil aggregates fueling conventional MB with energy but potentially also with nutrients and other essential minerals like metals (Fig. 5). Hence, with respect to soil methane sinks recent novel developments offer new views on functioning of high affinity MB but also point towards the potential of as yet non-methane

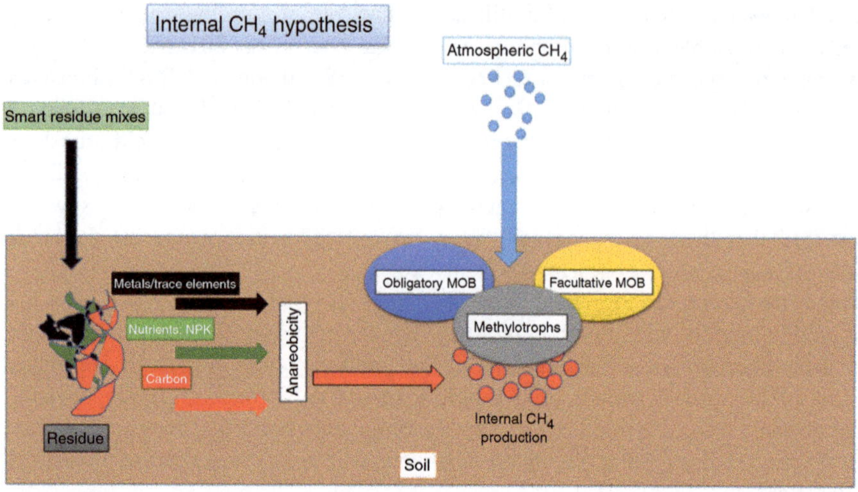

Fig. 5 Proposed mechanisms of residue-stimulated uptake of atmospheric CH_4 by agricultural soils. Addition of organic carbon can lead to anaerobic conditions mediated by the aggregate structure in the soil which leads to internal CH_4 production stimulating growth of obligatory and facultative MB or methylotrophs carrying sMMO genes. This stimulation maybe enhanced by nutrients (N, P, metals, trace elements) present in the residues. These stimulated cells will have a higher capacity to co-oxidize CH_4 from the atmosphere

sink habitats to be explored for their role in climate change mitigation scenarios by assessing the functioning of conventional MB and their role in atmospheric methane uptake. The latter will be necessary since climate change phenomena (lower rainfall, high temperature) may increase atmospheric methane consumption (Fest et al. 2017).

3.2.2 Wetlands

Due to their tight-aquatic terrestrial coupling and large organic matter accumulation, wetlands are hotspots of biogeochemical processing. The high methanogenic activity in their anoxic, carbon-rich soils has made them largest source of global atmospheric methane, emitting 142-284 Tg CH_4 per year (Kirschke et al. 2013). This provides an ideal habitat for anaerobic and aerobic methanotrophs, which both are abundant in wetlands, and can reduce CH_4 emissions by 10–90% (Segers 1998). The functioning of aerobic methanotrophs in wetlands has been reviewed extensively (Conrad 2007; Bodelier and Steenbergh 2014a, b; Dedysh 2009; Bridgham et al. 2013; Bodelier 2011) mainly also due to the relevance of these microbes in mitigation of methane emission from rice paddies (e.g. Bodelier et al. 2000; Kruger et al. 2002; Chowdhury and Dick 2013). MB abundance and community composition in wetlands largely depend on local environmental conditions, in part determined by the ecosystem type. We can distinguish several different types of wetlands, including peatlands (bogs and fens, arctic permafrost), salt marshes and estuaries, tropical (natural) wetlands including river floodplains, agricultural wetlands such as rice paddies, and geothermal wetlands. Methanotroph communities have been best described for temperate and (ant) arctic ecosystems, although some research has been done in the tropics, which also harbour large peatland areas. In peatlands, aerobic proteobacterial methanotrophs dominate, with members of the alphaproteobacteria dominating in oligotrophic, acidic (pH < 5) bogs, whereas in general gammaproteobacteria dominate in minerotrophic, nutrient-rich and only mildly acidic (pH > 5) fens (reviewed by Verbeke et al. 2018). Wetlands occur over a continuum of hydrological and geological conditions, and therefore often lie somewhere in between these two conditions, thus harbouring a mixture of alpha and gammaproteobacteria. Within the alphaproteobacteria occurring in bogs, *Methylocystis* strains have been found to be most active, while *Methylocella* and *Methylocapsa* species are also abundant (Verbeke et al. 2018). Within the gammaproteobacteria, *Methylobacter*, *Methylomonas* and *Methylomicrobium* species are the most abundant (Verbeke et al. 2018). Geothermal wetlands are characterized by high temperatures and tend to be very acidic (pH < 4) due to sulfur oxidation. Methanotrophs isolated from these wetlands have been found to belong to the *Verrucomicrobia*, and are currently described as the candidate cluster "*Methylacidiphilum*" and the candidate genus "*Methylacidimicrobium*" (Sharp et al. 2014).

Typically, aerobic MB are present at or above the oxic-anoxic interface of wetland soils (Reim et al. 2012; Vaksmaa et al. 2017b). However, in recent years MB have also been detected in association with *Sphagnum* mosses (Larmola et al. 2010; Kip et al. 2010), either occurring in air-filled pores within *Sphagnum* tissues, or associated to their roots. It was suggested that MB live in symbiosis with peat mosses (Raghoebarsing et al. 2005) providing the moss with CO_2 while receiving photosynthetic oxygen from the moss. Other studies reported on methane dependent nitrogen fixation by mosses suggesting an important role for aerobic MB in peat growth (Larmola et al. 2014), which was contradicted by other studies failing to observe an effect of methane addition on N_2 fixation by mosses (Kox et al. 2018). As yet conclusive evidence for mechanism of symbiosis between MB and peat mosses in missing (Ho and Bodelier 2015) as are the responsible microbes (Kox et al. 2018). A similar, potentially far reaching symbiosis is claimed for MB and rice roots, where diazotrophic MB in the rhizosphere fix nitrogen using methane as energy source and transferring this nitrogen to the rice plants (Ikeda et al. 2014). A gene similar to the one involved in plant-fungal symbiosis was proposed to be involved in the activation of rice-MB nitrogen transfer (Bao et al. 2014). However, besides the localization of type II MB proteins in rice root tissue (Bao et al. 2014) no direct proof for this symbiosis is available, which can be of great influence for the sustainable cultivation of rice emitting less methane while less nitrogen fertilizer is required. With respect to the latter, the recently discovered ability of some MB to denitrify (Dam et al. 2013; Kits et al. 2015) sheds new light on interactions between methane and nitrogen cycling in wetland systems and beyond. The recently described genome sequence of the denitrifying methanotroph *Methylomonas denitrificans* FGJ1 reveals genes involved in oxygen scavenging under hypoxic conditions, hypothesized to maintain oxygen provision for the *pmoA* under low oxygen conditions (Orata et al. 2018).

Hence, consumption of methane maybe providing wetland plants with nitrogen by diazotrophic MB but on the other hand may also lead to N loss by denitrifying MB. MB community composition will play a crucial role in this delicate balance. When it comes to wetlands, in many cases *Methylobacter* species are the dominant active MB present (Bodelier et al. 2013; Smith et al. 2018). Recently it has been shown that *Methylobacter* genomes retrieved from freshwater wetlands harbour genes for dissimilatory denitrification as well as genes for performing under hypoxic conditions, suggesting that the phenotype discovered in the described strain *Methylomonas denitrificans* is functional in nature and is more widespread among type I MB (Smith et al. 2018). The preference of low oxygen concentrations for growth and activity has been described as a niche differentiating factor for MB in other wetland (Reim et al. 2012) and lake sediments (Hernandez et al. 2015) with always *Methylobacte*r species being dominant and active MB at low oxygen (Oshkin et al. 2015). *Methylobacter* was also the most active and abundant MB species in artic peat (Tveit et al. 2013, 2014) as well as in the high methane fen sediment layers along an permafrost thaw gradient (Singleton et al. 2018). The latter study represents the most extensive metagenome and metatranscriptome analyses of MB communities to date. Permafost wetlands release high amounts of methane upon thawing which makes mitigation by MB of high environmental relevance. The investigated thawing gradient in a mire system in Northern Sweden revealed a clear niche differentiation of type IIa and b (USCα) MB in the permafrost palsa and in the acidic bog, while type I MB dominated the high methane deeper fen sediment layers (Singleton et al. 2018). The retrieved MB genomes revealed many features of relevance of MB survival and functioning in these habitats and also yield evidence for *Hypomicrobia* to belong to the type II MB families. The authors concluded that methane concentration was a strong selective factor shaping the observed niche differentiation. Fairly recent, a novel described MB, *Candidatus Methylospira mobilis* (type Ib) from an acidic peat bog, showed distinct chemotactic behaviour towards high methane and low oxygen, demonstrating the potential for MB to actively position themselves within the environmental gradients inherent to wetland habitats (Danilova et al. 2016).

The latter will be quite important given the most recent view on methane oxidation in wetland systems where aerobic MB do not have exclusive access to methane considering the presence of anaerobic methane oxidisers (AOM) which can mediated anaerobic methane consumption by various electron acceptors, including iron, sulphate, and manganese (Dean et al. 2018; Vaksmaa et al. 2017b; Welte et al. 2016). Anaerobic methane oxidation can be coupled to nitrite or nitrate reduction. This has been observed in the candidate species *Methylomirabilis oxyfera (nitrite-AOM)*, first enriched from nitrite-rich sediments (Ettwig et al. 2010) and the ANME-2D archaeal lineage "Candidatus *Methanoperedenaceae*", which has been recently enriched from rice-paddy soils (Haroon et al. 2013; Vaksmaa et al. 2017a). In rice paddies (Vaksmaa et al. 2017b) as well as in coastal wetlands (He et al. 2019), anaerobic methane oxidation rates as well as numbers can be substantial and indicate that a strong niche differentiation of AOM and MB in wetland system together forming effective filters for the methane produced by oxidizing from source to atmosphere. However, the effectiveness will strongly depend on the environmental gradients but also on the traits and competitive abilities of the respective AOM and MB present.

Sometimes, important controlling factors are overlooked in these interactions. In a range of wetland ditches a strong correlation with methane aerobic methane oxidation and phosphorus availability was found in the surface cm of the sediments which were dominated by type I MB (Veraart et al. 2015). Scanning the literature on these findings revealed strong positive effects of P on methane oxidation in arctic permafrost wetlands, also dominated by type I MB (Gray et al. 2014). Browsing MB genomes on traits relating to P uptake and utilization revealed a clear separation between type I and II MB, pointing towards a stronger reliance on P in type I MB, making P an important niche differentiating factor in wetland systems (Veraart et al. 2015).

3.2.3 Aquatic Systems

Aquatic systems are water bodies classified as freshwater, transitional, coastal or marine, coastal and marine. The marine ecosystem is the biggest covering almost 70% of the Earth surface but CH_4 released from them to the atmosphere is a small source with a mean value of 12 Tg CH_4 year^{-1} in comparison to other natural environments (Saunois et al. 2016). On the contrary, freshwater ecosystems (~4% Earth surface) can contribute up to 100 Tg CH_4 year^{-1} (Bastviken et al. 2011). Estimations or future trends in CH_4 emissions lack robustness due to the restricted number of measurements, different methodology used and also the little knowledge on the dynamics of CH_4 emissions in some systems (Reay et al. 2018). The release of CH_4 from aquatic systems to the atmosphere involves processes such as diffusion, ebullition (Walter et al. 2007), seepage, resuspension (Bussmann 2005) or bioturbation (Oliveira Junior et al. 2019) of the sediments. Typically, methane production occurs in sediments or anoxic bottom water layers. However, fairly recent it was discovered that oxic waters are supersaturated with CH_4 (Grossart et al. 2011; Karl et al. 2008). This phenomenon of the so called "methane-paradox" has spurred aquatic methane cycling research to pinpoint the possible abiotic (DelSontro et al. 2017; McGinnis et al. 2017) or biotic (Grossart et al. 2011; Tang et al. 2016; Yan et al. 2019) factors underlying this phenomenon (Yan et al. 2019). The large role of freshwater systems in global methane emission, combined with the unexpected high methane concentrations in oxic water columns has spurred attention for MB in aquatic systems.

Marine CH_4 oxidation can be performed by anaerobic and aerobic CH_4 oxidizers (Ruff et al. 2019). Anaerobic oxidation is driven by syntrophic interaction between archaea and $SO_4{}^{-2}$-reducing δ-Proteobacteria (Krukenberg et al. 2018). These consortia of ANME-archaea (ANME 1, 2 and 3) and sulfate-reducing bacteria (SRB) dominate methane oxidation in deep-sea sediments (Ruff et al. 2013; McGlynn et al. 2018; Durisch-Kaiser et al. 2005). This interaction depends on controlled gene expression to enhance electron transfer among the consortium members (Krukenberg et al. 2018; McGlynn et al. 2018). Aerobic MB in marine environments are comprised by five so called deep-sea clusters all associated with type I lineages (reviewed in Knief 2015). As yet, there is no clear evidence of the factors that drive niche differentiation of MB belonging to these clusters (Ruff et al. 2013). Despite the low amount of O_2 in deep regions of marine systems, aerobic methanotrophs from the order Methylococcales are present (Ruff et al. 2013; Tavormina et al. 2008). It has been recently described that ANME-archaea and Type I MB form a "methanotrophic microbiome" where Type I MB are the first to colonize the deep-sea sediments (Ruff et al. 2019). Regarding Type I, some lineages related to terrestrial environments are present in a deep-sea fan because of the inputs of soil derived organic matter (Bessette et al. 2017). The *pmoA* phylotypes found in the water-column differ from the ones inhabitant the sediments in terms of diversity and abundance. For instance, the planktonic MB members are distributed in three clades belonging to the order of Methylococcales whereas the benthic ones are less diverse (Tavormina et al. 2008). Similar to wetlands, MB can be found in Oxygen Minimum Zones (OMZ), regions in oceans and seas where O_2 drops down to nano or picomolar levels. These regions show an active CH_4 oxidation where Type I MB and NC10 (Chronopoulou et al. 2017) seemed to play a relevant role.

Freshwater ecosystems harbor environments such as lakes, rivers, ponds and reservoirs playing a relevant role in the C cycle as well as in the cycling of other nutrients. They also provide many services as water, electrical energy and food supply or can serve as recreational areas (Vollmer et al. 2016). Still, they are very vulnerable to climate change especially regarding the increase of temperature which alters the thermal structure and water chemistry of these systems (Yvon-Durocher et al. 2011). Among the freshwater systems, lakes are the most studied in relation to CH_4 dynamics. Lakes cover a small area (0.9%) of the Earth's surface but have a large contribution (6–16%) to the total global CH_4 budget (Tang et al. 2016; Bastviken et al. 2004). In comparison to other environments, freshwater systems hold "endemic" MB which we may call typical freshwater MB. These lineages comprise six sequence clusters within type Ia and Ib MB (Knief 2015; Dedysh and Knief 2018). Most of these clusters are ubiquitous in FW systems, except for the so called cluster 4a which is only detected so far in sediments while cluster 2 is more frequent in the water column (Knief 2015). First reports on freshwater MB communities focused on the sediment (Naguib and Overbeck 1970; Reeburgh and Heggie 1977; Cappenberg 1974). These studies, demonstrated clearly that sediments harbored much higher numbers of MB than the pelagic part of the lake (Zhou et al. 2018). In most pelagic zones of lakes, type I MB belonging to the genus *Methylobacter* are the dominating MB group (Blees et al. 2014; Biderre-Petit et al. 2011; Kojima et al. 2009), Recently a novel genus within the family of the *Methylococcaceae*, *Candidatus Methylomidiphilus alinensis*, was demonstrated to be the most abundant MB in a humic lake in Finland (Rissanen et al. 2018). Next to this, MB related the filamentous bacterium *Crenothrix polyspora* were reported to be the main CH_4 sink in two stratified lakes (Oswald et al. 2017), demonstrating the lack of knowledge on MB in freshwater habitats. In contrast, other reports found that Type II representatives may play a relevant role in the lacustrine CH_4 cycle. For instance, in an alkaline lake there was an almost equal abundance of type II MB, represented 2% of the whole microbial community, while type I accounted between 3% and 5% (Carini et al. 2005). In the oxic waters of a nutrient deprived deep lake (Zigah et al. 2015) type II was more abundant and active in nutrient than type I (Parks et al. 2018). Similar results were found in sub-arctic lakes (He et al. 2012).

Surprisingly, high abundance of type I MB have been detected in anoxic parts of the water column of lakes (Schubert et al. 2011) suggesting that they are either inactive or performing anaerobic methane oxidation. However, this is not the case as they were actively transcribing *pmoA* and no archaeal AOM were detectable (Blees et al. 2014; Rissanen et al. 2018; Milucka et al. 2015). This would suggest that they are interacting with other members of the microbial planktonic community or using other electron acceptors rather than O_2. Incubations with light increased the CH_4 oxidation in anoxic water samples dominated by Type I from boreal (Rissanen et al. 2018) and temperate lakes (Oswald et al. 2015). This implies an interaction between MB and phototrophs (oxygenic photosynthesis) as demonstrated recently in a number of lakes (Milucka et al. 2015; Oswald et al. 2016). Hence, the low amount of light penetrating into the deeper pars of lakes is sufficient for phototrophs to produce oxygen used by methanotrophs to oxidize methane aerobically. Recently, it was also reported that members of the NC10 phylum, i.e. *Candidatus Methylomirabilis oxyfera* can play an important role in methane oxidation. This bacterium links the C and N cycles by coupling CH_4 oxidation with NO_2^-/NO_3^- reduction (nitrate/nitrite-dependent anaerobic methane oxidation, n-damo). In a freshwater reservoir, clone libraries were comprised of type I, II MB with a dominance of members of the NC10 phylum. The latter dominated in deep waters of this system (Kojima et al. 2014). More recently, a bloom of NC10 "*Ca. Methylomirabilis limnetica*" reaching 27% of the whole bacterial community dominated the deepest water layers of a stratified lake (Graf et al. 2018). A similar strong link between nitrogen cycling and methane oxidation was found in 15 Indian dam-reservoirs. Using ^{15}N-incubations it was shown that CH_4 amendment boosted N loss from this systems with more than 10 times due to denitrification by aerobic MB (Naqvi et al. 2018).

Anaerobic CH_4 oxidation (AMO) also takes place in freshwater systems and plays a significant role in freshwater CH_4 dynamics. This process can happen simultaneously with the aerobic oxidation (Arp et al. 2018) or seasonally (Roland et al. 2017). CH_4 oxidizers different from the Proteobacterial were first described in anoxic water layers of Lake Plußsee (Eller et al. 2005). This study showed the existence of anaerobic methane-oxidizing archaea (ANME 1 and 2) and the role of the anaerobic CH_4 oxidation in a freshwater system. Different to their marine counterparts, these archaeal methanotrophs seemed not to depend on sulfate reducing partners, as there was no evidence of physical attachment to them. In a meromictic, temperate and eutrophic lake, CH_4 anaerobic oxidation was coupled with SO_4^{-2} reduction once NO_3^- is scarce or both acceptors may be used concomitantly (Roland et al. 2017). In a tropical lake, a new cluster of ANME (different from the marine members) was described and associated to SO_4^{-2} (Zigah et al. 2015).

3.2.4 Exotic Habitats

As already described in 1.3.1, MB have a very broad distribution. In recent years, however, a few "exotic" habitats have been described where MB are present and active. Termites are a significant global source of methane, contributing around 3–4% to the total global methane budget (IPCC 2014; Nauer et al. 2018). Methane produced in the termite guts will accumulate in their nests and leave through the typical "chimneys". The MB community was investigated in the termite nests as well as in these chimneys (Ho et al. 2013a, b) where the largest methane oxidation activity was found, higher even than reference soil surrounding the termite mounts. The termites in this study selected for a specific MB community, where mostly sequence associated with USCγ (upland cluster gamma), proposed to be affiliated with uptake of atmospheric methane, and type Ib affiliated with *Methylocaldum/ Methylococcus* were detected within the nests and chimneys. This selection maybe related to the "extreme" conditions in these habitats, of low methane with high through flow. Similarly, recent investigations found similar deviating MB communities in caves habitats (Pratscher et al. 2018). In this study, the first 16S rRNA sequence of USCα representatives was discovered which enabled a biogeographic survey in metagenome and amplicon sequencing databases. Extremely high abundance of USCα sequences were observed in subterranean habitats like caves worldwide but also in soil crusts and arctic cryosols (Pratscher et al. 2018), suggesting that these habitats should be considered as significant methane sinks. The genomic information revealed traits that facilitates growth and survival in biofilms and under dry and cold conditions again confirmed by involvement of USCα-related sequences in acidic arctic mineral cryosols (Lau et al. 2015). Alkaline cryosols seem to be populated by USCγ where a genome sequence has been recovered from recently (Edwards et al. 2017) which typically is related more to the order of the *Chromatiales* which does not contain any MB. Occupation of another pioneering habitat with extreme conditions by USCγ representatives was also demonstrated in glacier fore fields (Nauer et al. 2012) which took up atmospheric, with highest rates in oldest fields. Hence, the USC cluster MB seem to be the pioneering MB in low methane habitats, but *Methylobacter* species (related to *Methylobacter tundripaludum*) seem to fulfill a similar role in high methane habitats, acting under extreme conditions once activity become possible, like in polar waters beneath the ice-sheet (Michaud et al. 2017) or thawing permafrost (Singleton et al. 2018; Tveit et al. 2013, 2015) or permafrost thaw ponds (Crevecoeur et al. 2017).

A quite unexpected habitat for the presence of MB was described recently. Members of the *Methylocystaceae* (*Methylocystis*) as well as *Beijerinckiaceae* (*Methylocapsa*, *Methyferula*) were detected in decaying wood (Makipaa et al. 2018). Decaying wood is an extreme habitat with low pH, low nitrogen availability and high methanol availability due to activity of wood degrading fungi (van der Wal et al. 2013, 2015). Methylotrophs utilizing the methanol inside wood have already been described (Vorob'ev et al. 2009). However, methanotrophs so far not, which raises the question what substrate these MB are using. Off course they can grow on methanol or mixotrophic on acetate and methane. The origin of methane in wood is not clear yet. Fungi have been reported to be able to produce methane (Lenhart et al. 2012), although in low amounts it may be a source of methane for MB. The MB have been suggested to deliver nitrogen to the fungi in decaying wood, given the increase of *nifH* genes associated with MB with increasing decay stage of the wood (Makipaa et al. 2018). However, this still has to be verified.

4 Ecology

Many aspects of MB ecology have already been referred to or touched upon in the previous sections. Below we will focus mainly on more recently acquired insight in MB ecology and environmental functioning.

4.1 Controlling Abiotic Factors

Since the activity of MB depends to a great extent on the presence of methane, basically any factor that affects methane formation in the environment will also have implications for MB functioning. Hence, a complex set of often interacting physico-chemical factors (oxygen, pH, nutrients, metals, temperature, soil porosity, soil granulometry, etc.) can modulate MB survival and activity in various habitats. Numerous studies have been executed assessing controlling factors of MB and have been extensively reviewed (Hanson and Hanson 1996; Conrad 1996, 2007; Bodelier and Laanbroek 2004; Semrau et al. 2010; Bodelier and Steenbergh 2014a; Bridgham et al. 2013; Bodelier 2011; Tate 2015; DiSpirito et al. 2016). From its background as a potential inhibitor for methane oxidation in upland and wetland soils, nitrogen has been probably the most intensively studied (Bodelier and Laanbroek 2004; Bodelier 2011). The current state of the art regarding control of nitrogen compounds can still be summarized as depicted in Fig. 6.

Besides acting as inhibitor of the monooxygenase enzyme, nitrogen off course is essential nutrient for MB for which they may need to compete with other organisms. It has been elegantly demonstrated recently by a combination of stable isotope probing and high through put sequencing that MB successfully compete with ammonia oxidizers in rice soil (Zheng et al. 2014) and with archaeal and bacterial nitrifiers in volcanic grassland soils (Daebeler et al. 2014). In fact, for quite some time it is already known that MB can nitrify themselves (Bodelier and Frenzel 1999) and fairly recent that MB can execute dissimilatory denitrification (Kits et al. 2015). In case of accumulation of toxic intermediates of nitrification or denitrification (i.e. NO, nitrate, nitrite, NOx) MB possess many genes indicating the ability to deal with nitrosative stress (e.g. Dam et al. 2013; Orata et al. 2018) which has been revealed by genome analyses in recent years. What is far less well developed is the potential interaction between aerobic MB and anaerobic methane oxidizers (ANME,

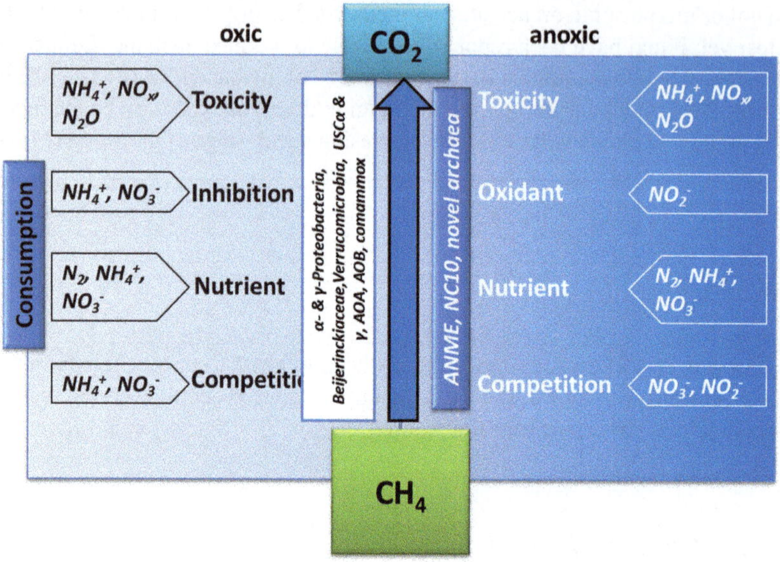

Fig. 6 Schematic overview of interactions between methane and nitrogen cycles under oxic and anoxic conditions and organisms involved. The mechanisms under which methane consumption is affected by nitrogen are as essential nutrient, as toxicant (nitrosative stress), by competitive substrate inhibition or by competition for nitrogen with other microbes. Under aerobic conditions interaction of aerobic MB will mainly take place with bacterial and archaeal nitrifiers, with comammox nitrifiers but also with heterotrophic bacteria. At the oxic/anoxic interfaces in e.g. wetlands intense interaction may occur with anaerobic MB, but also with denitrifiers producing toxic intermediates (adapted from Bodelier and Steenbergh 2014a)

NC10) (Kuypers et al. 2018). Potentially, competition for nitrate could occur at the oxic/anoxic interface in wetlands for example. There is no information yet on direct competition, but, anaerobic methane oxidizers of the genus *Methanoperedens* have been demonstrated to be able to cope with oxic conditions (Welte et al. 2016; Guerrero-Cruz et al. 2018), indicating that a niche overlap is possible with aerobic MB which are definitely equipped to survive anaerobic conditions and may even thrive at hypoxic conditions (see previous sections). Also unknown as yet is the interaction with the recently discovered Comammox nitrifiers belonging to the genus *Nitrospira* (van Kessel et al. 2015; Daims et al. 2015). These microbes catalyzing the oxidation from ammonium to nitrate in a single organisms also seem to have a broad distribution (Orellana et al. 2018) and also have a high affinity for ammonium (Kits et al. 2017) making potential strong competitors. Next to the myriad of potential novel and unknown interactions between N-cycling microbes and MB, work with known pure culture also revealed important novel info. Preference of combinations of carbon and nitrogen source differed significantly between different MB species (Tays et al. 2018), a fact that also need to be considered when trying to explain observations in the environment.

One of the most striking novel findings regarding controlling factors of MB functioning concerns metals as nicely reviewed recently (Semrau 2018). It was well known that many genes in the rather complex biosynthetic pathways that MB catalyze, require metals, especially copper. The pMMO needs copper for expression of the enzyme and in the absence of copper, a switch is made to express the sMMO, in the case of MB possessing both forms of the enzyme (DiSpirito et al. 2016; Semrau 2018). However, how MB acquire copper was only discovered recently. Representatives of the family of the *Methylocystaceae* can produce a chalkophore called methanobactin, which can mobilize copper in the environment, similar to siderophores for Iron (see (DiSpirito et al. 2016). Next to copper, methanobactin can bind a range of other metals, albeit with lower affinity (DiSpirito et al. 2016). Next to its role as chalkophore, methanobactin was also proposed to function as a signaling compound and as a detoxifying agent for oxygen radicals (see DiSpirito et al. 2016). The potential environmental relevance of methanobactin has been shown in a study where MB and denitrifying Pseudomonas species had to compete for copper (Chang et al. 2018). The competition was won by *Methylosinus trichosporium* leading to a reduction in the expression of the copper requiring *NosZ* gene, with the consequence that nitrous oxide (N_2O) is not reduced to N_2. Hence, expression of methanobactin in the environment could lead to negative effects for GHG emissions. Even more revolutionary than the discovery of methanobactin was the finding that Verrucomicrobial MB were dependent on rare earth metals (e.g. La, Ce) for the expression of MDH, catalyzing the conversion of methanol to formaldehyde in MB. This phenomenon was first described for Verrucomicrobial MB growing in Volcano mud pots (Pol et al. 2014). It appeared that next to a calcium dependent MDH (*mxaF*), MB can have a lanthanide dependent so called XoxF type MDH, which appeared to form several gene families (Keltjens et al. 2014). These findings have spurred a range of studies into the role of XoxF and lanthanides for the functioning of MB (see Semrau 2018). A wide environmental distribution (Chistoserdova 2016) and broad functional role of lanthanide dependent enzymes in microbes were suggested. In an elegant experiment with a range of MB and methylotrophs in a synthetic culture mix, it was demonstrated that the switch in MB from expressing mxaF of XoxF was determined by nitrogen source, oxygen but also by the respective species (Yu 2017). These findings bear the potential to explain environmental observations on MB functioning and distribution but also challenges to way to a full understanding of environmental methane consumption (Chistoserdova and Kalyuzhnaya 2018).

4.2 *Biotic Interactions*

Besides the rather complex interaction between MB and their abiotic environment, recent work has also shown that biotic associations between MB and other organisms maybe just as complex, but far less well studied. As already described in Sect. 3.2.2, MB have been reported to form a “symbiosis” with Sphagnum mosses. Next to plants, MB have been demonstrated to form a crucial component of a symbioses between deep-sea crabs and mussels near hydrothermal vents. Gammaproteobacterial MB were found to be part of the active epibiotic community in the setae of *Shinkaiacrosniere*, a deep sea dwelling crab found around hydrothermal vents (Watsuji et al. 2014). In a stable isotope labeling study, ^{13}C derived from $^{13}C–CH_4$ to $H^{13}CO_3^-$ (bicarbonate) could be retrieved from tissue of the *S. crosniere*, showing that carbon derived from MB or other epibionts was assimilated into the crab, and provided evidence that epibionts may also nutritionally support their host. With the deep-sea mussel *Bathymodiolus azoricus* mechanisms of symbiotic relationships with MB were elucidated recently using a proteomic approach (Ponnudurai et al. 2017). These mussels have gills which are colonized by gamma proteobacterial sulfur and methane oxidizers. The mussels provide CO_2 and tricarboxylic acid cycle intermediates to the autotrophic sulfur oxidisers while the microbes give vitamins and co-factors to the host (Ponnudurai et al. 2017). The MB involved in this symbiosis are related to type I MB but have as yet not brought into culture. Carbon is transferred from MB to the host very likely by digestion of the microbial cells by the host and subsequent uptake (Ponnudurai et al. 2017). What the advantage for the MB is not clear but it seems that providing a surface to attach to with a constant supply of methane would be a suitable niche to dwell in these energy limited deep-sea habitats. A similar symbiosis was also reported for deep-sea sponges, where MB made up 50–70% of the total microbiome of these hosts (Rubin-Blum et al. 2017).

What has been investigated far less intensive is the direct interaction between MB and other bacteria. It is well known that MB fix carbon from methane but that a substantial portion of this carbon is exuded (Kalyuzhnaya et al. 2013) and can possibly be used by other bacteria such as methylotrophs using the methanol (Zheng et al. 2014; Oshkin et al. 2015; Krause et al. 2017; Beck et al. 2013). This so called methane-based food web can give rise to many interactions as recently summarized (Ho et al. 2016a, b). By analyzing ^{13}C-labelling studies where the whole community was assessed in heavy-DNA fractions, it was shown that MB types (Ia, Ib, II) fed their own specific microbial community indicating that the composition of the compounds exuded differs (Ho et al. 2016a, b). There are also reports MB get something in return, as was demonstrated for *Methylovulum miyakonense* which got cobalamin from a *Rhizobium* species (Iguchi et al. 2011). An extensive screening study bringing numerous combinations of MB and heterotrophs in direct contact assays also pointed towards stimulation by provision of essential compounds for MB (Stock et al. 2013). Also, the composition and diversity of heterotrophic bacterial community was tested and proven to have a positive effect on methane oxidation in pure culture assays (Ho et al. 2014). When growing not in direct cell contact heterotroph was have also been demonstrated to influence MB activity and growth. In the presence of a *Hyphomicrobium* species that was encapsulated in a dialyses membrane, positive effects on growth and activity was observed on MB growing in a bioreactor (Jeong and Kim 2019). The first mechanistic information on possible modes of action came from experiments were MB and heterotrophs were physically separated allowing only for interaction via the atmosphere (Veraart et al. 2018). MB growth was strongly stimulated, but also inhibited depending on the heterotrophic species by means of volatile compounds. Analyses of these volatiles pointed towards sulfur containing compounds like dimethylsulfide (DMS), dimethyldisulfide (DMDS) and dimethyltrisulfide (DMTS) but addition of pure compounds to MB cultures did not yield the observed effects (Veraart et al. 2018). *Methylobacter* species used in these experiments also turned out to produce volatile compounds themselves. The terpenes produced may enable them to communicate with other microbes (Veraart et al. 2018; Schulz-Bohm et al. 2017) and may play a role in the positive diversity effect observed in a biodiversity-functioning experiment carried out with MB (Schnyder et al. 2018). In species mixtures of up to 10 MB species a positive relationship between diversity and activity as found (Schnyder et al. 2018). This effect could not be explained by the traits available as well as by the phylogenetic diversity. A phenotypic property like production of secondary metabolites was not included in the analyses and may be responsible for these observations. Very recent work shows that MB can also communicate with each other by means of quorum sensing (Puri et al. 2019).

A very nice example of biotic interactions is the "manipulation" of MB gene expression by methylotrophs. Presence of methylotrophs in incubations led to the expression of the Ca dependent mxaF MDH, which has a lower conversion efficiency compared XoxF MDH, leading to methanol accumulation and exudation into the medium (Krause et al. 2017). In this way the methylotrophs are manipulating the MB to have a continuous supply of substrate.

4.3 Life Strategies

The previous sections have demonstrated that MB make a very unique case in microbial ecology. The fact that it is possible to link the utilization of their primary substrate with the identity of the active species makes it possible to measure microbial behavior, a fact necessary to understand community functioning when looking at its composition at any given time. The combination of stable isotope probing using ^{13}C labelled CH_4 incorporation into taxonomically relevant molecules (e.g. DNA, RNA, PLFA, proteins) has revealed which MB species are active under which specific conditions (e.g., Zheng et al. 2014; Bodelier et al. 2013; Daebeler et al. 2014; Cai et al. 2016). Of great value in this respect as also the fact that many molecular detection tools were available that were based on functional genes related the methane processing (McDonald et al. 2008). Using these techniques, we could also assess resistance and resilience under environmental stress and linking this to community composition and functioning of individual species. It was shown that frequency of stress (drying; Ho et al. 2016a, b), environmental contingency (Krause et al. 2018), time after ecosystem restoration (Levine et al. 2011; Reumer et al. 2018) as well as stress exposure history (van Kruistum et al. 2018) all influenced the recovery of methane oxidation from these stresses. Remarkable was that in all cases a specific part of the community was responsible for the recovery of the activity likely depending on the habitat. In acidic habitats often type II MB were responsible for recovery of activity while in high methane neutral environments it was type Ia species, mostly *Methylobacter* related species that recovered. Type Ib MB usually were the first to be affected by stress.

The wealth of MB behavioral studies in a broad range of habitats combined with the biogeographical data available, has led to put MB in a macro ecological framework, designated to typify plant communities on the basis of their integrated traits within a triangular framework classifying them as either competitors, stress tolerators or ruderals (see Fig. 7).

This framework used for plants has been proposed to be used for microbial ecology (Ho et al. 2013a, b, 2017a; Krause et al. 2014). The major advantage is that organisms are assigned life strategies that comprise all their traits necessary to be present and function in a particular environment rather than classifying on a single or a few traits like *pmoA* or methane oxidation alone. Community composition at any given moment is the result of filtering by habitat characteristics, gaining or losing species with a match or no fit with the environmental conditions or stresses exposed to (see Ho et al. 2017a). The crude primary “habitat filtering” mode of action (i.e. pH, temperature) can eliminate species completely while also filtering at a finer level (e.g. nutrients, methane, oxygen) can select leading to changes in relative

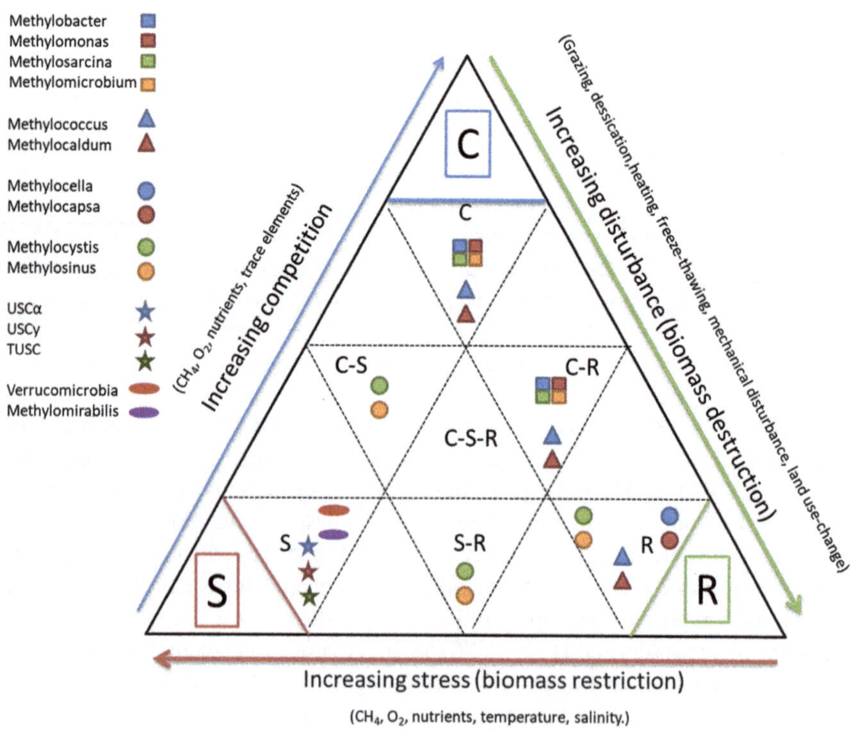

Fig. 7 Placement of known MB in the CSR life-strategy scheme. Adapted from Ho et al. (2013a, b)

abundance (e.g. type I vs Type II). This way of typifying MB communities may be a useful way of using community data in biogeochemical models by incorporating life strategies modules as representatives of aggregated functional traits of methane consuming or emitting habitats. The current developments in MB genomics (Orata et al. 2018; Singleton et al. 2018) and the ever growing range of novel isolates and biogeochemical knowledge will lead to a richer trait bases for MB. The challenge will be to integrate these data into approaches that will make it possible to use models generating data useful at policy level.

5 Synthesis and Outlook

Aerobic MB are incredibly important microbes in terms of our climate but also in our future sustainable society. Their role in greenhouse gas cycling and production of sustainable bio-based chemical is indisputable. For many years the knowledge on these microbes was lagging behind in microbiology. However, it is clear that the amount of novel organisms and even metabolism involved in methane cycling in the past decades has increased tremendously. We have a wealth of data on the environmental behavior of MB and in recent years also cultured MB are increasingly being sequenced yielding important novel insights into evolution, taxonomy, biochemistry of MB. Our view on methane oxidation as a monophyletic guild has been revised thoroughly and will be even more so in near future when more species outside of the currently known phyla will be able to show methane oxidation. The dichotomy of type I and II has been expanded to subdivisions and a type III group as has been turned into a phylogenomic framework which still has many ecological implications. As yet, most uncultivated *pmoA* lineages still are phylogenetically associated with phyla containing pure cultures of which genomes are becoming available. This is an ideal basis to use MB as model group in trait-based frameworks to use community composition and its inherent traits to predict environmental responses. These models are becoming available (Ebrahimi and Or 2018; Wieder et al. 2015) are waiting to be fed with digestible info maybe coming from the CSR framework applied on MB.

However, we also saw that many avenues in MB research are in the early stages. Especially, with respect to biotic interactions and how these interactions affect methane emission form natural environments is still a big challenge. At the same time, more researchers are starting to work with MB cultures which will yield novel and detailed mechanistic information needed to inform models but also to guide future environmental experiments. The degree of resistance or resilience of microbial communities to habitat fragmentation and climate are important topics for the future of our planet. Also, here MB communities may serve as models informing policy.

Nevertheless, the biggest challenge will be to integrate the stream of big data that will come from environmental genomics (e.g. Singleton et al. 2018) which will yield terra bases of genomic info. How to translate that into microbial traits will require verification will real life experiments with cultures in the lab. Getting "back to base" will be a challenge ahead for the coming decades.

Acknowledgements This publication is publication number 6717 of the Netherlands Institute of Ecology (NIOO-KNAW). This publication was supported by a grant of the Applied and Engineering Science division of the Netherlands Organization of Scientific Research (NWO-TTW) grant number 16475.

References

Arp J, Gotze S, Mukherji R, Mattern DJ, Garcia-Altares M, Klapper M, Brock DA, Brakhage AA, Strassmann JE, Queller DC, Bardl B, Willing K, Peschel G, Stallforth P (2018) Synergistic activity of cosecreted natural products from amoebae-associated bacteria. Proc Natl Acad Sci USA 115:3758–3763

Bao ZH, Okubo T, Kubota K, Kasahara Y, Tsurumaru H, Anda M, Ikeda S, Minamisawa K (2014) Metaproteomic identification of diazotrophic methanotrophs and their localization in root tissues of field-grown rice plants. Appl Environ Microbiol 80:5043–5052

Bastviken D, Cole J, Pace M, Tranvik L (2004) Methane emissions from lakes: dependence of lake characteristics, two regional assessments, and a global estimate. Global Biogeochem Cycles 18. https://doi.org/10.1029/2004GB002238

Bastviken D, Tranvik LJ, Downing JA, Crill PM, Enrich-Prast A (2011) Freshwater methane emissions offset the continental carbon sink. Science 331:50

Beck DAC, Kalyuzhnaya MG, Malfatti S, Tringe SG, Glavina Del Rio T, Ivanova N, Lidstrom ME, Chistoserdova L (2013) A metagenomic insight into freshwater methane-utilizing communities and evidence for cooperation between the Methylococcaceae and the Methylophilaceae. PeerJ 1:e23-e23

Bessette S, Moalic Y, Gautey S, Lesongeur F, Godfroy A, Toffin L (2017) Relative abundance and diversity of bacterial methanotrophs at the oxic-anoxic interface of the Congo deep-sea fan. Front Microbiol 8:715

Biderre-Petit C, Jezequel D, Dugat-Bony E, Lopes F, Kuever J, Borrel G, Viollier E, Fonty G, Peyret P (2011) Identification of microbial communities involved in the methane cycle of a freshwater meromictic lake. FEMS Microbiol Ecol 77:533–545

Blees J, Niemann H, Wenk CB, Zopfi J, Schubert CJ, Kirf MK, Veronesi ML, Hitz C, Lehmann MF (2014) Micro-aerobic bacterial methane oxidation in the chemocline and anoxic water column of deep south-Alpine Lake Lugano (Switzerland). Limnol Oceanogr 59:311–324

Bodelier PLE (2011) Interactions between nitrogenous fertilizers and methane cycling in wetland and upland soils. Curr Opin Environ Sustain 3:379–388

Bodelier PLE, Frenzel P (1999) Contribution of methanotrophic and nitrifying bacteria to CH_4 and NH_4^+ oxidation in the rhizosphere of rice plants as determined by new methods of discrimination. Appl Environ Microbiol 65:1826–1833

Bodelier PLE, Laanbroek HJ (2004) Nitrogen as a regulatory factor of methane oxidation in soils and sediments. FEMS Microbiol Ecol 47:265–277

Bodelier PLE, Steenbergh AK (2014a) Interactions between methane and the nitrogen cycle in light of climate change. Curr Opin Environ Sustain 9–10:26–36

Bodelier PLE, Steenbergh AK (2014b) Interactions between methane and nitrogen cycling; current metagenomic studies and future trends. In: Marco D (ed) Metagenomics of the microbial nitrogen cycle: theory, methods and applications. Caister Academic Press, Norfolk, pp 33–63

Bodelier PLE, Roslev P, Henckel T, Frenzel P (2000) Stimulation by ammonium-based fertilizers of methane oxidation in soil around rice roots. Nature 403:421–424

Bodelier PLE, Gillisen MJB, Hordijk K, Damste JSS, Rijpstra WIC, Geenevasen JA, Dunfield PF (2009) A reanalysis of phospholipid fatty acids as ecological biomarkers for methanotrophic bacteria. ISME J 3:606–617

Bodelier PLE, Meima-Franke M, Hordijk CA, Steenbergh AK, Hefting MM, Bodrossy L, von Bergen M, Seifert J (2013) Microbial minorities modulate methane consumption through niche partitioning. ISME J 7:2214–2228

Boschker HTS, Nold SC, Wellsbury P, Bos D, de Graaf W, Pel R, Parkes RJ, Cappenberg TE (1998) Direct linking of microbial populations to specific biogeochemical processes by C-13-labelling of biomarkers. Nature 392:801–805

Bowman JP, Sly LI, Nichols PD, Hayward AC (1993) Revised taxonomy of the methanotrophs – description of methylobacter gen-nov, emendation of methylococcus, validation of methylosinus and methylocystis species, and a proposal that the family methylococcaceae includes only the group-i methanotrophs. Int J Syst Bacteriol 43:735–753

Bridgham SD, Cadillo-Quiroz H, Keller JK, Zhuang QL (2013) Methane emissions from wetlands: biogeochemical, microbial, and modeling perspectives from local to global scales. Glob Change Biol 19:1325–1346

Bussmann I (2005) Methane release through resuspension of littoral sediment. Biogeochemistry 74:283–302

Cai YF, Zheng Y, Bodelier PLE, Conrad R, Jia ZJ (2016) Conventional methanotrophs are responsible for atmospheric methane oxidation in paddy soils. Nat Commun 7:11728

Cappenberg TE (1974) Interrelations between sulfate-reducing and methane-producing bacteria in bottom deposits of a fresh-water lake. II. Inhibition experiments. Antonie van Leeuwenhoek 40:297–306

Carini S, Bano N, LeCleir G, Joye SB (2005) Aerobic methane oxidation and methanotroph community composition during seasonal stratification in Mono Lake, California (USA). Environ Microbiol 7:1127–1138

Chang J, Gu WY, Park D, Semrau JD, DiSpirito AA, Yoon S (2018) Methanobactin from methylosinus trichosporium OB3b inhibits N_2O reduction in denitrifiers. ISME J 12:2086–2089

Chistoserdova L (2016) Lanthanides: new life metals? World J Microbiol Biotechnol 32:138

Chistoserdova L, Kalyuzhnaya MG (2018) Current trends in methylotrophy. Trends Microbiol 26:703–714

Chowdhury TR, Dick RP (2013) Ecology of aerobic methanotrophs in controlling methane fluxes from wetlands. Appl Soil Ecol 65:8–22

Chronopoulou PM, Shelley F, Pritchard WJ, Maanoja ST, Trimmer M (2017) Origin and fate of methane in the Eastern Tropical North Pacific oxygen minimum zone. ISME J 11:1386–1399

Conrad R (1996) Soil microorganisms as controllers of atmospheric trace gases (H_2, CO, CH_4, OCS, N_2O, and NO). Microbiol Rev 60:609

Conrad R (2007) Microbial ecology of methanogens and methanotrophs. Adv Agron 96:1–63

Crevecoeur SVW, Comte J, Matveev A, Lovejoy C (2017) Diversity and potential activity of methanotrophs in high methane-emitting permafrost thaw ponds. PLoS One 12:e0188223

Crombie AT, Murrell JC (2014) Trace-gas metabolic versatility of the facultative methanotroph Methylocella silvestris. Nature 510:148

Daebeler A, Bodelier PLE, Yan Z, Hefting MM, Jia ZJ, Laanbroek HJ (2014) Interactions between Thaumarchaea, Nitrospira and methanotrophs modulate autotrophic nitrification in volcanic grassland soil. ISME J 8:2397–2410

Daims H, Lebedeva EV, Pjevac P, Han P, Herbold C, Albertsen M, Jehmlich N, Palatinszky M, Vierheilig J, Bulaev A, Kirkegaard RH, von Bergen M, Rattei T, Bendinger B, Nielsen PH, Wagner M (2015) Complete nitrification by Nitrospira bacteria. Nature 528:504

Dam B, Dam S, Blom J, Liesack W (2013) Genome analysis coupled with physiological studies reveals a diverse nitrogen metabolism in methylocystis sp. strain SC2. PLoS One 8:e74767

Danilova OV, Suzina NE, Van De Kamp J, Svenning MM, Bodrossy L, Dedysh SN (2016) A new cell morphotype among methane oxidizers: a spiral-shaped obligately microaerophilic methanotroph from northern low-oxygen environments. ISME J 10:2734–2743

Dean JF, Middelburg JJ, Rockmann T, Aerts R, Blauw LG, Egger M, Jetten MS, Jong AE, Meisel OH, Rasigraf O (2018) Methane feedbacks to the global climate system in a warmer world. Rev Geophys 56:207–250

Dedysh SN (2009) Exploring methanotroph diversity in acidic northern wetlands: molecular and cultivation-based studies. Microbiology 78:655–669

Dedysh SN, Knief C (2018) Diversity and phylogeny of described aerobic methanotrophs. In: Kalyuzhnaya MG, Xing XH (eds) Methane biocatalysis: paving the way to sustainability. Springer, Cham, pp 17–42

Dedysh SN, Belova SE, Bodelier PLE, Smirnova KV, Khmelenina VN, Chidthaisong A, Trotsenko YA, Liesack W, Dunfield PF (2007) Methylocystis heyeri sp nov., a novel type II methanotrophic bacterium possessing 'signature' fatty acids of type I methanotrophs. Int J Syst Evol Microbiol 57:472–479

DelSontro T, del Giorgio PA, Prairie YT (2017) No longer a paradox: the interaction between physical transport and biological processes explains the spatial distribution of surface water methane within and across lakes. Ecosystems 21:1073–1087

DiSpirito AA, Semrau JD, Murrell JC, Gallagher WH, Dennison C, Vuilleumier S (2016) Methanobactin and the link between copper and bacterial methane oxidation. Microbiol Mol Biol Rev 80:387–409

Dunfield PF (2007) The soil methane sink. In: Reay DS, Hewitt CN, Smith KA, Grace J (eds) Greenhouse gas sinks. CABI, Wallingford, pp 152–170

Dunfield PF, Dedysh SN (2014) Methylocella: a gourmand among methanotrophs. Trends Microbiol 22:368–369

Dunfield PF, Yuryev A, Senin P, Smirnova AV, Stott MB, Hou S, Ly B, Saw JH, Zhou Z, Ren Y (2007) Methane oxidation by an extremely acidophilic bacterium of the phylum Verrucomicrobia. Nature 450:879-U18

Durisch-Kaiser E, Klauser L, Wehrli B, Schubert C (2005) Evidence of intense archaeal and bacterial methanotrophic activity in the Black Sea water column. Appl Environ Microbiol 71:8099–8106

Ebrahimi A, Or D (2018) On upscaling of soil microbial processes and biogeochemical fluxes from aggregates to landscapes. J Geophys Res Biogeosci 123:1526–1547

Edwards CROT, Miller JM, Wiggins JB, Wang W, Lee CK, Cary SC, Pointing SB, Lau MCY (2017) Draft genome sequence of uncultured upland soil cluster Gammaproteobacteria gives molecular insights into high-affinity methanotrophy. Genome Announc 5:e00047-17

Eller G, Känel L, Krüger M (2005) Cooccurrence of aerobic and anaerobic methane oxidation in the water column of lake Plußsee. Appl Environ Microbiol 71:8925–8928

Ettwig KF, Butler MK, Le Paslier D, Pelletier E, Mangenot S, Kuypers MM, Schreiber F, Dutilh BE, Zedelius J, de Beer D, Gloerich J, Wessels HJ, van Alen T, Luesken F, Wu ML, van de Pas-Schoonen KT, Op den Camp HJ, Janssen-Megens EM, Francoijs KJ, Stunnenberg H, Weissenbach J, Jetten MS, Strous M (2010) Nitrite-driven anaerobic methane oxidation by oxygenic bacteria. Nature 464:543–548

Farhan Ul Haque M, Crombie AT, Ensminger SA, Baciu C, Murrell JC (2018) Facultative methanotrophs are abundant at terrestrial natural gas seeps. Microbiome 6:118

Fest B, Hinko-Najera N, von Fischer JC, Livesley SJ, Arndt SK (2017) Soil methane uptake increases under continuous throughfall reduction in a temperate evergreen, Broadleaved Eucalypt Forest. Ecosystems 20:368–379

Graf JS, Mayr MJ, Marchant HK, Tienken D, Hach PF, Brand A, Schubert CJ, Kuypers MMM, Milucka J (2018) Bloom of a denitrifying methanotroph, 'Candidatus Methylomirabilis limnetica', in a deep stratified lake. Environ Microbiol 20:2598–2614

Gray ND, McCann CM, Christgen B, Ahammad SZ, Roberts JA, Graham DW (2014) Soil geochemistry confines microbial abundances across an arctic landscape; implications for net carbon exchange with the atmosphere. Biogeochemistry 120:307–317

Grossart HP, Frindte K, Dziallas C, Eckert W, Tang KW (2011) Microbial methane production in oxygenated water column of an oligotrophic lake. Proc Natl Acad Sci USA 108:19657–19661

Guerrero-Cruz S, Cremers G, van Alen TA, den Camp H, Jetten MSM, Rasigraf O, Vaksmaa A (2018) Response of the anaerobic methanotroph "Candidatus Methanoperedens nitroreducens" to oxygen stress. Appl Environ Microbiol 84:17

Hanson RS, Hanson TE (1996) Methanotrophic bacteria. Microbiol Rev 60:439

Haroon MF, Hu S, Shi Y, Imelfort M, Keller J, Hugenholtz P, Yuan Z, Tyson GW (2013) Anaerobic oxidation of methane coupled to nitrate reduction in a novel archaeal lineage. Nature 500:567

He D, Ren L, Wu Q (2012) Epiphytic bacterial communities on two common submerged macrophytes in Taihu Lake: diversity and host-specificity. Chin J Oceanol Limnol 30:237–247

He ZF, Wang JQ, Hu JJ, Yu HG, Jetten MSM, Liu H, Ren H, Zhang X, Hua M, Xu X, Zheng P, Hu B (2019) Regulation of coastal methane sinks by a structured gradient of microbial methane oxidizers. Environ Pollut 244:228–237

Hernandez ME, Beck DAC, Lidstrom ME, Chistoserdova L (2015) Oxygen availability is a major factor in determining the composition of microbial communities involved in methane oxidation. PeerJ 3:13

Heyer J, Galchenko VF, Dunfield PF (2002) Molecular phylogeny of type II methane-oxidizing bacteria isolated from various environments. Microbiology 148:2831–2846

Ho A, Bodelier PLE (2015) Diazotrophic methanotrophs in peatlands: the missing link? Plant Soil 389:419–423

Ho A, Erens H, Mujinya BB, Boeckx P, Baert G, Schneider B, Frenzel P, Boon N, Van Ranst E (2013a) Termites facilitate methane oxidation and shape the methanotrophic community. Appl Environ Microbiol 79:7234–7240

Ho A, Kerckhof FM, Luke C, Reim A, Krause S, Boon N, Bodelier PL (2013b) Conceptualizing functional traits and ecological characteristics of methane-oxidizing bacteria as life strategies. Environ Microbiol Rep 5:335–345

Ho ADRK, Thas O, De Neve J, Hoefman S, Vandamme P, Heylen K, Boon N (2014) The more, the merrier: heterotroph richness stimulates methanotrophic activity. ISME J 8:1945–1948

Ho A, Reim A, Kim SY, Meima-Franke M, Termorshuizen A, De Boer W, van der Putten WH, Bodelier PL (2015) Unexpected stimulation of soil methane uptake as emergent property of agricultural soils following bio-based residue application. Glob Change Biol 21:3864–3879

Ho A, Angel R, Veraart AJ, Daebeler A, Jia ZJ, Kim SY, Kerckhof FM, Boon N, Bodelier PL (2016a) Biotic interactions in microbial communities as modulators of biogeochemical processes: methanotrophy as a model system. Front Microbiol 7:1285

Ho A, van den Brink E, Reim A, Krause SMB, Bodelier PLE (2016b) Recurrence and frequency of disturbance have cumulative effect on methanotrophic activity, abundance, and community structure. Front Microbiol 6:1493

Ho A, Di Lonardo DP, Bodelier PLE (2017a) Revisiting life strategy concepts in environmental microbial ecology. FEMS Microbiol Ecol 93. https://doi.org/10.1093/femsec/fix006

Ho A, Ijaz UZ, Janssens TKS, Ruijs R, Kim SY, de Boer W, Termorshuizen A, Putten WH, Bodelier PL (2017b) Effects of bio-based residue amendments on greenhouse gas emission from agricultural soil are stronger than effects of soil type with different microbial community composition. Glob Change Biol Bioenergy 9:1707–1720

Ho A, Lee HJ, Reumer M, Meima-Franke M, Raaijmakers C, Zweers H, de Boer W, Van der Putten WH, Bodelier PLE (2019) Unexpected role of canonical aerobic methanotrophs in upland agricultural soils. Soil Biol Biochem 131:1–8

Iguchi H, Yurimoto H, Sakai Y (2011) Stimulation of methanotrophic growth in cocultures by cobalamin excreted by rhizobia. Appl Environ Microbiol 77:8509–8515

Ikeda S, Sasaki K, Okubo T, Yamashita A, Terasawa K, Bao Z, Liu D, Watanabe T, Murase J, Asakawa S, Eda S, Mitsui H, Sato T, Minamisawa K (2014) Low nitrogen fertilization adapts rice root microbiome to low nutrient environment by changing biogeochemical functions. Microbes Environ 29:50–59

in 't Zandt MH, de Jong AEE, Slomp CP, Jetten MSM (2018) The hunt for the most-wanted chemolithoautotrophic spookmicrobes. FEMS Microbiol Ecol 94. https://doi.org/10.1093/femsec/fiy064

IPCC (2014) Climate change 2013: the physical science basis: Working Group I contribution to the Fifth assessment report of the Intergovernmental Panel on Climate Change. Cambridge University Press, Cambridge

Jeong SY, Kim TG (2019) Development of a novel methanotrophic process with the helper microorganism Hyphomicrobium sp. NM3. J Appl Microbiol 126:534–544

Kalyuzhnaya MG, Yang S, Rozova ON, Smalley NE, Clubb J, Lamb A, Gowda GA, Raftery D, Fu Y, Bringel F, Vuilleumier S, Beck DA, Trotsenko YA, Khmelenina VN, Lidstrom ME (2013) Highly efficient methane biocatalysis revealed in a methanotrophic bacterium. Nat Commun 4:2785

Karl DM, Beversdorf L, Björkman KM, Church MJ, Martinez A, Delong EF (2008) Aerobic production of methane in the sea. Nat Geosci 1:473–478

Keltjens JT, Pol A, Reimann J, Op den Camp HJM (2014) PQQ-dependent methanol dehydrogenases: rare-earth elements make a difference. Appl Microbiol Biotechnol 98:6163–6183

Khadem AF, Pol A, Wieczorek A, Mohammadi SS, Francoijs KJ, Stunnenberg HG, Jetten MS, Op den Camp HJ (2011) Autotrophic methanotrophy in verrucomicrobia: methylacidiphilum fumariolicum SolV uses the Calvin-Benson-Bassham cycle for carbon dioxide fixation. J Bacteriol 193:4438–4446

Khadka R, Clothier L, Wang L, Lim CK, Klotz MG, Dunfield PF (2018) Evolutionary history of copper membrane monooxygenases. Front Microbiol 9:2493

Kip N, van Winden JF, Pan Y, Bodrossy L, Reichart G-J, Smolders AJP, Jetten MSM, Damsté JSS, Op den Camp HJM (2010) Global prevalence of methane oxidation by symbiotic bacteria in peat-moss ecosystems. Nat Geosci 3:617–621

Kirschke S, Bousquet P, Ciais P, Saunois M, Canadell JG, Dlugokencky EJ, Bergamaschi P, Bergmann D, Blake DR, Bruhwiler L, Cameron-Smith P, Castaldi S, Chevallier F, Feng L, Fraser A, Heimann M, Hodson EL, Houweling S, Josse B, Fraser PJ, Krummel PB, Lamarque J-F, Langenfelds RL, Le Quéré C, Naik V, O'Doherty S, Palmer PI, Pison I, Plummer D, Poulter B, Prinn RG, Rigby M, Ringeval B, Santini M, Schmidt M, Shindell DT, Simpson IJ, Spahni R, Steele LP, Strode SA, Sudo K, Szopa S, van der Werf GR, Voulgarakis A, van Weele M, Weiss RF, Williams JE, Zeng G (2013) Three decades of global methane sources and sinks. Nat Geosci 6:813–823

Kits KD, Klotz MG, Stein LY (2015) Methane oxidation coupled to nitrate reduction under hypoxia by the Gammaproteobacterium Methylomonas denitrificans, sp. nov. type strain FJG1. Environ Microbiol 17:3219–3232

Kits KDSC, Lebedeva EV, Han P, Bulaev A, Pjevac P, Daebeler A, Romano S, Albertsen M, Stein LY, Daims H, Wagner M (2017) Kinetic analysis of a complete nitrifier reveals an oligotrophic lifestyle. Nature 549:269–272

Knief C (2015) Diversity and habitat preferences of cultivated and uncultivated aerobic methanotrophic bacteria evaluated based on pmoA as molecular marker. Front Microbiol 6:1346

Knief C, Dunfield PF (2005) Response and adaptation of different methanotrophic bacteria to low methane mixing ratios. Environ Microbiol 7:1307–1317

Kojima H, Iwata T, Fukui M (2009) DNA-based analysis of planktonic methanotrophs in a stratified lake. Freshwater Biol 54:1501–1509

Kojima H, Tokizawa R, Kogure K, Kobayashi Y, Itoh M, Shiah FK, Okuda N, Fukui M (2014) Community structure of planktonic methane-oxidizing bacteria in a subtropical reservoir characterized by dominance of phylotype closely related to nitrite reducer. Sci Rep 4:5728

Kolb S (2009) The quest for atmospheric methane oxidizers in forest soils. Environ Microbiol Rep 1:336–346

Kolb S, Horn MA (2012) Microbial CH_4 and N_2O consumption in acidic wetlands. Front Microbiol 3:78

Kox MAR, Aalto SL, Penttila T, Ettwig KF, Jetten MSM, van Kessel M (2018) The influence of oxygen and methane on nitrogen fixation in subarctic Sphagnum mosses. AMB Exp 8:76

Krause S, Meima-Franke M, Hefting MM, Bodelier PLE (2013) Spatial patterns of methanotrophic communities along a hydrological gradient in a riparian wetland. FEMS Microbiol Ecol 86:59–70

Krause S, Le Roux X, Niklaus PA, Van Bodegom PM, Lennon JT, Bertilsson S, Grossart H-P, Philippot L, Bodelier PLE (2014) Trait-based approaches for understanding microbial biodiversity and ecosystem functioning. Front Microbiol 5:10

Krause SMB, Johnson T, Karunaratne YS, Fu YF, Beck DAC, Chistoserdova L, Lidstrom ME (2017) Lanthanide-dependent cross-feeding of methane-derived carbon is linked by microbial community interactions. Proce Natl Acad Sci USA 114:358–363

Krause SMB, Meima-Franke M, Veraart AJ, Ren GD, Ho A, Bodelier PLE (2018) Environmental legacy contributes to the resilience of methane consumption in a laboratory microcosm system. Sci Rep 8:8862

Kruger M, Eller G, Conrad R, Frenzel P (2002) Seasonal variation in pathways of CH_4 production and in CH_4 oxidation in rice fields determined by stable carbon isotopes and specific inhibitors. Glob Change Biol 8:265–280

Krukenberg V, Riedel D, Gruber-Vodicka HR, Buttigieg PL, Tegetmeyer HE, Boetius A, Wegener G (2018) Gene expression and ultrastructure of meso- and thermophilic methanotrophic consortia. Environ Microbiol 20:1651–1666

Kuypers MMM, Marchant HK, Kartal B (2018) The microbial nitrogen-cycling network. Nat Rev Microbiol 16:263–276

Larmola T, Tuittila E-S, Tiirola M, Nykänen H, Martikainen PJ, Yrjälä K, Tuomivirta T, Fritze H (2010) The role of Sphagnum mosses in the methane cycling of a boreal mire. Ecology 91:2356–2365

Larmola T, Leppanen SM, Tuittila ES, Aarva M, Merila P, Fritze H, Tiirola M (2014) Methanotrophy induces nitrogen fixation during peatland development. Proc Natl Acad Sci USA 111:734–739

Lau MCY, Stackhouse BT, Layton AC, Chauhan A, Vishnivetskaya TA, Chourey K, Ronholm J, Mykytczuk NC, Bennett PC, Lamarche-Gagnon G, Burton N, Pollard WH, Omelon CR, Medvigy DM, Hettich RL, Pfiffner SM, Whyte LG, Onstott TC (2015) An active atmospheric methane sink in high Arctic mineral cryosols. ISME J 9:1880–1891

Lenhart K, Bunge M, Ratering S, Neu TR, Schuttmann I, Greule M, Kammann C, Schnell S, Müller C, Zorn H, Keppler F (2012) Evidence for methane production by saprotrophic fungi. Nat Commun 3:1046

Levine UY, Teal TK, Robertson GP, Schmidt TM (2011) Agriculture's impact on microbial diversity and associated fluxes of carbon dioxide and methane. ISME J 5:1683–1691

Makipaa R, Leppanen SM, Munoz SS, Smolander A, Tiirola M, Tuomivirta T, Fritze H (2018) Methanotrophs are core members of the community in decayingdiazotroph Norway spruce logs. Soil Biol Biochem 120:230–232

Maurer D, Kolb S, Haumaier L, Borken W (2008) Inhibition of atmospheric methane oxidation by monoterpenes in Norway spruce and European beech soils. Soil Biol Biochem 40:3014–3020

McDonald IR, Bodrossy L, Chen Y, Murrell JC (2008) Molecular ecology techniques for the study of aerobic methanotrophs. Appl Environ Microbiol 74:1305–1315

McGinnis DF, Flury S, Tang KW, Grossart HP (2017) Porewater methane transport within the gas vesicles of diurnally migrating Chaoborus spp.: an energetic advantage. Sci Rep 7:44478

McGlynn SE, Chadwick GL, O'Neill A, Mackey M, Thor A, Deerinck TJ, Ellisman MH, Orphan VJ (2018) Subgroup characteristics of marine methane-oxidizing ANME-2 archaea and their syntrophic partners as revealed by integrated multimodal analytical microscopy. Appl Environ Microbiol 84:e00399-18

Menyailo OV, Abraham W-R, Conrad R (2010) Tree species affect atmospheric CH_4 oxidation without altering community composition of soil methanotrophs. Soil Biol Biochem 42:101–107

Michaud AB, Dore JE, Achberger AM, Christner BC, Mitchell AC, Skidmore ML, Vick-Majors TJ, Priscu JC (2017) Microbial oxidation as a methane sink beneath the West Antarctic Ice Sheet. Nat Geosci 10:582

Milucka J, Kirf M, Lu L, Krupke A, Lam P, Littmann S, Kuypers MM, Schubert CJ (2015) Methane oxidation coupled to oxygenic photosynthesis in anoxic waters. ISME J 9:1991
Naguib M, Overbeck J (1970) On methane oxidizing bacteria in fresh waters. I. Introduction to the problem and investigations on the presence of obligate methane oxidizers. Zeitschrift für allgemeine Mikrobiologie 10:17–36
Naqvi SWA, Lam P, Narvenkar G, Sarkar A, Naik H, Pratihary A, Shenoy DM, Gauns M, Kurian S, Damare S, Duret M, Lavik G, Kuypers MMM (2018) Methane stimulates massive nitrogen loss from freshwater reservoirs in India. Nat Commun 9:1265
Nauer PA, Dam B, Liesack W, Zeyer J, Schroth MH (2012) Activity and diversity of methane-oxidizing bacteria in glacier forefields on siliceous and calcareous bedrock. Biogeosciences 9:2259–2274
Nauer PA, Hutley LB, Arndt SK (2018) Termite mounds mitigate half of termite methane emissions. Proc Natl Acad Sci USA 115:13306–13311
Nazaries L, Tate KR, Ross DJ, Singh J, Dando J, Saggar S, Baggs EM, Millard P, Murrell JC, Singh BK (2011) Response of methanotrophic communities to afforestation and reforestation in New Zealand. ISME J 5:1832–1836
Nazaries L, Karunaratne SB, Delgado-Baquerizo M, Campbell CD, Singh BK (2018) Environmental drivers of the geographical distribution of methanotrophs: insights from a national survey. Soil Biol Biochem 127:264–279
Nguyen AD, Hwang IY, Lee OK, Hur DH, Jeon YC, Hadiyati S, Kim MS, Yoon SH, Jeong H, Lee EY (2018) Functional analysis of methylomonas sp. DH-1 genome as a promising biocatalyst for bioconversion of methane to valuable chemicals. Catalysts 8:117
Nisbet EG, Dlugokencky EJ, Bousquet P (2014) Methane on the rise-again. Science 343:493–495
Oliveira Junior ES, Temmink RJM, Buhler BF, Souza RM, Resende N, Spanings T, Muniz CC, Lamers LPM, Kosten S (2019) Benthivorous fish bioturbation reduces methane emissions, but increases total greenhouse gas emissions. Freshw Biol 64:197–207
Orata FD, Kits KD, Stein LY (2018) Complete genome sequence of methylomonas denitrificans strain FJG1, an obligate aerobic methanotroph that can couple methane oxidation with denitrification. Microbiol Resour Announ 6. https://doi.org/10.1128/genomeA.00276-18
Orellana LH, Chee-Sanford JC, Sanford RA, Loffler FE, Konstantinidis KT (2018) Year-round shotgun metagenomes reveal stable microbial communities in agricultural soils and novel ammonia oxidizers responding to fertilization. Appl Environ Microbiol 84. https://doi.org/10.1128/AEM.01646-17
Osborne CD, Haritos VS (2018) Horizontal gene transfer of three co-inherited methane monooxygenase systems gave rise to methanotrophy in the proteobacteria. Mol Phylogenet Evol 129:171–181
Oshkin IY, Beck DAC, Lamb AE, Tchesnokova V, Benuska G, McTaggart TL, Kalyuzhnaya MG, Dedysh SN, Lidstrom ME, Chistoserdova L (2015) Methane-fed microbial microcosms show differential community dynamics and pinpoint taxa involved in communal response. ISME J 9:1119–1129
Oswald K, Milucka J, Brand A, Littmann S, Wehrli B, Kuypers MM, Schubert CJ (2015) Light-dependent aerobic methane oxidation reduces methane emissions from seasonally stratified lakes. PLoS One 10:e0132574
Oswald K, Milucka J, Brand A, Hach P, Littmann S, Wehrli B, Kuypers MMM, Schubert CJ (2016) Aerobic gammaproteobacterial methanotrophs mitigate methane emissions from oxic and anoxic lake waters. Limnol Oceanogr 61:S101–SS18
Oswald K, Graf JS, Littmann S, Tienken D, Brand A, Wehrli B, Albertsen M, Daims H, Wagner M, Kuypers MM, Schubert CJ, Milucka J (2017) Crenothrix are major methane consumers in stratified lakes. ISME J 11:2124–2140
Parks DH, Chuvochina M, Waite DW, Rinke C, Skarshewski A, Chaumeil PA, Hugenholtz P (2018) A standardized bacterial taxonomy based on genome phylogeny substantially revises the tree of life. Nat Biotechnol 36:996

Pieja AJ, Morse MC, Cal AJ (2017) Methane to bioproducts: the future of the bioeconomy? Curr Opin Chem Biol 41:123–131

Pol A, Heijmans K, Harhangi HR, Tedesco D, Jetten MSM, den Camp H (2007) Methanotrophy below pH1 by a new Verrucomicrobia species. Nature 450:874–U17

Pol A, Barends TRM, Dietl A, Khadem AF, Eygensteyn J, Jetten MS, Op den Camp HJ (2014) Rare earth metals are essential for methanotrophic life in volcanic mudpots. Environ Microbiol 16:255–264

Ponnudurai R, Kleiner M, Sayavedra L, Petersen JM, Moche M, Otto A, Becher D, Takeuchi T, Satoh N, Dubilier N, Schweder T, Markert S (2017) Metabolic and physiological interdependencies in the Bathymodiolus azoricus symbiosis. ISME J 11:463–477

Pratscher J, Dumont MG, Conrad R (2011) Assimilation of acetate by the putative atmospheric methane oxidizers belonging to the USC alpha clade. Environ Microbiol 13:2692–2701

Pratscher J, Vollmers J, Wiegand S, Dumont MG, Kaster AK (2018) Unravelling the identity, metabolic potential and global biogeography of the atmospheric methane-oxidizing upland soil cluster alpha. Environ Microbiol 20:1016–1029

Puri AW, LiuD SAL, Yu Z, Pesesky MW, Greenberg PE, Lidstrom ME (2019) Interspecies chemical signaling in a methane-oxidizing bacterial community. Appl Environ Microbiol 85 (7):e02702-18

R Development Core Team (2008) R: a language and environment for statistical computing. R Foundation for Statistical Computing, Vienna, Austria. ISBN: 3-900051-07-0, http://www.R-project.org

Raghoebarsing AA, Smolders AJP, Schmid MC, Rijpstra WIC, Wolters-Arts M, Derksen J, Jetten MS, Schouten S, Sinninghe Damsté JS, Lamers LP, Roelofs JG, Op den Camp HJ, Strous M (2005) Methanotrophic symbionts provide carbon for photosynthesis in peat bogs. Nature 436:1153–1156

Rahman MT, Crombie A, Chen Y, Stralis-Pavese N, Bodrossy L, Meir P, McNamara NP, Murrell JC (2010) Environmental distribution and abundance of the facultative methanotroph Methylocella. ISME J 5(6):1061

Reay DS, Smith P, Christensen TR, James RH, Clark H (2018) Methane and global environmental change. Annu Rev Environ Resour 43:165–192

Reeburgh WS, Heggie DT (1977) Microbial methane consumption reactions and their effect on methane distributions in freshwater and marine environments. Limnol Oceanogr 22:1–9

Reim A, Luke C, Krause S, Pratscher J, Frenzel P (2012) One millimetre makes the difference: high-resolution analysis of methane-oxidizing bacteria and their specific activity at the oxic-anoxic interface in a flooded paddy soil. ISME J 6:2128–2139

Reumer M, Harnisz M, Lee HJ, Reim A, Grunert O, Putkinen A, Fritze H, Bodelier PLE, Ho A (2018) Impact of peat mining and restoration on methane turnover potential and methane-cycling microorganisms in a Northern bog. Appl Environ Microbiol 84:e02218-17

Rissanen AJ, Saarenheimo J, Tiirola M, Peura S, Aalto SL, Karvinen A, Nykänen H (2018) Gammaproteobacterial methanotrophs dominate methanotrophy in aerobic and anaerobic layers of boreal lake waters. Aquat Microb Ecol 81:257–276

Roland FAE, Darchambeau F, Morana C, Bouillon S, Borges AV (2017) Emission and oxidation of methane in a meromictic, eutrophic and temperate lake (Dendre, Belgium). Chemosphere 168:756–764

Rubin-Blum M, Antony CP, Borowski C, Sayavedra L, Pape T, Sahling H, Bohrmann G, Kleiner M, Redmond MC, Valentine DL, Dubilier N (2017) Short-chain alkanes fuel mussel and sponge Cycloclasticus symbionts from deep-sea gas and oil seeps. Nat Microbiol 2:17093

Ruff SE, Arnds J, Knittel K, Amann R, Wegener G, Ramette A, Boetius A (2013) Microbial communities of deep-sea methane seeps at Hikurangi continental margin (New Zealand). PLoS One 8:e72627

Ruff SE, Felden J, Gruber-Vodicka HR, Marcon Y, Knittel K, Ramette A, Boetius A (2019) In situ development of a methanotrophic microbiome in deep-sea sediments. ISME J 13:197–213

Saunois M, Bousquet P, Poulter B, Peregon A, Ciais P, Canadell JG, Dlugokencky EJ, Etiope G, Bastviken D, Houweling S, Janssens-Maenhout G, Tubiello FN, Castaldi S, Jackson RB, Alexe M, Arora VK, Beerling DJ, Bergamaschi P, Blake DR, Brailsford G, Brovkin V, Bruhwiler L, Crevoisier C, Crill P, Covey K, Curry C, Frankenberg C, Gedney N, Höglund-Isaksson L, Ishizawa M, Ito A, Joos F, Kim HS, Kleinen T, Krummel P, Lamarque JF, Langenfelds R, Locatelli R, Machida T, Maksyutov S, McDonald KC, Marshall J, Melton JR, Morino I, Naik V, O'Doherty S, Parmentier FJW, Patra PK, Peng C, Peng S et al (2016) The global methane budget 2000–2012. Earth Syst Sci Data 8:697–751

Schnyder E, Bodelier PLE, Hartmann M, Henneberger R, Niklaus PA (2018) Positive diversity-functioning relationships in model communities of methanotrophic bacteria. Ecology 99:714–723

Schubert CJ, Vazquez F, Losekann-Behrens T, Knittel K, Tonolla M, Boetius A (2011) Evidence for anaerobic oxidation of methane in sediments of a freshwater system (Lago di Cadagno). FEMS Microbiol Ecol 76:26–38

Schulz-Bohm K, Martin-Sanchez L, Garbeva P (2017) Microbial volatiles: small molecules with an important role in infra- and inter-kingdom interactions. Front Microbiol 8:2484

Segers R (1998) Methane production and methane consumption: a review of processes underlying wetland methane fluxes. Biogeochemistry 41:23–51

Semrau JD (2018) Metals and methanotrophy. Appl Environ Microbiol 84:e02289-17

Semrau JD, DiSpirito AA, Yoon S (2010) Methanotrophs and copper. FEMS Microbiol Rev 34:496–531

Sharp CE, Smirnova AV, Graham JM, Stott MB, Khadka R, Moore TR, Grasby SE, Strack M, Dunfield PF (2014) Distribution and diversity of Verrucomicrobia methanotrophs in geothermal and acidic environments. Environ Microbiol 16:1867–1878

Singleton CM, McCalley CK, Woodcroft BJ, Boyd JA, Evans PN, Hodgkins SB, Chanton JP, Frolking S, Crill PM, Saleska SR, Rich VI, Tyson GW (2018) Methanotrophy across a natural permafrost thaw environment. ISME J 12:2544–2558

Smith GJ, Angle JC, Solden LM, Borton MA, Morin TH, Daly RA, Johnston MD, Stefanik KC, Wolfe R, Gil B, Wrighton KC (2018) Members of the genus methylobacter are inferred to account for the majority of aerobic methane oxidation in oxic soils from a freshwater wetland. Mbio 9. https://doi.org/10.1128/mBio.00815-18

Söhngen NL (1906) Ueber Bakterien, welche Methan als Kohlenstoffnahrung und Energiequelle gebrauchen. Centralbl Bakteriol Parasitenk Infektionskr Hyg Abt II 15:513–517

Stock M, Hoefman S, Kerckhof FM, Boon N, De Vos P, De Baets B, Heylen K, Waegeman W (2013) Exploration and prediction of interactions between methanotrophs and heterotrophs. Res Microbiol 164:1045–1054

Stoecker K, Bendinger B, Schoning B, Nielsen PH, Nielsen JL, Baranyi C, Toenshoff ER, Daims H, Wagner M (2006) Cohn's Crenothrix is a filamentous methane oxidizer with an unusual methane monooxygenase. Proc Natl Acad Sci USA 103:2363–2367

Tang KW, McGinnis DF, Ionescu D, Grossart H-P (2016) Methane production in oxic lake waters potentially increases aquatic methane flux to air. Environ Sci Technol Lett 3:227–233

Tate KR (2015) Soil methane oxidation and land-use change – from process to mitigation. Soil Biol Biochem 80:260–272

Tavormina PL, Ussler W 3rd, Orphan VJ (2008) Planktonic and sediment-associated aerobic methanotrophs in two seep systems along the North American margin. Appl Environ Microbiol 74:3985–3995

Tays C, Guarnieri MT, Sauvageau D, Stein LY (2018) Combined effects of carbon and nitrogen source to optimize growth of proteobacterial methanotrophs. Front Microbiol 9:2239

Thompson LR, Sanders JG, McDonald D, Amir A, Ladau J, Locey KJ, Prill RJ, Tripathi A, Gibbons SM, Ackermann G, Navas-Molina JA, Janssen S, Kopylova E, Vázquez-Baeza Y, González A, Morton JT, Mirarab S, Zech Xu Z, Jiang L, Haroon MF, Kanbar J, Zhu Q, Jin Song S, Kosciolek T, Bokulich NA, Lefler J, Brislawn CJ, Humphrey G, Owens SM, Hampton-Marcell J, Berg-Lyons D, McKenzie V, Fierer N, Fuhrman JA, Clauset A, Stevens RL, Shade A, Pollard KS, Goodwin KD, Jansson JK, Gilbert JA, Knight R (2017) A communal catalogue reveals Earth's multiscale microbial diversity. Nature 551:457

Tveit A, Schwacke R, Svenning MM, Urich T (2013) Organic carbon transformations in high-Arctic peat soils: key transformations and microorganisms. ISME J 7:299–311
Tveit AT, Urich T, Svenning MM (2014) Metatranscriptomic analysis of Arctic peat soil microbiota. Appl Environ Microbiol 80:5761–5772
Tveit AT, Urich T, Frenzel P, Svenning MM (2015) Metabolic and trophic interactions modulate methane production by Arctic peat microbiota in response to warming. Proc Natl Acad Sci USA 112:E2507–E2E16
Vaksmaa A, Guerrero-Cruz S, van Alen TA, Cremers G, Ettwig KF, Lüke C, Jetten MSM (2017a) Enrichment of anaerobic nitrate-dependent methanotrophic 'Candidatus Methanoperedens nitroreducens' archaea from an Italian paddy field soil. Appl Microbiol Biotechnol 101:7075–7084
Vaksmaa A, van Alen TA, Ettwig KF, Lupotto E, Vale G, Jetten MSM, Luke C (2017b) Stratification of diversity and activity of methanogenic and methanotrophic microorganisms in a nitrogen-fertilized Italian paddy soil. Front Microbiol 8:15
van der Wal A, Geydan TD, Kuyper TW, de Boer W (2013) A thready affair: linking fungal diversity and community dynamics to terrestrial decomposition processes. FEMS Microbiol Rev 37:477–494
van der Wal A, Ottosson E, de Boer W (2015) Neglected role of fungal community composition in explaining variation in wood decay rates. Ecology 96:124–133
van Kessel M, Speth DR, Albertsen M, Nielsen PH, Op den Camp HJM, Kartal B, Jetten MS, Lucker S (2015) Complete nitrification by a single microorganism. Nature 528:555
van Kruistum H, Bodelier PLE, Ho A, Meima-Franke M, Veraart AJ (2018) Resistance and recovery of methane-oxidizing communities depends on stress regime and history; a microcosm study. Front Microbiol 9:1714
van Teeseling MCF, Pol A, Harhangi HR, van der Zwart S, Jetten MSM, den Camp H (2014) Expanding the verrucomicrobial methanotrophic world: description of three novel species of methylacidimicrobium gen. nov. Appl Environ Microbiol 80:6782–6791
Veraart AJ, Steenbergh AK, Ho A, Kim SY, Bodelier PLE (2015) Beyond nitrogen: the importance of phosphorus for CH4 oxidation in soils and sediments. Geoderma 259:337–346
Veraart AJ, Garbeva P, van Beersum F, Ho A, Hordijk CA, Meima-Franke M, Zweers AJ, Bodelier PLE (2018) Living apart together-bacterial volatiles influence methanotrophic growth and activity. ISME J 12:1163–1166
Verbeke TJ, Dedysh SN, Dunfield PF (2018) Methanotrophy in acidic soils, including Northern Peatlands. In: McGenity TJ (ed) Microbial communities utilizing hydrocarbons and lipids: members, metagenomics and ecophysiology. Springer, Cham, pp 1–25
Versantvoort W, Guerrero-Cruz S, Speth DR, Frank J, Gambelli L, Cremers G, van Alen T, Jetten MSM, Kartal B, Op den Camp HJM, Reimann J (2018) Comparative genomics of candidatus methylomirabilis species and description of Ca. methylomirabilis lanthanidiphila. Front Microbiol 9:1672
Vigliotta G, Nutricati E, Carata E, Tredici SM, De Stefano M, Pontieri P, Massardo DR, Prati MV, De Bellis L, Alifano P (2007) Clonothrix fusca Roze 1896, a filamentous, sheathed, methanotrophic gamma-proteobacterium. Appl Environ Microbiol 73:3556–3565
Vollmer D, Regan HM, Andelman SJ (2016) Assessing the sustainability of freshwater systems: a critical review of composite indicators. Ambio 45:765–780
Vorob'ev AV, de Boer W, Folman LB, Bodelier PLE, Doronina NV, Suzina NE, Trotsenko YA, Dedysh SN (2009) Methylovirgula ligni gen. nov., sp nov., an obligately acidophilic, facultatively methylotrophic bacterium with a highly divergent mxaF gene. Int J Syst Evol Microbiol 59:2538–2545
Vorobev AV, Baani M, Doronina NV, Brady AL, Liesack W, Dunfield PF, Dedysh SN (2011) Methyloferula stellata gen. nov., sp nov., an acidophilic, obligately methanotrophic bacterium that possesses only a soluble methane monooxygenase. Int J Syst Evol Microbiol 61:2456–2463
Walter KM, Smith LC, Chapin FS 3rd (2007) Methane bubbling from northern lakes: present and future contributions to the global methane budget. Philos Trans A Math Phys Eng Sci 365:1657–1676

Wang JJ, Krause S, Muyzer G, Meima-Franke M, Laanbroek HJ, Bodelier PLE (2012) Spatial patterns of iron- and methane-oxidizing bacterial communities in an irregularly flooded, riparian wetland. Front Microbiol 3:13

Waring CL, Hankin SI, Griffith DWT, Kertesz MA, Kobylski V, Wilson NL, Coleman NV, Kettlewell G, Zlot R, Bosse M, Bell G (2017) Seasonal total methane depletion in limestone caves. Sci Rep 7:8314

Watsuji TO, Yamamoto A, Takaki Y, Ueda K, Kawagucci S, Takai K (2014) Diversity and methane oxidation of active epibiotic methanotrophs on live Shinkaia crosnieri. ISME J 8:1020–1031

Welte CU, Rasigraf O, Vaksmaa A, Versantvoort W, Arshad A, Op den Camp HJ, Jetten MS, Luke C, Reimann J (2016) Nitrate- and nitrite-dependent anaerobic oxidation of methane. Environ Microbiol Rep 8:941–955

Whittenbury R, Philips KC, Wilkinson JF (1970) Enrichment, isolation and some properties of methane-utilizing bacteria. J Gen Microbiol 61:205–218

Wieczorek AS, Drake HL, Kolb S (2011) Organic acids and ethanol inhibit the oxidation of methane by mire methanotrophs. FEMS Microbiol Ecol 77:28–39

Wieder WR, Allison SD, Davidson EA, Georgiou K, Hararuk O, He Y, Hopkins F, Luo Y, Smith MJ, Sulman B, Todd-Brown K, Wang Y-P, Xia J, Xu X (2015) Explicitly representing soil microbial processes in Earth system models. Glob Biogeochem Cycles 29:1782–1800

Yan X, Xu X, Ji M, Zhang Z, Wang M, Wu S, Wang G, Zhang C, Liu H (2019) Cyanobacteria blooms: a neglected facilitator of CH4 production in eutrophic lakes. Sci Total Environ 651:466–474

Yu ZCL (2017) Communal metabolism of methane and the rare earth element switch. J Bacteriol 199:e00328-17

Yvon-Durocher G, Montoya J, Woodward G, Jones J, Trimmer M (2011) Warming increases the proportion of primary production emitted as methane from freshwater mesocosms. Glob Change Biol 17:1225–1234

Zhao R, Wang HM, Cheng XY, Yun Y, Qiu X (2018) Upland soil cluster γ dominates the methanotroph communities in the karst Heshang Cave. FEMS Microbiol Ecol 94. https://doi.org/10.1093/femsec/fiy192

Zheng Y, Huang R, Wang BZ, Bodelier PLE, Jia ZJ (2014) Competitive interactions between methane- and ammonia-oxidizing bacteria modulate carbon and nitrogen cycling in paddy soil. Biogeosciences 11:3353–3368

Zhou X, Jin F, Lu C, Baoyin T, Jia Z (2018) Shifts in the community composition of methane-cycling microorganisms during lake shrinkage. Geoderma 311:9–14

Zigah PK, Oswald K, Brand A, Dinkel C, Wehrli B, Schubert CJ (2015) Methane oxidation pathways and associated methanotrophic communities in the water column of a tropical lake. Limnol Oceanogr 60:553–572

Enrichment and Isolation of Aerobic and Anaerobic Methanotrophs

Sung-Keun Rhee, Samuel Imisi Awala, and Ngoc-Loi Nguyen

Contents

Abstract Methanotrophs mediate methane oxidation to produce CO_2 which is facilitated by methane monooxygenase. Owing to this unique nature, methanotrophs are required in many industrial processes and have environmental applications. For this purpose, diverse aerobic proteobacterial and verrucomicrobial methanotrophs with different traits have been isolated and characterized. Members of the "NC10" phylum of bacteria and Archaea are involved in methane oxidation in anoxic conditions. These methanotrophs have been isolated from various environments including extreme environments, and thus, have diverse physiological and biochemical properties as regards to the required optimum pH, oxygen, temperature, salinity, and electron acceptors. To obtain novel methanotrophs with high potential of application, new cultivation technologies have been developed with the aid of advanced methodologies for monitoring the identity and activity of methanotrophs. In this chapter, we summarize the factors to be considered in the cultivation of methanotrophs and technologies developed for enrichment and isolation of diverse aerobic and anaerobic methanotrophs from various environments.

S.-K. Rhee (✉) · S. I. Awala · N.-L. Nguyen
Department of Microbiology, Chungbuk National University, Cheongju, Republic of Korea
e-mail: rhees@cbnu.ac.kr; siawala@futa.edu.ng

E. Y. Lee (ed.), *Methanotrophs*, Microbiology Monographs 32,
https://doi.org/10.1007/978-3-030-23261-0_2

1 Introduction

Methanotrophs, i.e., methane-oxidizing organisms, can remove methane (CH_4) from the environment under oxic and anoxic conditions and have been found in both the domains of bacteria and archaea. From the time of the first isolation of methanotrophs reported in 1906 (Söhngen 1906), the diversity and growth characteristics of methanotrophic cultures have significantly increased due to the new enrichments and isolates obtained from a wide variety of environments. Most of the aerobic methanotrophic bacteria that have been isolated belong to the subclasses Gammaproteobacteria or type I methanotrophs and Alphaproteobacteria or type II methanotrophs (Hanson and Hanson 1996), primarily according to their use of the ribulose monophosphate pathway or serine pathways for formaldehyde assimilation and arrangement of intracellular membranes (ICMs) throughout the cell. Extremely acidophilic methanotrophic members belonging to the phylum Verrucomicrobia or type III methanotrophs were isolated from hot springs and found to be widely distributed in acidic geothermal environments (Carere et al. 2017; Dunfield et al. 2007). These thermoacidophiles use carbon dioxide as the carbon source via the Calvin-Benson-Bassham (CBB) cycle.

Anaerobic oxidation of methane (AOM) has been demonstrated to be coupled to the reduction of electron acceptors such as sulfate, nitrate, nitrite, iron, manganese (Ettwig et al. 2010, 2016; Haroon et al. 2013). Methanotrophic bacterial representatives of the candidate phylum NC10 occur in anoxic habitats and have an intra-aerobic pathway of CH_4 oxidation (Ettwig et al. 2010; He et al. 2016). In the process, oxygen is generated from nitrite through the dismutation of nitric oxide (NO) and used as a substrate for methane monooxygenase (MMO)-dependent methane oxidation. Besides, methane can be oxidized by anaerobic methane-oxidizing archaea (ANME) in the phylum of Euryarchaeota. So far, three types of ANME (ANME-1, ANME-2, and ANME-3) that are phylogenetically related to different methanogenic archaea have been identified for AOM depending on the use of sulfate as the terminal electron acceptor (Boetius et al. 2000; Knittel et al. 2005), whereas ANME-2d is shown to be responsible for nitrate driven AOM (Haroon et al. 2013).

Methanotrophs play an important role in regulating methane emission and have been researched extensively for many years due to their role in industrial applications. Methanotrophs can be used for removal of greenhouse gases, degrading pollutants, biological denitrification, and recovery of metals and sulfur compounds (Kalyuzhnaya et al. 2015; Strong et al. 2015). Owing to their potential in industrial applications, various enrichment and isolation technologies have been developed for rapid recovery of targeted methanotrophs or novel methanotrophs (Hoefman et al. 2012; Kim et al. 2018).

There are diverse, uncultivated methanotrophs as demonstrated by molecular ecology techniques employing genes involved in methane oxidation (Knief 2015; Lau et al. 2007). The initial step of methane oxidation from methane to methanol by methanotrophic bacteria is catalyzed by MMOs that are most commonly used as a biomarker for the identification of methanotrophic bacteria (Knief 2015). Thus, isolation of methanotrophs from natural resources remain challenging due to the limitations of current approaches. For example, unknown growth factors might be required for stimulation of growth of clades of uncultivated methanotrophs. Slow-growing uncultivated methanotrophs might be diluted out when fast-growing strains are selectively enriched during typical liquid enrichment cultures. Technically, tightly integrated satellite bacteria reduce the efficiency of isolating diverse methanotrophs from enrichment cultures. It is the goal of this chapter to provide current knowledge of cultivation and isolation of methanotrophs to help obtain new methanotroph resources from the environment.

2 Cultivation Condition of Aerobic Methanotrophs

Diverse aerobic methanotrophs have been obtained from various habitats having lower or higher temperature, pH, oxygen, and salinity (Knief 2015; Trotsenko and Khmelenina 2002). The methanotrophs isolated from these habitats show physiological diversities which are relevant for potential industrial applications.

2.1 Physiological Diversity of Aerobic Methanotrophs

From natural environments characterized by permanently low temperatures (10 °C or below), high concentration of methane, and low-oxygen tension, methanotrophic enrichments can be obtained by incubating under low-temperature and static conditions coupled with a reduction of oxygen concentration in the headspace (under 10%). Psychrophilic methanotrophs such as *Methylobacter psychrophilus* (Omel'chenko et al. 1996) and *Methylomonas scandinavica* (Kalyuzhnaya et al. 1999) were isolated from tundra soil. *Methylosphaera hansonii* (Bowman et al. 1997) was isolated from surface sediments of Antarctic lakes. Recently, cold-tolerant methanotrophs including *Methylosoma difficile* (Rahalkar et al. 2007), *Methyloglobulus morosus* (Deutzmann et al. 2014), *Methyloprofundus sediment* (Tavormina et al. 2015) and *Methylovulum psychrotolerans* (Oshkin et al. 2016) were isolated from marine and terrestrial sediments with low in situ temperature. A spiral-shaped microaerophilic methanotroph '*Candidatus* Methylospira mobilis' was cultivated from a lake (Danilova et al. 2016). Most of strains isolated from the environment with permanently low temperatures were found to grow under wider range of temperatures. With possible adaptation to their habitats comprising high concentrations of methane and low oxygen tensions, aerobic methanotrophs of Gammaproteobacteria are assumed to play a central role in methane oxidation in hypoxic and even anoxic environments where they are frequently detected (Chistoserdova 2015; Martinez-Cruz et al. 2017; Oswald et al. 2016; Rissanen et al. 2018).

Thermotolerant or moderately thermophilic methanotrophs of the class Gammaproteobacteria were isolated from thermal environments including marine sediment, hot spring or aquifer, and tropical soils. The genera *Methylococcus* (Bodrossy et al. 1995), *Methylocaldum* (Bodrossy et al. 1997), and *Methylomarinovum* (Hirayama et al. 2014) were considered to be thermotolerant or moderately thermophilic growing at temperatures up to 50 °C. The genus *Methylothermus* (Bodrossy et al. 1999; Hirayama et al. 2011; Tsubota et al. 2005), isolated from hot springs, was found to include the most thermophilic gammaproteobacterial methanotrophs with optimal growth at 55–60 °C. Media with low-salt contents were frequently used for enrichments of thermophilic methanotrophs from terrestrial origins (Hirayama et al. 2011, 2014; Islam et al. 2016). Supplementation of growth factors to the growth medium was found to be necessary (Danilova et al. 2016; Hirayama et al. 2011).

Halotolerant/halophilic methanotrophs have been isolated from saline environments. From marine environments, *Methylomicrobium japanense* (Kalyuzhnaya et al. 2008), *Methylomicrobium pelagicum* (Bowman et al. 1995), *Methylobacter marinus* (Bowman et al. 1993), *Methylomarinum vadi* (Hirayama et al. 2013), *Methylomarinovum caldicuralii* (Hirayama et al. 2014), *Methyloprofundus sediment* (Tavormina et al. 2015), and *Methylocaldum marinum* (Takeuchi et al. 2014) were isolated. From soda lakes, *Methylosphaera hansonii* (Bowman et al. 1997), *Methylohalobius crimeensis* (Heyer et al. 2005), *Methylomicrobium* spp. (Kaluzhnaya et al. 2001) and *Methylocystis* sp. B3 (Eshinimaev et al. 2008) also were isolated. The species of *Methylomicrobium alcaliphium* and *Methylomicrobium kenyense* are alkaliphilic and halophilic methanotrophs which were isolated from soda lakes (Kalyuzhnaya et al. 2008). The modified nitrate mineral salts (NMS) medium with supplementation of NaCl was successfully used to enrich and isolate halophilic methanotrophic bacteria which have an obligate requirement for NaCl. *Methylohalobius* species possess the most halophilic strains with the highest salt tolerance, growing optimally at 1–1.5 M NaCl and tolerating NaCl concentrations of up to 2.5 M (14.6% w/v). For isolation of alkaliphilic methanotrophs from soda lakes, the salts contents of mineral media should be adjusted to pH 9–10 using $NaHCO_3$/Na_2CO_3 to match the pH profile of the source samples (Table 1).

Acidophilic or acid-tolerant methanotrophs have been isolated using mild to strong acidic media from peat ecosystems, floodplain of the river, forest soils and geothermal habitats. Proteobacterial species of *Methylocella* (Dedysh et al. 2000; 2004; Dunfield et al. 2003), *Methylocaps*a (Dedysh et al. 2002), *Methylocystis* (Belova et al. 2013; Dedysh et al. 2007), *Methylomonas*, *Methylovulum* (Danilova et al. 2013; Kip et al. 2011; Oshkin et al. 2016), and *Methylocaldum* (Islam et al. 2016) were isolated from these environments. Thermo-acidophilic verrucomicrobial methanotrophs can grow in a wide range of temperatures (22.5–81.6 °C) but only in extremely acidic pH (1.8–5.0) (Sharp et al. 2014). The verrucomicrobial methanotrophs were obtained from geothermal habitats in strongly acidified mineral salt media. H_2SO_4 was used for acidification of media for growing methanotrophs. Enrichment culture experiments showed that a better growth was observed in a medium supplemented with NH_4Cl than with KNO_3. The media used for isolating moderate acidophiles were low-salt mineral media with using a phosphate buffer.

Table 1 Mineral media used for cultivation of aerobic methanotrophic bacteria

Mineral media	Composition (g per liter distilled water)	Target group of methanotrophs	Environmental associations
Nitrate mineral salts (NMS) medium (Bowman et al. 1993; Whittenbury et al. 1970)	KNO_3, 1; $MgSO_4.7H_2O$, 1; $Na_2HPO_4.12H_2O$, 0.717; KH_2PO_4, 0.272; $CaCl_2.6H_2O$, 0.2; ferric ammonium EDTA, 0.004; 0.1% (v/v) of TE 1 solution (Table 2); pH 6.8.	Neutrophilic methanotrophs	Terrestrial (soil, fresh water, sediments, groundwater)
Ammonium mineral salts (AMS) medium (Whittenbury et al. 1970)	NH_4Cl, 0.5; $MgSO_4.7H_2O$, 1; $Na_2HPO_4.12H_2O$, 0.717; KH_2PO_4, 0.272; $CaCl_2.6H_2O$, 0.2; ferric ammonium EDTA, 0.005; 0.1% (v/v) of TE 1 solution (Table 2); pH 6.8.	Neutrophilic methanotrophic	Terrestrial (soil, fresh water, sediments, groundwater) environments
Dilute nitrate mineral salts (DNMS) medium (Dunfield et al. 2003)	NMS medium diluted 1:5 with distilled water, 1 mM of NaH_2PO_4-Na_2HPO_4 buffer, pH 5.5–7.0.	Mild acidophiles and neutrophiles	Terrestrial (soil, fresh water, sediments) environments
Medium 10 (Kip et al. 2011)	NH_4Cl, 0.5; $MgSO_4.7H_2O$, 0.1; $Na_2HPO_4.12H_2O$, 0.7; KH_2PO_4, 0.3; $CaCl_2.6H_2O$, 0.1; $FeSO_4.7H_2O$; 0.005; 0.1% (v/v) of TE 2 solution (Table 2); pH 7.0.	Neutrophilic methanotrophic	Terrestrial (soil, fresh water, sediments) environments
Medium M1 (Dedysh et al. 1998)	KNO_3, 1; $MgSO_4.7H_2O$, 0.1; KH_2PO_4, 0.2; $CaCl_2.6H_2O$, 0.02; 0.1% (v/v) of TE 1 solution (Table 2), pH 4.5 to 5.0.	Mildly acidophilic methanotrophs	Acidic peat, sphagnum moss soils, tundra soils
Medium M2 (Dedysh et al. 1998)	KNO_3, 0.25; $MgSO_4.7H_2O$, 0.05; KH_2PO_4, 0.1; $CaCl_2.6H_2O$, 0.01; NaCl, 0.02; 0.1% (v/v) of TE 1 solution (Table 2), pH 5.5.	Mildly acidophilic methanotrophs	Mildly acidic soils and freshwater wetlands
Medium N (Kip et al. 2011)	KNO_3, 0.001; NH_4Cl, 0.008; NaCl, 0.003; $CaCl_2.6H_2O$, 0.0018; $MgSO_4.7H_20$, 0.010; KH_2PO_4, 0.00175; Na_2SiO_3, 0.002; $AlCl_3$, 0.001; 0.2% (v/v) of 10x TE 1 solution (Table 2).	Mildly acidophilic methanotrophs	Acidic wetlands and forest soils
Medium V41 (Dunfield et al. 2007)	NH_4Cl, 0.1; KH_2PO_4, 0.015; $Na_2HPO_4.12H_2O$, 0.01; Yeast extract, 0.01; $FeNH_4$-EDTA, 0.003; 0.3% (v/v) of TE 2 and 0.1% (v/v) of TE 5 (Table 2), pH 5.5.	Thermo-acidophilic methanotrophs	Geothermal soils (volcanic soil, thermal mud pod)

(continued)

Table 1 (continued)

Mineral media	Composition (g per liter distilled water)	Target group of methanotrophs	Environmental associations
Medium 3.9C10.2 (Sharp et al. 2014)	NH_4Cl, 0.2; $MgSO_4.7H_2O$, 0.02; KH_2PO_4, 0.05; $CaCl_2.6H_2O$, 0.01; Fe-EDTA powder, 0.005; 0.3% (v/v) of TE 3 solution (Table 2), pH 3.9.	Thermal-acido-philic methanotrophs	Geothermal areas (volcanic soils, thermal mud pod)
Medium S (Heyer et al. 2005)	NH_4NO_3, 0.5; $Na_2HPO_4.12H_2O$, 0.5; KH_2PO_4, 0.1; $FeSO_4.7H_2O$, 0.005; NaCl, 66.0; Na_2SO_4, 10.45; Na_2CO_3, 0.10; $NaHCO_3$, 0.06; NaBr, 0.28; $MgCl_2$, 20.89; $CaCl_2$, 2.21; 0.1% (v/v) of TE 1 solution (Table 2), pH 7.5.	Neutrophilic halophiles	Hyper saline lake sediment
Medium MJmet (Hirayama et al. 2011)	$MgSO_4.7H_2O$, 0.34, $MgCl_2.6H_2O$, 0.42; $CaCl_2$, 0.08; KCl, 0.033, NH_4Cl, 0.25; $NaNO_3$, 0.25; NaCl, 30; K_2HPO_4; 0.014, $NaHCO_3$, 0.5; $NiCl_2.6H_2O$, 0.00005, $Na_2SeO_3.5H_2O$, 0.00005, Na_2WO_4, 0.00001, Fe $(NH_4)_2(SO_4)_2.6H_2O$, 0.002; 0.1% (v/v) of TE 1 solution and 0.1% (v/v) of VS1 (Table 3), pH 6.6 to 7.0.	Thermophilic methanotrophs	Marine hydrother-mal system, sedi-ment, mud, hot spring and aquifer.
Medium SL (Sorokin et al. 2000)	Na_2CO_3, 23; $NaHCO_3$, 7; NaCl, 5; KH_2PO_4, 0.5; KNO_3, 0.5; $MgSO_4.7H_2O$, 0.12; 0.1% (v/v) of TE 4 solution (Table 2), pH 10.1.	Alkaliphilic methanotrophs	Soda lake sediment

For isolation of *Methylocella*, *Methylocapsa*, and *Methyloferula*, the low-salt medium was prepared by diluting NMS medium to five to ten-fold or designed based on the salt contents of the environment.

For obtaining methanotrophs, it is important to choose or design appropriate culture media (Table 1) and growth conditions based on the characteristics of the methanotrophs, closely mimicking natural conditions. Most known species of neutrophilic and mesophilic methanotrophs can be enriched, isolated and cultivated in a nitrate or ammonium mineral salts (NMS/AMS) medium proposed by Whittenbury et al. (1970) from freshwater and low salinity terrestrial environments. Modified media include ammonium nitrate mineral salts (ANMS) medium with addition of both nitrate and ammonia or nitrogen source-free medium for targeting nitrogen-fixing methanotrophs. The AMS medium is less common in practice than NMS, due to high concentrations of ammonia which is inhibitory to the methanotrophs as it

Table 2 Trace element stock solutions

Trace element (TE)	Composition (g per liter distilled water)	Applicability
TE 1 (Bowman 2006)	Na_2EDTA, 0.5; $FeSO_4.7H_2O$, 0.2; H_3BO_3, 0.03; $ZnSO_4.7H_2O$, 0.01; $MnCl_2.4H_2O$, 0.03; $CoCl_2.6H_2O$, 0.02; $CuSO_4.5H_2O$, 0.03; $NiCl_2.6H_2O$, 0.002; $Na_2MoO_4.2H_2O$, 0.003.	Most methanotrophs
TE 2 (Heyer et al. 2002)	$ZnSO_4.7H_2O$, 0.44; $CuSO_4.5H_2O$, 0.22; $MnSO_4.2H_2O$, 0.17; $Na_2MoO_4.2H_2O$, 0.06; H_3BO_3, 0.10; $CoCl_2.6H_2O$, 0.08.	Most methanotrophs
TE 3 (Carere et al. 2017)	Na_2EDTA, 2.06; $FeSO_4.7H_2O$, 1.54; $ZnSO_4.7H_2O$, 0.44; $CuSO_4.5H_2O$, 0.20; $MnCl_2.4H_2O$, 0.19; $Na_2MoO_4.2H_2O$, 0.06; H_3BO_3, 0.10; $CoCl_2.6H_2O$, 0.08; $CeCl_3.7H_2O$ 0.1; $LaCl_3.7H_2O$ 0.1.	Methanotrophs from geothermal area
TE 4 (Sorokin et al. 2000)	Na_2EDTA, 5.0; $FeSO_4.7H_2O$, 2.0; $CuSO_4.5H_2O$, 0.1; $ZnSO_4.7H_2O$, 0.1; $MnCl_2.4H_2O$, 0.03; $CoCl_2.6H_2O$, 0.2; $NiCl_2.6H_2O$, 0.02; $Na_2MoO_4.2H_2O$, 0.03.	Alkaliphilic methanotrophs
TE 5 (Dunfield et al. 2007)	Nitrilotriacetic acid, 1.5; $Fe(NH_4)_2(SO_4)_2{\cdot}6H_2O$, 0.2; Na_2SeO_3, 0.2; $CoCl_2{\cdot}6H_2O$, 0.1; $MnSO_4{\cdot}2H_2O$, 0.1; $Na_2MoO_4{\cdot}2H_2O$, 0.1; $Na_2WO_4.2H_2O$, 0.1; $ZnSO_4.7H_2O$, 0.1; $AlCl_3.6H_2O$, 0.04; $NiCl_2.6H_2O$, 0.025; H_3BO_3, 0.01; $CuSO_4.5H_2O$, 0.01.	Methanotrophs from geothermal area

competitively inhibits MMO. The diluted NMS/AMS medium or alternative medium have been applied for obtaining various methanotrophs (Islam et al. 2015; Nguyen et al. 2017; Sharpe et al. 2007).

Environmental factors including methane supply, oxygen level, temperature, pH, salinity, trace metals and growth-stimulating factors can influence growth of methanotrophs. A high methane concentration at the initial enrichment step appears to be selective towards rapid growing methanotrophs. To increase cultivability of diverse methanotrophs, media have been modified for promoting specific traits of methanotroph. For example, low levels of copper and absence of copper in media ensures selective advantage to methanotrophs with high affinity to copper and soluble MMO-containing methanotrophs (Semrau et al. 2018). Augmentation of soil extracts, trace elements (TE) (Table 2), vitamin solutions (Table 3), or yeast extract was used to support the growth of fastidious methanotrophic bacteria (Islam et al. 2008; Tsubota et al. 2005). Replacement of solidifying agents (agar, agarose or noble agar) by alternative gelling agents (gellan gum, phytagel or silica gel) was found to increase selectivity and efficiency in obtaining methanotroph colonies. In case of isolation of verrucomicrobial methanotrophs from geothermal environments, growth was strictly dependent on the presence of lanthanide (e.g., cerium and lanthanum) in the media (Carere et al. 2017). Lanthanides were also found to promote methanol oxidation and thus was selective for the methanotrophs with lanthanide-requiring Xox-type methanol dehydrogenases (Xox-MeDH) (Krause et al. 2017).

Table 3 Vitamin stock solution

Vitamin stock (VS)	Composition (g per liter distilled water)	Applicability
VS 1 (Balch et al. 1979)	Folic acid, 0.02; biotin, 0.02; pyridoxine-HCl, 0.1; riboflavin, 0.05; thiamine-HC1 × $2H_2O$, 0.05; nicotinamide, 0.05; Ca-pantothenate, 0.05; vitamin B12, 0.001; p-aminobenzoic acid, 0.05.	Marine and terrestrial methanotrophs.
VS 2 (Danilova et al. 2016)	d-biotin, 0.1; folic acid, 0.1; pyridoxine HCl, 0.5; thiamine HCl, 0.25; riboflavin, 0.25; nicotinic acid, 0.25; DL-calcium pantothenate, 0.25; vitamin B12, 0.05; p-aminobenzoic acid, 0.25; lipoic acid, 0.25; 1,4-naphthaquinone, 0.2; nicotinamide, 0.5; hemin, 0.05.	Microaerophilic methanotrophs

2.2 *Enrichment and Isolation of Aerobic Methanotrophs*

Methanotrophic bacteria can be isolated from samples of various environments directly or after enrichment. Using direct isolation approach diverse methanotrophs can be obtained from environmental samples. To increase chances of successful isolation of methanotrophs, the second approach uses prior enrichment of methanotrophic communities which is followed by several techniques to obtain pure cultures.

2.2.1 Direct Isolation from Environmental Samples

Methanotrophic isolates can be directly obtained from environmental samples when fast-growing methanotrophs, which are relatively abundant, are targeted (Dunfield et al. 2007). The sample of interest can be serially diluted or directly plated on the surface of the mineral salt medium solidified by high-purified agar (Dunfield et al. 2003; Kip et al. 2011; Lidstrom 1988) or the polycarbonate membrane which can be placed on liquid media, soil substrate, or extracts of habitats (Pol et al. 2007; Svenning et al. 2003) (Fig. 1). After inoculation, the plates are incubated under a methane-containing atmosphere in gas-tight jars or desiccators. In floating membrane systems, the non-sterile soil suspension is used as the medium and nutrients from the soil can diffuse through the membrane to stimulate colony formation. The colonies are suspended, serially diluted and inoculated on a solid medium or the floating membrane and incubated under the same conditions. After several transfers of the diluted colonies, pure cultures can be obtained. In contrast to the plate method, the soil substrate membrane system developing by Svenning et al. (2003) is less time-consuming for isolation of more diverse isolates. Growth of methanotrophs can be challenging due to the necessity of having gas-tight systems in which methane can be provided to promote growth.

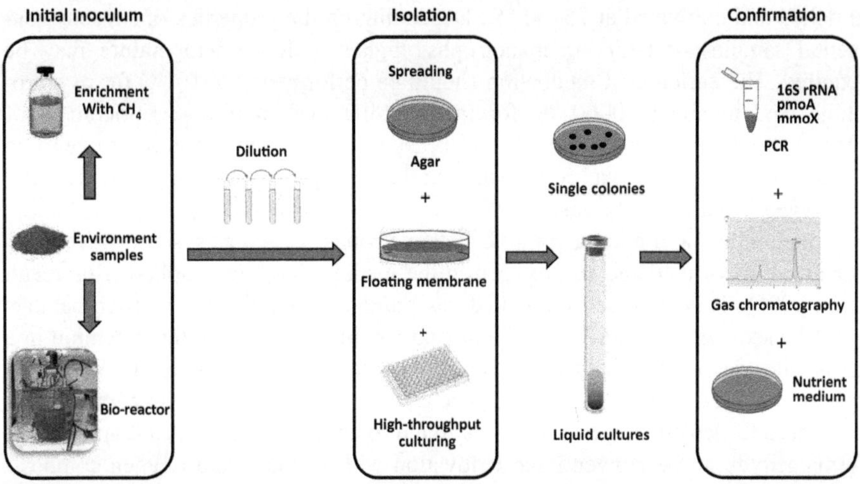

Fig. 1 A schematic overview of the workflows for the enrichment and isolation of aerobic methanotrophs

2.2.2 Enrichment Cultures

Due to relatively low abundance (<4% of total bacterial cells) of methane-oxidizing bacteria in the microbial community in most habitats (Lau et al. 2007), an enrichment procedure under a methane atmosphere is often carried out before isolation steps (Fig. 1). First, incubation of samples with methane under in situ conditions is clearly more suitable for isolation of environmentally relevant methanotrophs. Despite technical difficulties of incubation of samples in situ with a gaseous substrate, raw environmental samples such as soil can be incubated in the laboratory under defined conditions that mimic the natural site. Second, the samples are used to inoculate a liquid mineral media and are incubated under certain conditions in the laboratory which can be followed by serial dilution onto plates. Majority of the easily available methanotroph strains were isolated using this cultivation approach. A small amount of sample is inoculated into the medium in a sterile serum bottle which contains methane (5–50%, v/v) and CO_2 (5%, v/v) in the headspace. For routine cultivation, the ratio of methane and air most often used is a 1:4. In some cases, methanotrophs are greatly stimulated in the presence of CO_2, which is involved in primary and intermediate metabolism of these bacteria (Trotsenko and Murrell 2008). Verrucomicrobial methanotrophs obtain the necessary carbon by CO_2 fixation via the CBB cycle (Khadem et al. 2011). Some proteobacterial methanotrophs such as Type I (*Methylococcus* and *Methylocaldum*) and Type II (*Methyloferula*, *Methylocella* and *Methylocapsa*) also possess complete gene sets for autotrophic CO_2 fixation by the CBB cycle. After inoculation, the bottles are incubated on a rotary shaker (100–120 rpm) at an optimal growth temperature. Most known species of methanotrophs including those of *Methylocystis*, *Methylosinus*, *Methylomonas*, *Methylobacter*, and *Methylomicrobium* should be enriched and cultivated at 25–30 °C. Depending on the properties of raw environmental samples or target methanotrophs, higher or lower temperature may be required. The enrichment incubation should be performed at 2–10 °C for psychrophilic, and at about 40–60 °C for thermophilic methanotrophs. Enrichment of methanotrophs that have the ability to fix N_2 can be carried out in a nitrogen-free medium coupled with a gas composition in the headspace comprising of under 10% of oxygen in static conditions.

For enrichment of microaerophilic methanotrophs, a gradient system with opposing gradients of methane and oxygen in the medium has been applied (Bussmann et al. 2006). This approach mimicked gas profiles of many methanotroph habitats e.g. lake sediments, landfill soils. Semisolid medium mixed with the inoculum in a glass tube was filled with air from above and with methane from below. A gradient system was used only for initial enrichment and switched to a traditional cultivation procedure for isolation. An isolate was found to accumulate significant amounts of formaldehyde in the conventional cultivation with methane and oxygen compositions, which may explain its preferential enrichment in a gradient cultivation system.

The enrichment usually takes one to several weeks, depending on the sample and incubation conditions used. Gas chromatography (GC) is used to monitor the concentration of methane in the culture vials and turbidity needs to be monitored daily. In some cases, a pellicle of growth in the vials indicates development of target methanotrophs leading to the decrease of the methane concentration. Suspensions sampled from these enrichment cultures can be examined using phase-contrast microscopy. Due to the presence of intracytoplasmic membranes, cells of methanotrophs usually look somewhat darker than cells of non-methanotrophic satellites.

2.2.3 Enrichment of Methanotrophs in Bioreactors

A recently demonstrated enrichment technique utilizing a large-scale cultivation system aimed to select the target methanotrophs from an environmental sample under controlled conditions. Bioreactor systems were applied to select the fast growing methanotrophs (Kim et al. 2018), to enrich acid-adapted methanotrophs (Nguyen et al. 2018) and grow the pure cultures under controlled conditions. In order to cultivate methanotrophs at a large scale (e.g. >1 L) in continuous culture reactors, it is necessary to continuously add methane and oxygen (in the form of sterile air or oxygen) and remove gases. It is recommended to include a dissolved oxygen probe in the system to monitor oxygen levels in solution. The basic set-up of a bioreactor system should include a reactor fitted with a fermenter lid containing a stirrer, a dissolved oxygen probe, a pH probe, acid and base in-flow tubes for pH control, temperature detector, gas line, feed in-flow tube, sampling line, and effluent-withdraw line. If the methane oxidation and community composition of the methanotroph in chemostat are achieved, samples can be taken and further cultivated and characterized in batch cultures.

2.3 Isolation Techniques

Methane-oxidizing enrichment cultures represent complex microbial communities developed through association mechanisms between methanotrophs and non-methanotrophs, such as cross-feeding intermediates of methane metabolisms to non-methanotrophs such as methanol, formate, acetate, succinate, and other organic acids (Krause et al. 2017; Oshkin et al. 2015) and supplying growth factors (e.g., cobalamin, vitamin B_{12}) for methane oxidation from non-methanotrophs (Iguchi et al. 2011). Separation of methanotrophic cells from these associated bacteria is the most challenging and time-consuming task of the isolation procedure (Fig. 1).

After one or several passages of methanotrophic enrichment in liquid medium supplemented with CH_4 in the headspace gas, the isolation of pure cultures was followed by repeated single-colony picking on agar plate (Belova et al. 2013; Heyer et al. 2005; Ogiso et al. 2012). After serial dilutions (10^{-2}, 10^{-4}, and 10^{-6}), the enrichment cultures were spread onto mineral-salt agar plates. Plates were incubated in gas-tight jars in the same CH_4/air mixture used for the enrichment culture. Control setup without methane is recommended to be included in this case to detect heterotroph colonies co-cultured with methanotrophs. NMS agar slopes in tubes with methane injected in a head gas-phase also can be applied for development of methanotrophic colonies (Ogiso et al. 2012). Once colonies were apparent, single colonies were selected and continuously re-streaked onto fresh medium to obtain pure cultures. Methanotrophy of the isolates was confirmed for methane oxidation in liquid cultures by GC analysis.

Another modification of spreading-and-single-colony-picking technique can be performed on floating membrane (Hojberg et al. 1997) which was applied for isolation of the thermo-acidophilic verrucomicrobial methanotrophs (Pol et al. 2007). The floating membrane was used as a general technique for isolation of methanotrophs sensitive to organic matters like agar and restriction of development of colonies of methylotrophic and heterotrophic satellite bacteria (Nguyen et al. 2017). The enrichment culture of methanotrophs was diluted and filtered through 0.2 μm polycarbonate filters and then the filters were floated on mineral salt medium in petri dishes and incubated in closed jars in a methane atmosphere. Colonies were selected and repeatedly re-streaked on the membranes until uniform pure colonies were obtained. Most of the colonies were verified as methanotrophs.

For isolation of slow-growing, non-colony-forming but numerically dominant strains (Dedysh et al. 2002), serial dilution in liquid medium was an optional procedure. However, these processes to obtain pure cultures are tedious and time-consuming. Cultures of methanotroph communities were serially diluted nine- to tenfold in vials filled with the mineral salt medium. After dilution, the vials were sealed, and methane was injected into the headspace by using a syringe and a sterile filter (0.22 μm). The vials were then incubated under the same incubation conditions. The cultures with turbidity were examined by phase-contrast microscopy. The cultures containing a large proportion of morphologically uniform cells were selected and the process of serial dilutions was repeated until a culture consisting of morphologically uniform cells was obtained. In contrast to the serial dilution method, the utilization of miniaturized extinction culturing methods in 96-well plates (Hoefman et al. 2012) has to be exploited significantly for methanotrophic isolation and could be ideal approaches for isolation of fast-growth methanotrophs.

The methanotroph culture purity was evaluated by (i) colony morphology, (ii) phase-contrast microscopy, and (iii) absence of growth in complex organic media (standard undiluted and tenfold-diluted Luria-Bertani agar, Reasoner's 2A, or nutrient agar) and the same mineral media used for methanotroph isolation supplemented with 0.05% (w/v) glucose, fructose, or sucrose and 0.005% (w/v) yeast extract. Microscopic examination of uniform cell morphology in culture suspension is required to confirm culture purity. No growth should be observed on plates when an aliquot (0.05–0.1 mL) of a methanotroph culture is spread onto plates with nutrient media and incubated for 2–4 weeks without methane. Some non-methanotrophic satellites (e.g. facultative methylotrophs) may not develop colonies on these media and may escape detection. It should be noted here that some methanotrophs are facultative, i.e., some strains of methanotrophs of *Methylocystis* species (Belova et al. 2011) and Beijerinckiaceae can grow on short-chain organic acids and ethanol (Dunfield et al. 2003). However, to date no methanotrophs have been shown to grow on sugars-rich media. If it is suspected that novel isolates are facultative, it is necessary to be much more thorough to ensure culture purity.

2.4 *Monitoring Activity*

For successful enrichment, isolation and cultivation of methanotrophs monitoring activity and abundance of methanotrophs are necessary. Growth dynamics on methane can be monitored by measuring the methane concentration in headspace of liquid cultures. The genes encoding of MMO and 16S rRNA gene of methanotrophs can be used as biomarkers to determine diverse methanotrophs in environmental samples as well as cultures.

2.4.1 Measurement of Methane Oxidation Dynamics

Growth of methanotrophs can be determined by measuring the turbidity (OD_{600}) of cultures according to incubation time. Control experiments without CH_4 need to be carried out under similar conditions. These experiments are required to confirm the ability of the isolated strain to grow on methane and are usually conducted using liquid cultures. An increase in optical density (OD_{600}) should be accompanied by a decline in CH_4 concentration in the headspace, while no growth should be observed in the same medium in the absence of methane. Methane concentration is measured using a GC equipped with a flame ionization detector (FID) and an appropriate column. The optical density is measured on a spectrophotometer. The procedure used is as follows: (1) Preparation of 25-mL test tubes filled to 20% capacity with the liquid mineral medium. (2) Inoculation of the vial with 10% inoculum. (2) Sealing the vials and adding methane (10%, vol/vol) to the headspace by using a syringe and a sterile filter (0.22 μm). (3) Inclusion of uninoculated controls of the medium as blanks to check for methane leakage and as sterility controls, and also including inoculated medium with no added methane. (4) Incubation of vials on a rotary shaker (100–150 rpm) or in static conditions at an optimal growth temperature. (5) Taking gas samples and culture aliquots once every 1–2 days for determination of methane concentrations by GC and OD_{600} by spectrophotometer.

2.4.2 Molecular Identification

New isolates can be routinely identified based on the analysis of genes encoding the structural subunits of the MMO and 16S rRNA gene sequence. All aerobic methanotrophs contain one or both of two potential MMO, particulate (pMMO) and/or soluble (sMMO) MMO. The pMMO is encoded by a *pmoCAB* operon while sMMO is encoded by a more complex set of genes including *mmoXYBZDC*, and some methanotrophs contain additionally *pxmABC*, the homolog of *pmoCAB*. The *pmoA* and *mmoX* genes encoding the β-subunit of pMMO and the α-subunit of the sMMO, respectively, are commonly used as functional markers for methane-oxidizing bacteria. Primers and polymerase chain reaction (PCR) conditions to identify methanotrophs have been summarized in Table 4. Colony PCR of the *pmoA* and *mmoX* genes may also be used as an early screen of methanotrophs to identify target colonies. Other genes encoding subunits of methanol hydrogenase and di-nitrogenase, i.e., MxaF (Lau et al. 2013; McDonald et al. 2008), NifH (Auman et al. 2001), and PxmA (Tavormina et al. 2011), respectively, are also used for identification of methanotrophs. The 16S rRNA gene can be amplified and sequenced with the primers 27F/1492R (Weisburg et al. 1991). The *pmoA* gene can be amplified using the primer A186/mb661 (Costello and Lidstrom 1999) for proteobacterial methanotrophs and V170/V631 (Sharp et al. 2012) or LVpmoAf/LVpmoAr (Sharp et al. 2014) for verrucomicrobial methanotroph. For some Type II methanotrophs which possess only mmoX (*Methylocella* and *Methyloferula*), the mmoX gene can be amplified using mmoXA/mmoXD (Auman et al. 2000) or mmoX206f/mmoX886r (Hutchens et al. 2004).

3 Enrichment of Anaerobic Methanotrophs

Based on culture dependent and independent studies, anaerobic methane oxidizers have been shown to be widely distributed in both marine sediments and terrestrial environments. Due to thermodynamic constraints and oxygen sensitivity, cultivation of anaerobic methanotrophs are much more fastidious than that of aerobic methanotrophs. To successfully enrich diverse anaerobic methanotrophs, several key factors should be considered.

Table 4 Primers targeting MMO-encoding genes

Primer	Sequence (5′ → 3′)	PCR conditions					
		Denaturation		Annealing		Extension	
		Temp. (°C)	Time (s)	Temp. (°C)	Time (s)	Temp. (°C)	Time (s)
Proteobacterial *pmoA* gene (Costello and Lidstrom 1999)							
A189	GGNGACTGGGACTTCTGG	94	60	60	60	72	60
mb661	CCGGMGCAACGTCYTTACC						
Proteobacterial *pxmA* gene (Tavormina et al. 2011)							
pxmA230F	GGCARTGGTGGCCNTTGGT	95	60	59	60	72	60
pxmA732R	TGGCGAACCATTTACCGATGTAC						
Proteobacterial mmoX gene (Auman et al. 2000; Hutchens et al. 2004)							
mmoXA	ACCAAGGARCARTTCAAG	92	30	60	60	72	60
mmoXB	TGGCACTCRTARCGCTC						
mmoX206f	ATCGCBAARGAATAYGCSCG	94	60	55	60	72	60
mmoX886r	ACCCANGGCTCGACYTTGAA						
Thermophilic verrucomicrobial pmoA gene (Sharp et al. 2012)							
V170	GGATWGATTGGAAAGATMG	94	60	56	45	75	45
V631	GCAAARCTYCTCATYGTWCC						
Thermophilic verrucomicrobial pmoA gene (Sharp et al. 2014)							
LVpmoAf	GGRTKGACTGGAAAGAYCG	94	60	66	45	72	45
LVpmoAr	GCGAARCTYCGCATCG TTCC						
General bacterial 16S rDNA (Weisburg et al. 1991)							
27F	AGA GTT TGA TCC TGG CTC AG	95	30	50	30	72	60
1492R	GGT TAC CTT GTT ACG ACT T						

3.1 Factors for Anaerobic Methane Oxidation (AOM)

The major controlling factors for the distribution of anaerobic methane-oxidizing bacteria and/or archaea are the availability of methane and terminal electron acceptors which can support the AOM. In addition to electron acceptors, other environmental parameters such as oxygen, temperature, salinity, and pH also play decisive roles in the occurrence of AOM.

3.1.1 Electron Acceptors

AOM has been demonstrated to be coupled to the reduction of electron acceptors such as sulfate, nitrate, nitrite, iron, and manganese (Ettwig et al. 2010, 2016; Haroon et al. 2013). These different types of electron acceptors play critical roles in selecting the type of microorganisms responsible for AOM (Cui et al. 2015). For instance, AOM coupled to the reduction of sulfate is a process mediated by a consortium of anaerobic methanotrophic archaea (ANME) and sulfate reducing bacteria (SRB) (Knittel and Boetius 2009). So far, three types of ANME that are phylogenetically related to different methanogenic archaea have been identified for AOM depending on the use of sulfate as the terminal electron acceptor (ANME-1, ANME-2, ANME-3) (Boetius et al. 2000; Knittel et al. 2005). Elemental sulfur can presumably be used directly by ANME archaea (Milucka et al. 2012). Whereas ANME-2d is shown to be responsible for nitrate-driven AOM (Haroon et al. 2013). This process performed by the ANME-2d, named '*Ca.* Methanoperedens nitroreducens', occurs without a partner organism via reverse methanogenesis with nitrate as the terminal electron acceptor, using genes for nitrate reduction that have been laterally transferred from a bacterial donor (Haroon et al. 2013).

In addition to nitrate, iron, and manganese have been reported to serve as alternative electron acceptors for AOM in a freshwater enrichment culture of '*Ca.* M. nitroreducens' (Ettwig et al. 2016). Furthermore, nitrite reduction can be coupled to methane oxidation by a single bacterial species of the NC10 phylum, '*Ca.* Methylomirabilis oxyfera' (Ettwig et al. 2010). The nitrite-dependent methane-oxidizing bacterium, '*Ca.* M. oxyfera', likely produces molecular oxygen intracellularly from NO, which is fed into a canonic aerobic methane oxidation pathway (Ettwig et al. 2010), resembling that of aerobic methanotrophic bacteria.

Therefore, it is important to carefully select the electron acceptor of interest when trying to establish a successful AOM enrichment culture. Furthermore, collection of source materials from environments where methane and the appropriate electron acceptor (sulfate, nitrate, nitrite, iron, manganese) coexist is also critical in establishing a successful enrichment culture (Meulepas et al. 2009). The main niche for anaerobic methanotrophs on earth is the so-called sulfate-methane transition zone in the seabed (Reeburgh 2007). Gaseous hydrocarbons such as methane migrate via diffusion into sulfate-rich sediment layers. Here microorganisms oxidize methane with sulfate as electron acceptor, resulting in defined sulfate-methane interfaces (Reeburgh and Heggie 1977). To date, all successful enrichment of sulfate-dependent methane oxidizers used inoculum from sites that have a long history of methane exposure and show elevated concentrations of sulfide as well as inorganic carbon (i.e., the products of sulfate-dependent methane oxidation) (Holler et al. 2009, 2011; Meulepas et al. 2009). Besides sulfate, other sulfur compounds could also be used as electron acceptors for AOM. Thiosulfate and sulfite are more thermodynamically favorable than sulfate (Meulepas et al. 2009) and elemental sulfur can presumably be used directly by anaerobic methanotrophic archaea (ANME) (Milucka et al. 2012).

Successful establishment of a nitrate/nitrite-dependent anaerobic methane oxidation (N-DAMO) enrichment culture also requires the collection of inoculums from environments where methane and nitrate/nitrite coexist simultaneously. Thus, methanogenic sludge successfully served as an appropriate inoculum to enrich N-DAMO microorganisms (Luesken et al. 2011). Similarly, the high amount of methane produced by methanogens in paddy soil, coupled with the accumulation of nitrogenous components due to nitrogen fertilizer treatment, also makes paddy soil a suitable environment for the N-DAMO reaction (Conrad 2009). Several studies have reported the abundance of '*Ca.* M. oxyfera'-like bacteria and '*Ca.* M. nitroreducens' ANME-2d in paddy soil (Hu et al. 2014; Vaksmaa et al. 2017; Wang et al. 2012).

3.1.2 Temperature

AOM has been found to occur across a wide range of in situ temperatures from <20 °C (at cold methane seeps) to 90 °C (in hydrothermal environments) (Holler et al. 2011; Wankel et al. 2012). According to the temperature in which they occur, following classes of AOM are defined; psychrophilic (<20 °C), mesophilic (20–50 °C) and thermophilic (>50 °C) AOM. Sulfate-dependent AOM has been demonstrated in the psychrophilic (<20 °C), mesophilic (20–50 °C) and thermophilic temperature range (60 °C), whereas nitrite/nitrate dependent-AOM has only been described in the mesophilic range (20–35 °C) (Ettwig et al. 2010; Haroon et al. 2013; He et al. 2015a, c; Vaksmaa et al. 2017; Wang et al. 2012). To date, thermophilic AOM coupled to sulfate reduction has only been described in the hydrothermal seep sediments in Guaymas Basin (Gulf of Mexico) (Holler et al. 2011) and Middle Valley (Juan de Fuca Ridge) (Wankel et al. 2012) which can support high rates of AOM.

During the cultivation process, the incubation temperature chosen should be close to that of the sampling site. Enrichment cultures initiated with material from cold seep environment (in situ temperature 4–12 °C) were successful when incubated at temperatures between 5 and 15 °C (Girguis et al. 2005; Kruger et al. 2008; Meulepas et al. 2009). When materials from hydrothermally heated environments with steep temperature gradients were used for hydrocarbon-dependent sulfate reduction, temperatures between 37 and 60 °C were successfully used (Holler et al. 2011).

Furthermore, the storage temperature of environmental samples prior to setup is also important. Samples collected from geothermal environments can be stored at room temperature, whereas those collected from low temperature environments can be stored at 4 °C. In this way, samples can be stored for at least 2 months without losing substantial microbial activity.

3.1.3 pH

Another important environmental parameter to be considered during the enrichment process of ANME is the pH. For optimal AOM, in vitro pH should mimic that of the environment where the samples were collected. The in situ pH measurements in marine sediments with high AOM activities indicate only small variations in pH value, between 7.5 and 7.9 in the AOM zone (de Beer et al. 2006). In vitro study on the pH optimum of sulfate-driven AOM with different ANME populations also showed a maximum ca. pH 7.8, which is in the typical pH range for marine sediments (Knittel and Boetius 2009). Most N-DAMO cultures have been incubated in a pH range of 6.0–7.6 (Ettwig et al. 2010; Haroon et al. 2013; He et al. 2015b, c; Hu et al. 2014; Vaksmaa et al. 2017; Wang et al. 2012). A study on the effect of pH showed that N-DAMO cultures can grow between pH 6.0 and 9.0 (He et al. 2015c). However, a pH range of 7.0–8.0 was recommended as N-DAMO activity was inhibited at pH 9 (He et al. 2015c). During enrichment, it is also important to measure and control the pH, as processes such as denitrification increases alkalinity (Rittmann and MacCarty 2001). The pH can be controlled by, (1) adding HCl in the influent to balance the base production, (2) adding CO_2 gas in the bioreactor to control the pH with carbonic acid (Tang et al. 2011) and (3) using a mixture of CH_4–CO_2 gas in the bioreactor. The added CO_2 will serve not just to lower the pH value, but also as a direct carbon source of N-DAMO bacteria that have been shown to fix CO_2 via the CBB cycle (Rasigraf et al. 2014).

3.1.4 Salinity

In addition, successful enrichment of ANMEs from marine, saline lakes, fresh water or brackish habitats requires the adjustment of the salinity of the artificial medium to that of the in situ environment originally sampled. For sediment samples collected from marine environments, artificial seawater medium with marine salinity should be used. He et al. (2015c) reported absence of NaCl as the optimal salinity for N-DAMO cultures enriched from fresh water sediment. These fresh water N-DAMO were inhibited at high NaCl concentrations, probably due to increased osmotic pressure caused by high salt concentration (Jin et al. 2012). High salinity could also damage important metabolic enzymes in the periplasm, such as nitrite reductase (Nir), NO dismutase (NOD) and methanol dehydrogenase (MDH) (Wu et al. 2011; Zhao et al. 2013). However, the discovery of several 16S rRNA and *pmoA* genes of '*Ca.* M. oxyfera'-like bacteria from saline environments (Chen et al. 2014; Shen et al. 2014; Yang et al. 2012) and successful enrichment of '*Ca.* M. oxyfera'-like bacteria from marine samples (He et al. 2015b) show that N-DAMO is well adapted to survive beyond freshwater environments. These results show that it is important to mimic the in situ environment salinity during the enrichment culture setup. For instance, successful enrichment of N-DAMO has been achieved with samples obtained from marine (He et al. 2015b), freshwater (Ettwig et al. 2010; He et al. 2015c) and minerotrophic peatland (Zhu et al. 2012) using artificial medium with similar salinity to the environment or filtered anoxic water from the natural environment.

3.2 *Anaerobic Cultivation*

Cultures of ANME are defined as microbial enrichments obtained and maintained with methane as the sole energy source, an electron acceptor of interest (e.g. sulfate, nitrate, and nitrite), and with carbon dioxide as the sole carbon source. In addition to these essential elements, variations of environmental parameters such as temperature, pH, and salinity not only improve the growth of ANME, but also aid in purification attempts by increasing the selective pressure on the ANME of interest. Most conventional in vitro enrichment techniques involve the use of gas tight serum bottles, or the continuous bioreactor system.

3.2.1 Cultivation in Gas Tight Serum Bottles

The use of gas tight serum bottles for in vitro incubation is of great advantage due to the low cost of setup and maintenance. This method also gives room for testing the microbial activities, kinetics of the metabolic reactions and maintaining large numbers of ANME cultures. For this cultivation method, the incubation pressure should not exceed 0.25 MPa (Beal et al. 2009; Blumenberg et al. 2005; Holler et al. 2011; Meulepas et al. 2009).

Another noteworthy advantage this method provides is flexibility. With gas tight serum bottles, one can set up several experiments at the same time. Likewise, the ease of controlling and manipulating different environmental conditions for the batch incubation-based experiments makes it a widely used technique for cultivation of slow-growing anaerobic methanotrophs (Cassarini 2017). Another advantage this cultivation technique provides is that anaerobic condition can still be maintained during sampling. The use of the batch culture cultivation technique, however, requires that one pays critical attention to some factors which are keys to the enrichment of ANME. Some of these factors include the low solubility of methane and the accumulation of toxic metabolites such as sulfide and nitrite during batch incubations (Cassarini 2017). The techniques for establishing enrichment cultures for ANME using the batch culture method is well described by Laso-Perez et al. (2018).

A modified batch incubation system which made use of a glass tube connected through a needle to a syringe and placed in a pressure-proof cylinder was used for the cultivation of ANME-2 (Nauhaus et al. 2007). During the incubation period, the cells of the AOM consortia, which was tracked using FISH, increased exponentially (Nauhaus et al. 2007). Similarly, Haroon et al. (2013) used a modified batch incubation approach, which allowed the intermittent replacement of supernatant by fresh medium (i.e., fed batch system) once a month to successfully enrich a denitrifying ANME-2d, '*Ca.* M. nitroreducens', at abundances of about 78%.

3.2.2 Cultivation in Bioreactor

Continuous flow incubation columns are designed for nutrient provision and removal of end products at environmentally relevant rates (Girguis et al. 2003). In order to better understand the anaerobic methane oxidation process, Girguis et al. (2003) developed a continuous flow incubation system that was able to mimic in situ conditions and support growth and metabolism of ANME. In this system, 0.2 μm filtered seawater was pumped into a gas tight polyvinylchloride (PVC) manifold. The seawater was then reduced with hydrogen sulfide (510 μM) and saturated with methane (1.5 mM) in a conditioning column (4 h at 0.5 MPa). During operation, the manifold and the seawater reservoir were maintained at 0.2 MPa and 5 °C.

In a bid to attain high methane concentration, continuous bioreactors designed to withstand hydrostatic pressure of up to 44.5 MPa have been used (Deusner et al. 2010). The bioreactor was configured to allow for flexibility, such that it could be operated in batch, fed-batch or continuous mode. Using this bioreactor for the incubation of sediments from the Black Sea, AOM rate was found to increase six-fold when methane partial pressure increased from 0.2 to 6 MPa.

Sauer et al. (2012) employed a similar high-pressured incubation system (operated at 60 MPa and 120 °C) to incubate sediment samples from the Isis Mud volcano. The reactor can be configured to operate in batch or continuous mode, permitts sub-sampling during incubation without pressure loss. Despite the possibility of operating high-pressured reactors at high dissolved methane concentrations, operation and maintenance of such systems are cumbersome. Likewise, the establishment of such systems, requires the implementation of various safety criteria (Cassarini 2017).

3.3 Monitoring AOM

Throughout the enrichment process, the development of the microbial community is monitored using sequencing of amplicons of 16S rRNA and functional genes or metagenomics analysis. FISH is applied to enumerate clades of interest and to discover spatial associations of ANME and other cells in cell aggregates.

3.3.1 16S rRNA and Functional Gene Indicators of AOM

Monitoring of molecular markers such as the 16S rRNA and functional genes including the subunit A of the methyl-CoM reductase (*mcrA*), subunit A of the particulate methane monooxygenase (*pmoA*) and dissimilatory (bi)sulfite reductase (*dsrAB*) genes are important in assignment of identity to AOM in the microbial community as well as monitoring metabolic activities from natural environments and enrichment cultures (Hu et al. 2014; Lee et al. 2013; Yanagawa et al. 2011). Phylogenetic and functional gene analyses have contributed to the metabolic characterization of AOM consortia. Studies using quantitative PCR to determine the abundance of 16S rRNA (Hu et al. 2014) and key functional genes such as *mcrA* (reverse methanogenesis related) in ANME (Lee et al. 2013; Yanagawa et al. 2011) and *dsrA* (sulfite reductase related) in SRB (Lee et al. 2013) and *pmoA* (particulate methane monooxygenase related) in '*Ca. M. oxyfera*' (He et al. 2016; Shen et al. 2014) have been used to enumerate and distinguish microbes within the AOM community. Specific primers to quantify ANME types and anaerobic bacterial methanotrophs have also been designed (Miyashita et al. 2009; Vaksmaa et al. 2017; Wang et al. 2017; Zhou et al. 2014).

The use of the gene-based approach permits the detection and quantification of functional genes and phylotypes of interest (Evans et al. 2019; Wang et al. 2019). However, this approach can sometimes be misleading. For example, the detection of DNA/RNA might be from cells transported from neighboring environments and does not necessarily represent active cells. Hence, to establish the presence of metabolically active cells in an environment, it is important to complement the gene-based method with other techniques such as in situ activity measurement of the biomass (Cassarini 2017).

3.3.2 Activity and Kinetics of AOM

The common approach used to identify the occurrence of AOM is by direct measurement of the profile of indicator chemicals such as CH_4, CO_2, SO_4^{2-}, S^{2-}, NO_2^-, NO_3^-, and NH_4^+ in environments and cultures (Cord-Ruwisch 1985; Hu et al. 2011). Measurement of the chemical profiles though, does not ensure whether the CO_2, S^{2-}, N_2 production is due to AOM or not. Therefore, other complementary methods such as in vitro incubation with stable isotopes or radioisotopes (e.g. $^{13}CH_4$ and $^{14}CH_4$) and profile measurement of labeled carbon are used for estimation of the AOM rate. $^{13}CH_4$ oxidation and $^{13}CO_2$ production were monitored in a batch ANME culture (Knittel and Boetius 2009; Reeburgh 2007). Analysis of the microbial community ensures the presence of ANME and establishes the link between the identity of the ANME and the AOM activity.

3.3.3 Others

Fluorescence in situ hybridization (FISH) has been a useful tool in detecting and enumerating microorganisms associated with AOM microscopically, and in revealing spatial associations of ANME and other cells in cell aggregates (Haroon et al. 2013; He et al. 2015a; Roalkvam et al. 2012). For explicit AOM illustration, it is essential to combine FISH with other approaches such as activity measurements, quantification of ANME cells (by cell count or qPCR) or quantification FISH, isotope probing or mass spectrometric detection of metabolites. Combination of FISH with secondary ion mass spectrometry (SIMS) (Orphan and House 2009) allowed the archaeal component of an AOM consortium to be linked with its function. The isotopic signature of cells can be analyzed using SIMS in order to understand the mechanism of anaerobic methane oxidation along with the syntrophic associations between the AOM consortia and intermediate substrates transferred between the two microorganisms (Orphan and House 2009). Similarly, the use of nanometer-scale secondary ion mass spectrometry (NanoSIMS) in combination with FISH allows for the observation of single cell morphology and quantitative analysis of the elemental and isotopic signature of cells with high sensitivity and precision (Behrens et al. 2008; Musat et al. 2008; Polerecky et al. 2012). This combined technique was used by Dekas et al. (2009) to study nitrogen fixation by ANME-2d. The process involves incubating highly enriched microbial communities with isotopic labeled substrates and monitoring the fate of the substrates using NanoSIMS equipment.

Together with FISH, measurement of specific lipid biomarkers and stable carbon isotopes (Hinrichs and Boetius 2003; Reeburgh 2007; Valentine and Reeburgh 2000), have provided direct evidence of AOM, at potential AOM sites (Orphan and House 2009; Reeburgh 2007). Biomarkers are used to differentiate between archaeal and bacterial cells. Phospholipids, fatty acids with an ester linkage are usually common in bacteria and eukarya (Niemann and Elvert 2008). The lipid profile of the bacterial anaerobic methanotrophs '*Ca.* M. oxyfera', is dominated by 10-methylhexadecanoic acid (Kool et al. 2012). ANME-1 are found to have isoprenoidal dialkyl glycerol tetraethers on their lipid profile, while ANME-2 and ANME-3 mostly contain phosphate-based polar derivatives of archaeol and hydroxyarchaeol (Niemann and Elvert 2008; Rossel et al. 2008). While detection of lipid biomarkers provides information on the microorganism's identity, the carbon isotopic composition of the biomarkers provides information on the carbon source and/or metabolic pathway of the microbes (Hinrichs and Boetius 2003). Different ANME groups produce specific ratios of strongly ^{13}C-depleted membrane lipids (Rossel et al. 2008). Carbon isotope measurements by Kool et al. (2014) showed that '*Ca.* M. oxyfera' lipids are not depleted in ^{13}C due to the utilization of CO_2 for lipid and biomass production rather than methane.

4 Conclusion

This chapter is to summarize current knowledge of cultivation and isolation of methanotrophs which help to obtain new methanotroph resources for industrial processes and environmental applications. First, to obtain methanotroph strains, it is important to choose or design appropriate culture media and growth conditions to closely mimic natural conditions or to target specific methanotroph with desired properties. Diverse methanotrophs have been obtained from various habitats with appropriate manipulation of environmental factors including methane supply, oxygen level, temperature, pH, salinity, trace metals and growth-stimulating factors. Most of the strains isolated from permanently low temperatures (10 °C or below) with possible adaptation to their habitats with high concentrations of methane and low oxygen tensions were found to grow under wider ranges of temperature. Media with low-salt content were used for enrichment of thermophilic methanotrophs such as *Methylothermus* from terrestrial origins. The modified NMS medium with supplementation of NaCl was successfully used to enrich and isolate halophilic methanotrophic bacteria such as *Methylohalobius*. Thermo-acidophilic verrucomicrobial methanotrophs obtained from geothermal habitats in low-salt media acidified by H_2SO_4 can grow in a wide range of temperatures (22.5–81.6 °C) but only in extremely acidic pH (pH 1.8–5.0).

Due to relatively low abundance of methane-oxidizing bacteria in the microbial community in most habitats (Lau et al. 2007), an enrichment procedure under a methane atmosphere is often carried out before isolation steps. While the enrichment is routinely performed using serum bottles, bioreactor systems can be used to select the fast-growing methanotrophs or to select methanotrophs with specific traits under the controlled conditions. After one or several passages of methanotrophic enrichment in liquid medium, the isolation of pure cultures of aerobic methanotrophs was followed by repeated single-colony picking on agar plates or floating membrane filters. Modern technologies for monitoring methane oxidation and evaluation of culture purity were employed for successful enrichment and isolation of the methanotrophs.

AOM has been demonstrated to be coupled to the reduction of electron acceptors such as sulfate, nitrate, nitrite, iron, manganese. So far, three types of ANME that are phylogenetically related to different methanogenic archaea have been identified for AOM depending on the use of sulfate as the terminal electron acceptor (ANME-1, ANME-2, and ANME-3). ANME-2d is shown to be responsible for nitrate-driven AOM. Nitrite reduction can be coupled to methane oxidation by members of the NC10 phylum. Collection of source materials from environments where methane and the appropriate electron acceptor coexist is critical in establishing a successful enrichment culture. To date, all successful enrichment of sulfate-dependent methane oxidizers used inoculum from sites that have a long history of methane exposure and show elevated concentrations of sulfide as well as inorganic carbon. Successful in vitro enrichment techniques used gas-tight serum bottles or continuous bioreactor systems to maintain an anoxic atmosphere.

Despite the successful enrichment and isolation of various aerobic and anaerobic methanotrophs, there might be hidden diversities of methanotrophs in the environment. In the environment, *pmoA*-like genes can only be detectable using PCR-based molecular ecology approaches. Our current knowledge on diversity of ANMEs is largely dependent on the limited strains that are cultivated and thus ANMEs with new methane-activation and transformation mechanisms using diverse electron acceptors are possibly present in nature. Thus, the technologies summarized here should be improved and expanded to retrieve the hidden uncultivated methanotrophs as resources for industrial applications.

Acknowledgements This work was supported by the National Research Foundation of Korea (NRF) grants (NRF-2015-R1A4A1041869; C1 Gas Refinery Program, NRF-2015-M3D3A1A01064881) which are funded by the Ministry of Science, ICT and Future Planning.

References

Auman AJ, Speake CC, Lidstrom ME (2001) *nifH* sequences and nitrogen fixation in type I and type II methanotrophs. Appl Environ Microbiol 67(9):4009–4016

Auman AJ, Stolyar S, Costello AM, Lidstrom ME (2000) Molecular characterization of methanotrophic isolates from freshwater lake sediment. Appl Environ Microbiol 66 (12):5259–5266

Balch WE, Fox GE, Magrum LJ, Woese CR, Wolfe RS (1979) Methanogens: reevaluation of a unique biological group. Microbiol Rev 43(2):260–296

Beal EJ, House CH, Orphan VJ (2009) Manganese- and iron-dependent marine methane oxidation. Science 325(5937):184–187. https://doi.org/10.1126/science.1169984

Behrens S, Lösekann T, Pett-Ridge J, Weber PK, Ng W-O, Stevenson BS, Hutcheon ID, Relman DA, Spormann AM (2008) Linking microbial phylogeny to metabolic activity at the single-cell level by using enhanced element labeling-catalyzed reporter deposition fluorescence In situ hybridization (EL-FISH) and NanoSIMS. Appl Microbiol Biotechnol 74(10):3143–3150. https://doi.org/10.1128/aem.00191-08

Belova SE, Baani M, Suzina NE, Bodelier PLE, Liesack W, Dedysh SN (2011) Acetate utilization as a survival strategy of peat-inhabiting *Methylocystis* spp. Environ Microbiol Rep 3(1):36–46. https://doi.org/10.1111/j.1758-2229.2010.00180.x

Belova SE, Kulichevskaya IS, Bodelier PL, Dedysh SN (2013) *Methylocystis bryophila* sp. nov., a facultatively methanotrophic bacterium from acidic *Sphagnum* peat, and emended description of the genus *Methylocystis* (ex Whittenbury et al. 1970) Bowman et al. 1993. Int J Syst Evol Microbiol 63(Pt 3):1096–1104. https://doi.org/10.1099/ijs.0.043505-0

Blumenberg M, Seifert R, Nauhaus K, Pape T, Michaelis W (2005) In vitro study of lipid biosynthesis in an anaerobically methane-oxidizing microbial mat. Appl Environ Microbiol 71(8):4345–4351. https://doi.org/10.1128/AEM.71.8.4345-4351.2005

Bodrossy L, Holmes EM, Holmes AJ, Kovacs KL, Murrell JC (1997) Analysis of 16S rRNA and methane monooxygenase gene sequences reveals a novel group of thermotolerant and thermophilic methanotrophs, *Methylocaldum* gen. nov. Arch Microbiol 168(6):493–503

Bodrossy L, Kovács KL, McDonald IR, Murrell JC (1999) A novel thermophilic methane-oxidising γ-Proteobacterium. FEMS Microbiol Lett 170(2):335–341. https://doi.org/10.1111/j.1574-6968.1999.tb13392.x

Bodrossy L, Murrell JC, Dalton H, Kalman M, Puskas LG, Kovacs KL (1995) Heat-tolerant methanotrophic bacteria from the hot water effluent of a natural gas field. Appl Environ Microbiol 61(10):3549–3555

Boetius A, Ravenschlag K, Schubert CJ, Rickert D, Widdel F, Gieseke A, Amann R, Jorgensen BB, Witte U, Pfannkuche O (2000) A marine microbial consortium apparently mediating anaerobic oxidation of methane. Nature 407(6804):623–626. https://doi.org/10.1038/35036572

Bowman J (2006) The methanotrophs - The families *Methylococcaceae* and *Methylocystaceae*. In: Dworkin M, Falkow S, Rosenberg E, Schleifer K-H, Stackebrandt E (eds) The Prokaryotes, Proteobacteria: alpha and beta subclasses, vol 5. Springer, New York, NY, pp 266–289. https://doi.org/10.1007/0-387-30745-1_15

Bowman JP, McCammon SA, Skerratt JH (1997) *Methylosphaera hansonii* gen. nov., sp. nov., a psychrophilic, group I methanotroph from Antarctic marine-salinity, meromictic lakes. Microbiology 143(Pt 4):1451–1459. https://doi.org/10.1099/00221287-143-4-1451

Bowman JP, Sly LI, Nichols PD, Hayward AC (1993) Revised taxonomy of the methanotrophs: description of *Methylobacter* gen. nov., emendation of *Methylococcus*, validation of *Methylosinus* and *Methylocystis* species, and a proposal that the family *Methylococcaceae* includes only the group I methanotrophs. Int J Syst Evol Microbiol 43(4):735–753. https://doi.org/10.1099/00207713-43-4-735

Bowman JP, Sly LI, Stackebrandt E (1995) The phylogenetic position of the family *Methylococcaceae*. Int J Syst Bacteriol 45(1):182–185. https://doi.org/10.1099/00207713-45-1-182

Bussmann I, Rahalkar M, Schink B (2006) Cultivation of methanotrophic bacteria in opposing gradients of methane and oxygen. FEMS Microbiol Ecol 56(3):331–344. https://doi.org/10.1111/j.1574-6941.2006.00076.x

Carere CR, Hards K, Houghton KM, Power JF, McDonald B, Collet C, Gapes DJ, Sparling R, Boyd ES, Cook GM, Greening C, Stott MB (2017) Mixotrophy drives niche expansion of verrucomicrobial methanotrophs. ISME J 11(11):2599–2610. https://doi.org/10.1038/ismej.2017.112

Cassarini C (2017) Anaerobic oxidation of methane coupled to the reduction of different sulfur compounds in bioreactors. Université Paris-Est, Taylor & Francis

Chen J, Zhou ZC, Gu JD (2014) Occurrence and diversity of nitrite-dependent anaerobic methane oxidation bacteria in the sediments of the South China Sea revealed by amplification of both 16S rRNA and *pmoA* genes. Appl Microbiol Biotechnol 98(12):5685–5696. https://doi.org/10.1007/s00253-014-5733-4

Chistoserdova L (2015) Methylotrophs in natural habitats: current insights through metagenomics. Appl Microbiol Biotechnol 99(14):5763–5779. https://doi.org/10.1007/s00253-015-6713-z

Conrad R (2009) The global methane cycle: recent advances in understanding the microbial processes involved. Environ Microbiol Rep 1(5):285–292. https://doi.org/10.1111/j.1758-2229.2009.00038.x

Cord-Ruwisch R (1985) A quick method for the determination of dissolved and precipitated sulfides in cultures of sulfate-reducing bacteria. J Microbiol Methods 4(1):33–36. https://doi.org/10.1016/0167-7012(85)90005-3

Costello AM, Lidstrom ME (1999) Molecular characterization of functional and phylogenetic genes from natural populations of methanotrophs in lake sediments. Appl Environ Microbiol 65 (11):5066–5074

Cui M, Ma A, Qi H, Zhuang X, Zhuang G (2015) Anaerobic oxidation of methane: an "active" microbial process. Microbiology 4(1):1–11. https://doi.org/10.1002/mbo3.232

Danilova OV, Kulichevskaya IS, Rozova ON, Detkova EN, Bodelier PL, Trotsenko YA, Dedysh SN (2013) *Methylomonas paludis* sp. nov., the first acid-tolerant member of the genus *Methylomonas*, from an acidic wetland. Int J Syst Evol Microbiol 63(Pt 6):2282–2289. https://doi.org/10.1099/ijs.0.045658-0

Danilova OV, Suzina NE, Van De Kamp J, Svenning MM, Bodrossy L, Dedysh SN (2016) A new cell morphotype among methane oxidizers: a spiral-shaped obligately microaerophilic methanotroph from northern low-oxygen environments. ISME J 10(11):2734–2743. https://doi.org/10.1038/ismej.2016.48

de Beer D, Sauter E, Niemann H, Kaul N, Foucher J-P, Witte U, Schlüter M, Boetius A (2006) In situ fluxes and zonation of microbial activity in surface sediments of the Håkon Mosby Mud Volcano. Limnol Oceanogr 51(3):1315–1331. https://doi.org/10.4319/lo.2006.51.3.1315

Dedysh SN, Belova SE, Bodelier PL, Smirnova KV, Khmelenina VN, Chidthaisong A, Trotsenko YA, Liesack W, Dunfield PF (2007) *Methylocystis heyeri* sp. nov., a novel type II methanotrophic bacterium possessing 'signature' fatty acids of type I methanotrophs. Int J Syst Evol Microbiol 57(Pt 3):472–479. https://doi.org/10.1099/ijs.0.64623-0

Dedysh SN, Berestovskaya YY, Vasylieva LV, Belova SE, Khmelenina VN, Suzina NE, Trotsenko YA, Liesack W, Zavarzin GA (2004) *Methylocella tundrae* sp. nov., a novel methanotrophic bacterium from acidic tundra peatlands. Int J Syst Evol Microbiol 54(Pt 1):151–156. https://doi.org/10.1099/ijs.0.02805-0

Dedysh SN, Khmelenina VN, Suzina NE, Trotsenko YA, Semrau JD, Liesack W, Tiedje JM (2002) *Methylocapsa acidiphila* gen. nov., sp. nov., a novel methane-oxidizing and dinitrogen-fixing acidophilic bacterium from *Sphagnum* bog. Int J Syst Evol Microbiol 52(Pt 1):251–261. https://doi.org/10.1099/00207713-52-1-251

Dedysh SN, Liesack W, Khmelenina VN, Suzina NE, Trotsenko YA, Semrau JD, Bares AM, Panikov NS, Tiedje JM (2000) Methylocella palustris gen. nov., sp. nov., a new methane-oxidizing acidophilic bacterium from peat bogs, representing a novel subtype of serine-pathway methanotrophs. Int J Syst Evol Microbiol 50(Pt 3):955–969. https://doi.org/10.1099/00207713-50-3-955

Dedysh SN, Panikov NS, Tiedje JM (1998) Acidophilic methanotrophic communities from Sphagnum peat bogs. Appl Environ Microbiol 64(3):922–929

Dekas AE, Poretsky RS, Orphan VJ (2009) Deep-sea archaea fix and share nitrogen in methane-consuming microbial consortia. Science 326(5951):422–426. https://doi.org/10.1126/science.1178223

Deusner C, Meyer V, Ferdelman TG (2010) High-pressure systems for gas-phase free continuous incubation of enriched marine microbial communities performing anaerobic oxidation of methane. Biotechnol Bioeng 105(3):524–533. https://doi.org/10.1002/bit.22553

Deutzmann JS, Hoppert M, Schink B (2014) Characterization and phylogeny of a novel methanotroph, *Methyloglobulus morosus* gen. nov., spec. nov. Syst Appl Microbiol 37 (3):165–169. https://doi.org/10.1016/j.syapm.2014.02.001

Dunfield PF, Khmelenina VN, Suzina NE, Trotsenko YA, Dedysh SN (2003) *Methylocella silvestris* sp. nov., a novel methanotroph isolated from an acidic forest cambisol. Int J Syst Evol Microbiol 53(Pt 5):1231–1239. https://doi.org/10.1099/ijs.0.02481-0

Dunfield PF, Yuryev A, Senin P, Smirnova AV, Stott MB, Hou S, Ly B, Saw JH, Zhou Z, Ren Y, Wang J, Mountain BW, Crowe MA, Weatherby TM, Bodelier PL, Liesack W, Feng L, Wang L, Alam M (2007) Methane oxidation by an extremely acidophilic bacterium of the phylum *Verrucomicrobia*. Nature 450(7171):879–882. https://doi.org/10.1038/nature06411

Eshinimaev B, Khmelenina VN, Trotsenko Iu A (2008) First isolation of a type II methanotroph from a soda lake. Mikrobiologiia 77(5):704–707

Ettwig KF, Butler MK, Le Paslier D, Pelletier E, Mangenot S, Kuypers MM, Schreiber F, Dutilh BE, Zedelius J, de Beer D, Gloerich J, Wessels HJ, van Alen T, Luesken F, Wu ML, van de Pas-Schoonen KT, Op den Camp HJ, Janssen-Megens EM, Francoijs KJ, Stunnenberg H, Weissenbach J, Jetten MS, Strous M (2010) Nitrite-driven anaerobic methane oxidation by oxygenic bacteria. Nature 464(7288):543–548. https://doi.org/10.1038/nature08883

Ettwig KF, Zhu B, Speth D, Keltjens JT, Jetten MSM, Kartal B (2016) Archaea catalyze iron-dependent anaerobic oxidation of methane. Proc Natl Acad Sci U S A 113(45):12792–12796. https://doi.org/10.1073/pnas.1609534113

Evans PN, Boyd JA, Leu AO, Woodcroft BJ, Parks DH, Hugenholtz P, Tyson GW (2019) An evolving view of methane metabolism in the Archaea. Nat Rev Microbiol 17(4):219–232. https://doi.org/10.1038/s41579-018-0136-7

Girguis PR, Cozen AE, DeLong EF (2005) Growth and population dynamics of anaerobic methane-oxidizing archaea and sulfate-reducing bacteria in a continuous-flow bioreactor. Appl Environ Microbiol 71(7):3725–3733. https://doi.org/10.1128/AEM.71.7.3725-3733.2005

Girguis PR, Orphan VJ, Hallam SJ, DeLong EF (2003) Growth and methane oxidation rates of anaerobic methanotrophic archaea in a continuous-flow bioreactor. Appl Environ Microbiol 69 (9):5472–5482

Hanson RS, Hanson TE (1996) Methanotrophic bacteria. Microbiol Rev 60(2):439–471

Haroon MF, Hu S, Shi Y, Imelfort M, Keller J, Hugenholtz P, Yuan Z, Tyson GW (2013) Anaerobic oxidation of methane coupled to nitrate reduction in a novel archaeal lineage. Nature 500(7464):567

He Z, Cai C, Shen L, Lou L, Zheng P, Xu X, Hu B (2015a) Effect of inoculum sources on the enrichment of nitrite-dependent anaerobic methane-oxidizing bacteria. Appl Microbiol Biotechnol 99(2):939–946. https://doi.org/10.1007/s00253-014-6033-8

He Z, Cai C, Wang J, Xu X, Zheng P, Jetten MS, Hu B (2016) A novel denitrifying methanotroph of the NC10 phylum and its microcolony. Sci Rep 6:32241. https://doi.org/10.1038/srep32241

He Z, Geng S, Cai C, Liu S, Liu Y, Pan Y, Lou L, Zheng P, Xu X, Hu B (2015b) Anaerobic oxidation of methane coupled to nitrite reduction by halophilic marine NC10 bacteria. Appl Environ Microbiol 81(16):5538–5545. https://doi.org/10.1128/AEM.00984-15

He Z, Geng S, Shen L, Lou L, Zheng P, Xu X, Hu B (2015c) The short- and long-term effects of environmental conditions on anaerobic methane oxidation coupled to nitrite reduction. Water Res 68:554–562. https://doi.org/10.1016/j.watres.2014.09.055

Heyer J, Berger U, Hardt M, Dunfield PF (2005) *Methylohalobius crimeensis* gen. nov., sp. nov., a moderately halophilic, methanotrophic bacterium isolated from hypersaline lakes of Crimea. Int J Syst Evol Microbiol 55(Pt 5):1817–1826. https://doi.org/10.1099/ijs.0.63213-0

Heyer J, Galchenko VF, Dunfield PF (2002) Molecular phylogeny of type II methane-oxidizing bacteria isolated from various environments. Microbiology 148(Pt 9):2831–2846. https://doi.org/10.1099/00221287-148-9-2831

Hinrichs K-U, Boetius A (2003) The anaerobic oxidation of methane: new insights in microbial ecology and biogeochemistry. In: Wefer G, Billett D, Hebbeln D, Jørgensen BB, Schlüter M, van Weering TCE (eds) Ocean margin systems. Springer, Berlin, pp 457–477. https://doi.org/10.1007/978-3-662-05127-6_28

Hirayama H, Abe M, Miyazaki M, Nunoura T, Furushima Y, Yamamoto H, Takai K (2014) *Methylomarinovum caldicuralii* gen. nov., sp. nov., a moderately thermophilic methanotroph isolated from a shallow submarine hydrothermal system, and proposal of the family Methylothermaceae fam. nov. Int J Syst Evol Microbiol 64(Pt 3):989–999. https://doi.org/10.1099/ijs.0.058172-0

Hirayama H, Fuse H, Abe M, Miyazaki M, Nakamura T, Nunoura T, Furushima Y, Yamamoto H, Takai K (2013) *Methylomarinum vadi* gen. nov., sp. nov., a methanotroph isolated from two distinct marine environments. Int J Syst Evol Microbiol 63(Pt 3):1073–1082. https://doi.org/10.1099/ijs.0.040568-0

Hirayama H, Suzuki Y, Abe M, Miyazaki M, Makita H, Inagaki F, Uematsu K, Takai K (2011) *Methylothermus subterraneus* sp. nov., a moderately thermophilic methanotroph isolated from a terrestrial subsurface hot aquifer. Int J Syst Evol Microbiol 61(Pt 11):2646–2653. https://doi.org/10.1099/ijs.0.028092-0

Hoefman S, van der Ha D, De Vos P, Boon N, Heylen K (2012) Miniaturized extinction culturing is the preferred strategy for rapid isolation of fast-growing methane-oxidizing bacteria. Microb Biotechnol 5(3):368–378. https://doi.org/10.1111/j.1751-7915.2011.00314.x

Hojberg O, Binnerup SJ, Sorensen J (1997) Growth of silicone-immobilized bacteria on polycarbonate membrane filters, a technique to study microcolony formation under anaerobic conditions. Appl Environ Microbiol 63(7):2920–2924

Holler T, Wegener G, Knittel K, Boetius A, Brunner B, Kuypers MM, Widdel F (2009) Substantial (13) C/(12) C and D/H fractionation during anaerobic oxidation of methane by marine consortia enriched in vitro. Environ Microbiol Rep 1(5):370–376. https://doi.org/10.1111/j.1758-2229.2009.00074.x

Holler T, Widdel F, Knittel K, Amann R, Kellermann MY, Hinrichs KU, Teske A, Boetius A, Wegener G (2011) Thermophilic anaerobic oxidation of methane by marine microbial consortia. ISME J 5(12):1946–1956. https://doi.org/10.1038/ismej.2011.77

Hu BL, Shen LD, Lian X, Zhu Q, Liu S, Huang Q, He ZF, Geng S, Cheng DQ, Lou LP, Xu XY, Zheng P, He YF (2014) Evidence for nitrite-dependent anaerobic methane oxidation as a previously overlooked microbial methane sink in wetlands. Proc Natl Acad Sci U S A 111 (12):4495–4500. https://doi.org/10.1073/pnas.1318393111

Hu S, Zeng RJ, Keller J, Lant PA, Yuan Z (2011) Effect of nitrate and nitrite on the selection of microorganisms in the denitrifying anaerobic methane oxidation process. Environ Microbiol Rep 3(3):315–319. https://doi.org/10.1111/j.1758-2229.2010.00227.x

Hutchens E, Radajewski S, Dumont MG, McDonald IR, Murrell JC (2004) Analysis of methanotrophic bacteria in Movile Cave by stable isotope probing. Environ Microbiol 6 (2):111–120

Iguchi H, Yurimoto H, Sakai Y (2011) Stimulation of methanotrophic growth in cocultures by cobalamin excreted by rhizobia. Appl Environ Microbiol 77(24):8509–8515. https://doi.org/10.1128/aem.05834-11

Islam T, Jensen S, Reigstad LJ, Larsen O, Birkeland NK (2008) Methane oxidation at 55 degrees C and pH 2 by a thermoacidophilic bacterium belonging to the *Verrucomicrobia* phylum. Proc Natl Acad Sci U S A 105(1):300–304. https://doi.org/10.1073/pnas.0704162105

Islam T, Larsen O, Torsvik V, Ovreas L, Panosyan H, Murrell JC, Birkeland NK, Bodrossy L (2015) Novel methanotrophs of the family Methylococcaceae from different geographical regions and habitats. Microorganisms 3(3):484–499. https://doi.org/10.3390/microorganisms3030484

Islam T, Torsvik V, Larsen Ø, Bodrossy L, Øvreås L, Birkeland N-K (2016) Acid-tolerant moderately thermophilic methanotrophs of the class Gammaproteobacteria isolated from tropical topsoil with methane seeps. Front Microbiol 7(851):1–12. https://doi.org/10.3389/fmicb.2016.00851

Jin R-C, Yang G-F, Yu J-J, Zheng P (2012) The inhibition of the anammox process: a review. Chem Eng J 197:67–79. https://doi.org/10.1016/j.cej.2012.05.014

Kaluzhnaya M, Khmelenina V, Eshinimaev B, Suzina N, Nikitin D, Solonin A, Lin JL, McDonald I, Murrell C, Trotsenko Y (2001) Taxonomic characterization of new alkaliphilic and alkalitolerant methanotrophs from soda lakes of the Southeastern Transbaikal region and description of *Methylomicrobium buryatense* sp.nov. Syst Appl Microbiol 24(2):166–176. https://doi.org/10.1078/0723-2020-00028

Kalyuzhnaya MG, Khmelenina V, Eshinimaev B, Sorokin D, Fuse H, Lidstrom M, Trotsenko Y (2008) Classification of halo(alkali)philic and halo(alkali)tolerant methanotrophs provisionally assigned to the genera *Methylomicrobium* and *Methylobacter* and emended description of the genus *Methylomicrobium*. Int J Syst Evol Microbiol 58(Pt 3):591–596. https://doi.org/10.1099/ijs.0.65317-0

Kalyuzhnaya MG, Khmelenina VN, Kotelnikova S, Holmquist L, Pedersen K, Trotsenko YA (1999) *Methylomonas scandinavica* sp. nov., a new methanotrophic psychrotrophic bacterium isolated from deep igneous rock ground water of Sweden. Syst Appl Microbiol 22(4):565–572. https://doi.org/10.1016/S0723-2020(99)80010-1

Kalyuzhnaya MG, Puri AW, Lidstrom ME (2015) Metabolic engineering in methanotrophic bacteria. Metab Eng 29:142–152. https://doi.org/10.1016/j.ymben.2015.03.010

Khadem AF, Pol A, Wieczorek A, Mohammadi SS, Francoijs KJ, Stunnenberg HG, Jetten MS, Op den Camp HJ (2011) Autotrophic methanotrophy in verrucomicrobia: Methylacidiphilum fumariolicum SolV uses the Calvin-Benson-Bassham cycle for carbon dioxide fixation. J Bacteriol 193(17):4438–4446. https://doi.org/10.1128/JB.00407-11

Kim J, Kim DD, Yoon S (2018) Rapid isolation of fast-growing methanotrophs from environmental samples using continuous cultivation with gradually increased dilution rates. Appl Microbiol Biotechnol 102(13):5707–5715. https://doi.org/10.1007/s00253-018-8978-5

Kip N, Ouyang W, van Winden J, Raghoebarsing A, van Niftrik L, Pol A, Pan Y, Bodrossy L, van Donselaar EG, Reichart GJ, Jetten MS, Damste JS, Op den Camp HJ (2011) Detection, isolation, and characterization of acidophilic methanotrophs from *Sphagnum* mosses. Appl Environ Microbiol 77(16):5643–5654. https://doi.org/10.1128/AEM.05017-11

Knief C (2015) Diversity and habitat preferences of cultivated and uncultivated aerobic methanotrophic bacteria evaluated based on *pmoA* as molecular marker. Front Microbiol 6:1346. https://doi.org/10.3389/fmicb.2015.01346

Knittel K, Boetius A (2009) Anaerobic oxidation of methane: progress with an unknown process. Annu Rev Microbiol 63:311–334. https://doi.org/10.1146/annurev.micro.61.080706.093130

Knittel K, Losekann T, Boetius A, Kort R, Amann R (2005) Diversity and distribution of methanotrophic archaea at cold seeps. Appl Environ Microbiol 71(1):467–479. https://doi.org/10.1128/AEM.71.1.467-479.2005

Kool DM, Talbot HM, Rush D, Ettwig K, Sinninghe Damsté JS (2014) Rare bacteriohopanepolyols as markers for an autotrophic, intra-aerobic methanotroph. Geochim Cosmochim Acta 136:114–125. https://doi.org/10.1016/j.gca.2014.04.002

Kool DM, Zhu B, Rijpstra WI, Jetten MS, Ettwig KF, Sinninghe Damste JS (2012) Rare branched fatty acids characterize the lipid composition of the intra-aerobic methane oxidizer "*Candidatus* Methylomirabilis oxyfera". Appl Environ Microbiol 78(24):8650–8656. https://doi.org/10.1128/AEM.02099-12
Krause SMB, Johnson T, Samadhi Karunaratne Y, Fu Y, Beck DAC, Chistoserdova L, Lidstrom ME (2017) Lanthanide-dependent cross-feeding of methane-derived carbon is linked by microbial community interactions. Proc Natl Acad Sci U S A 114(2):358–363. https://doi.org/10.1073/pnas.1619871114
Kruger M, Wolters H, Gehre M, Joye SB, Richnow HH (2008) Tracing the slow growth of anaerobic methane-oxidizing communities by (15)N-labelling techniques. FEMS Microbiol Ecol 63(3):401–411. https://doi.org/10.1111/j.1574-6941.2007.00431.x
Laso-Perez R, Krukenberg V, Musat F, Wegener G (2018) Establishing anaerobic hydrocarbon-degrading enrichment cultures of microorganisms under strictly anoxic conditions. Nat Protoc 13(6):1310–1330. https://doi.org/10.1038/nprot.2018.030
Lau E, Ahmad A, Steudler PA, Cavanaugh CM (2007) Molecular characterization of methanotrophic communities in forest soils that consume atmospheric methane. FEMS Microbiol Ecol 60(3):490–500. https://doi.org/10.1111/j.1574-6941.2007.00308.x
Lau E, Fisher MC, Steudler PA, Cavanaugh CM (2013) The methanol mehydrogenase gene, *mxaF*, as a functional and phylogenetic marker for proteobacterial methanotrophs in natural environments. PLoS One 8(2):e56993. https://doi.org/10.1371/journal.pone.0056993
Lee J-W, Kwon KK, Azizi A, Oh H-M, Kim W, Bahk J-J, Lee D-H, Lee J-H (2013) Microbial community structures of methane hydrate-bearing sediments in the Ulleung Basin, East Sea of Korea. Mar Pet Geol 47:136–146. https://doi.org/10.1016/j.marpetgeo.2013.06.002
Lidstrom ME (1988) Isolation and characterization of marine methanotrophs. Antonie Leeuwenhoek 54(3):189–199
Luesken FA, Zhu B, van Alen TA, Butler MK, Diaz MR, Song B, Op den Camp HJ, Jetten MS, Ettwig KF (2011) pmoA primers for detection of anaerobic methanotrophs. Appl Environ Microbiol 77(11):3877–3880. https://doi.org/10.1128/AEM.02960-10
Martinez-Cruz K, Leewis MC, Herriott IC, Sepulveda-Jauregui A, Anthony KW, Thalasso F, Leigh MB (2017) Anaerobic oxidation of methane by aerobic methanotrophs in sub-Arctic lake sediments. Sci Total Environ 607-608:23–31. https://doi.org/10.1016/j.scitotenv.2017.06.187
McDonald IR, Bodrossy L, Chen Y, Murrell JC (2008) Molecular ecology techniques for the study of aerobic methanotrophs. Appl Environ Microbiol 74(5):1305–1315. https://doi.org/10.1128/AEM.02233-07
Meulepas RJ, Jagersma CG, Gieteling J, Buisman CJ, Stams AJ, Lens PN (2009) Enrichment of anaerobic methanotrophs in sulfate-reducing membrane bioreactors. Biotechnol Bioeng 104(3):458–470. https://doi.org/10.1002/bit.22412
Milucka J, Ferdelman TG, Polerecky L, Franzke D, Wegener G, Schmid M, Lieberwirth I, Wagner M, Widdel F, Kuypers MMM (2012) Zero-valent sulphur is a key intermediate in marine methane oxidation. Nature 491:541. https://doi.org/10.1038/nature11656
Miyashita A, Mochimaru H, Kazama H, Ohashi A, Yamaguchi T, Nunoura T, Horikoshi K, Takai K, Imachi H (2009) Development of 16S rRNA gene-targeted primers for detection of archaeal anaerobic methanotrophs (ANMEs). FEMS Microbiol Lett 297(1):31–37. https://doi.org/10.1111/j.1574-6968.2009.01648.x
Musat N, Halm H, Winterholler B, Hoppe P, Peduzzi S, Hillion F, Horreard F, Amann R, Jorgensen BB, Kuypers MM (2008) A single-cell view on the ecophysiology of anaerobic phototrophic bacteria. Proc Natl Acad Sci U S A 105(46):17861–17866. https://doi.org/10.1073/pnas.0809329105
Nauhaus K, Albrecht M, Elvert M, Boetius A, Widdel F (2007) In vitro cell growth of marine archaeal-bacterial consortia during anaerobic oxidation of methane with sulfate. Environ Microbiol 9(1):187–196. https://doi.org/10.1111/j.1462-2920.2006.01127.x

Nguyen N-L, Yu W-J, Gwak J-H, Kim S-J, Park S-J, Herbold CW, Kim J-G, Jung M-Y, Rhee S-K (2018) Genomic insights into the acid adaptation of novel methanotrophs enriched from acidic forest soils. Front Microbiol 9:1982. https://doi.org/10.3389/fmicb.2018.01982

Nguyen NL, Yu WJ, Yang HY, Kim JG, Jung MY, Park SJ, Roh SW, Rhee SK (2017) A novel methanotroph in the genus *Methylomonas* that contains a distinct clade of soluble methane monooxygenase. J Microbiol 55(10):775–782. https://doi.org/10.1007/s12275-017-7317-3

Niemann H, Elvert M (2008) Diagnostic lipid biomarker and stable carbon isotope signatures of microbial communities mediating the anaerobic oxidation of methane with sulphate. Org Geochem 39(12):1668–1677. https://doi.org/10.1016/j.orggeochem.2007.11.003

Ogiso T, Ueno C, Dianou D, Huy TV, Katayama A, Kimura M, Asakawa S (2012) *Methylomonas koyamae* sp. nov., a type I methane-oxidizing bacterium from floodwater of a rice paddy field. Int J Syst Evol Microbiol 62(Pt 8):1832–1837. https://doi.org/10.1099/ijs.0.035261-0

Omel'chenko MV, Vasil'eva LV, Zavarzin GA, Savel'eva ND, Lysenko AM, Mityushina LL, Khmelenina VN, Trotsenko YA (1996) A novel psychrophilic methanotroph of the genus *Methylobacter*. Microbiology 65:339–343

Orphan VJ, House CH (2009) Geobiological investigations using secondary ion mass spectrometry: microanalysis of extant and paleo-microbial processes. Geobiology 7(3):360–372. https://doi.org/10.1111/j.1472-4669.2009.00201

Oshkin IY, Beck DA, Lamb AE, Tchesnokova V, Benuska G, McTaggart TL, Kalyuzhnaya MG, Dedysh SN, Lidstrom ME, Chistoserdova L (2015) Methane-fed microbial microcosms show differential community dynamics and pinpoint taxa involved in communal response. ISME J 9 (5):1119–1129. https://doi.org/10.1038/ismej.2014.203

Oshkin IY, Belova SE, Danilova OV, Miroshnikov KK, Rijpstra WI, Sinninghe Damste JS, Liesack W, Dedysh SN (2016) *Methylovulum psychrotolerans* sp. nov., a cold-adapted methanotroph from low-temperature terrestrial environments, and emended description of the genus *Methylovulum*. Int J Syst Evol Microbiol 66(6):2417–2423. https://doi.org/10.1099/ijsem.0.001046

Oswald K, Milucka J, Brand A, Hach P, Littmann S, Wehrli B, Kuypers MMM, Schubert CJ (2016) Aerobic gammaproteobacterial methanotrophs mitigate methane emissions from oxic and anoxic lake waters. Limnol Oceanogr 61(S1):S101–S118. https://doi.org/10.1002/lno.10312

Pol A, Heijmans K, Harhangi HR, Tedesco D, Jetten MS, Op den Camp HJ (2007) Methanotrophy below pH 1 by a new Verrucomicrobia species. Nature 450(7171):874–878. https://doi.org/10.1038/nature06222

Polerecky L, Adam B, Milucka J, Musat N, Vagner T, Kuypers MM (2012) Look@NanoSIMS—a tool for the analysis of nanoSIMS data in environmental microbiology. Environ Microbiol 14 (4):1009–1023. https://doi.org/10.1111/j.1462-2920.2011.02681.x

Rahalkar M, Bussmann I, Schink B (2007) *Methylosoma difficile* gen. nov., sp. nov., a novel methanotroph enriched by gradient cultivation from littoral sediment of Lake Constance. Int J Syst Evol Microbiol 57(Pt 5):1073–1080. https://doi.org/10.1099/ijs.0.64574-0

Rasigraf O, Kool DM, Jetten MS, Sinninghe Damste JS, Ettwig KF (2014) Autotrophic carbon dioxide fixation via the Calvin-Benson-Bassham cycle by the denitrifying methanotroph "*Candidatus* Methylomirabilis oxyfera". Appl Environ Microbiol 80(8):2451–2460. https://doi.org/10.1128/AEM.04199-13

Reeburgh WS (2007) Oceanic methane biogeochemistry. Chem Rev 107(2):486–513. https://doi.org/10.1021/cr050362v

Reeburgh WS, Heggie DT (1977) Microbial methane consumption reactions and their effect on methane distributions in freshwater and marine environments. Limnol Oceanogr 22(1):1–9. https://doi.org/10.4319/lo.1977.22.1.0001

Rissanen AJ, Saarenheimo J, Tiirola M, Peura S, Aalto SL, Karvinen A, Nykänen H (2018) Gammaproteobacterial methanotrophs dominate methanotrophy in aerobic and anaerobic layers of boreal lake waters. Aquat Microb Ecol 81(3):257–276

Rittmann BE, MacCarty PL (2001) Environmental biotechnology: principles and applications. McGraw-Hill, Boston, MA

Roalkvam I, Dahle H, Chen Y, Jorgensen SL, Haflidason H, Steen IH (2012) Fine-scale community structure analysis of ANME in Nyegga sediments with high and low methane flux. Front Microbiol 3:216. https://doi.org/10.3389/fmicb.2012.00216

Rossel PE, Lipp JS, Fredricks HF, Arnds J, Boetius A, Elvert M, Hinrichs K-U (2008) Intact polar lipids of anaerobic methanotrophic archaea and associated bacteria. Org Geochem 39 (8):992–999. https://doi.org/10.1016/j.orggeochem.2008.02.021

Söhngen NL (1906) Ueber Bakterien, welche Methan als Kohlenstoffnahrung und Energiequelle gebrauchen. Centralbl Bakteriol Parasitenk Infektionskr Hyg Abt II 15:513–517

Sauer P, Glombitza C, Kallmeyer J (2012) A system for incubations at high gas partial pressure. Front Microbiol 3:25. https://doi.org/10.3389/fmicb.2012.00025

Semrau JD, DiSpirito AA, Gu W, Yoon S (2018) Metals and methanotrophy. Appl Environ Microbiol 84(6). https://doi.org/10.1128/AEM.02289-17

Sharp CE, Smirnova AV, Graham JM, Stott MB, Khadka R, Moore TR, Grasby SE, Strack M, Dunfield PF (2014) Distribution and diversity of Verrucomicrobia methanotrophs in geothermal and acidic environments. Environ Microbiol 16(6):1867–1878. https://doi.org/10.1111/1462-2920.12454

Sharp CE, Stott MB, Dunfield PF (2012) Detection of autotrophic verrucomicrobial methanotrophs in a geothermal environment using stable isotope probing. Front Microbiol 3:303. https://doi.org/10.3389/fmicb.2012.00303

Sharpe PL, Dicosimo D, Bosak MD, Knoke K, Tao L, Cheng Q, Ye RW (2007) Use of transposon promoter-probe vectors in the metabolic engineering of the obligate methanotroph *Methylomonas* sp. strain 16a for enhanced C40 carotenoid synthesis. Appl Environ Microbiol 73(6):1721

Shen LD, Liu S, Zhu Q, Li XY, Cai C, Cheng DQ, Lou LP, Xu XY, Zheng P, Hu BL (2014) Distribution and diversity of nitrite-dependent anaerobic methane-oxidising bacteria in the sediments of the Qiantang River. Microb Ecol 67(2):341–349. https://doi.org/10.1007/s00248-013-0330-0

Sorokin DY, Jones BE, Gijs Kuenen J (2000) An obligate methylotrophic, methane-oxidizing Methylomicrobium species from a highly alkaline environment. Extremophiles 4(3):145–155. https://doi.org/10.1007/s007920070029

Strong PJ, Xie S, Clarke WP (2015) Methane as a resource: can the methanotrophs add value? Environ Sci Technol 49(7):4001–4018. https://doi.org/10.1021/es504242n

Svenning MM, Wartiainen I, Hestnes AG, Binnerup SJ (2003) Isolation of methane oxidising bacteria from soil by use of a soil substrate membrane system. FEMS Microbiol Ecol 44 (3):347–354. https://doi.org/10.1016/S0168-6496(03)00073-4

Takeuchi M, Kamagata Y, Oshima K, Hanada S, Tamaki H, Marumo K, Maeda H, Nedachi M, Hattori M, Iwasaki W, Sakata S (2014) *Methylocaldum marinum* sp. nov., a thermotolerant, methane-oxidizing bacterium isolated from marine sediments, and emended description of the genus Methylocaldum. Int J Syst Evol Microbiol 64(Pt 9):3240–3246. https://doi.org/10.1099/ijs.0.063503-0

Tang Y, Zhou C, Ziv-El M, Rittmann BE (2011) A pH-control model for heterotrophic and hydrogen-based autotrophic denitrification. Water Res 45(1):232–240. https://doi.org/10.1016/j.watres.2010.07.049

Tavormina PL, Hatzenpichler R, McGlynn S, Chadwick G, Dawson KS, Connon SA, Orphan VJ (2015) *Methyloprofundus sedimenti* gen. nov., sp. nov., an obligate methanotroph from ocean sediment belonging to the 'deep sea-1' clade of marine methanotrophs. Int J Syst Evol Microbiol 65(Pt 1):251–259. https://doi.org/10.1099/ijs.0.062927-0

Tavormina PL, Orphan VJ, Kalyuzhnaya MG, Jetten MS, Klotz MG (2011) A novel family of functional operons encoding methane/ammonia monooxygenase-related proteins in gammaproteobacterial methanotrophs. Environ Microbiol Rep 3(1):91–100. https://doi.org/10.1111/j.1758-2229.2010.00192.x

Trotsenko YA, Khmelenina VN (2002) Biology of extremophilic and extremotolerant methanotrophs. Arch Microbiol 177(2):123–131. https://doi.org/10.1007/s00203-001-0368-0

Trotsenko YA, Murrell JC (2008) Metabolic aspects of aerobic obligate methanotrophy. Adv Appl Microbiol 63:183–229. https://doi.org/10.1016/S0065-2164(07)00005-6

Tsubota J, Eshinimaev B, Khmelenina VN, Trotsenko YA (2005) *Methylothermus thermalis* gen. nov., sp. nov., a novel moderately thermophilic obligate methanotroph from a hot spring in Japan. Int J Syst Evol Microbiol 55(Pt 5):1877–1884. https://doi.org/10.1099/ijs.0.63691-0

Vaksmaa A, Guerrero-Cruz S, van Alen TA, Cremers G, Ettwig KF, Luke C, Jetten MSM (2017) Enrichment of anaerobic nitrate-dependent methanotrophic '*Candidatus* Methanoperedens nitroreducens' archaea from an Italian paddy field soil. Appl Microbiol Biotechnol 101 (18):7075–7084. https://doi.org/10.1007/s00253-017-8416-0

Valentine DL, Reeburgh WS (2000) New perspectives on anaerobic methane oxidation. Environ Microbiol 2(5):477–484

Wang S, Liu Y, Liu G, Huang Y, Zhou Y (2017) A new primer to amplify *pmoA* gene from NC10 bacteria in the sediments of Dongchang Lake and Dongping Lake. Curr Microbiol 74 (8):908–914. https://doi.org/10.1007/s00284-017-1260-8

Wang Y, Wegener G, Hou J, Wang F, Xiao X (2019) Expanding anaerobic alkane metabolism in the domain of Archaea. Nat Microbiol 4(4):595–602. https://doi.org/10.1038/s41564-019-0364-2

Wang Y, Zhu G, Harhangi HR, Zhu B, Jetten MS, Yin C, Op den Camp HJ (2012) Co-occurrence and distribution of nitrite-dependent anaerobic ammonium and methane-oxidizing bacteria in a paddy soil. FEMS Microbiol Lett 336(2):79–88. https://doi.org/10.1111/j.1574-6968.2012.02654.x

Wankel SD, Adams MM, Johnston DT, Hansel CM, Joye SB, Girguis PR (2012) Anaerobic methane oxidation in metalliferous hydrothermal sediments: influence on carbon flux and decoupling from sulfate reduction. Environ Microbiol 14(10):2726–2740. https://doi.org/10.1111/j.1462-2920.2012.02825.x

Weisburg WG, Barns SM, Pelletier DA, Lane DJ (1991) 16S ribosomal DNA amplification for phylogenetic study. J Bacteriol 173(2):697–703

Whittenbury R, Phillips KC, Wilkinson JF (1970) Enrichment, isolation and some properties of methane-utilizing bacteria. J Gen Microbiol 61(2):205–218. https://doi.org/10.1099/00221287-61-2-205

Wu ML, Ettwig KF, Jetten MS, Strous M, Keltjens JT, van Niftrik L (2011) A new intra-aerobic metabolism in the nitrite-dependent anaerobic methane-oxidizing bacterium "*Candidatus* Methylomirabilis oxyfera". Biochem Soc Trans 39(1):243–248. https://doi.org/10.1042/BST0390243

Yanagawa K, Sunamura M, Lever MA, Morono Y, Hiruta A, Ishizaki O, Matsumoto R, Urabe T, Inagaki F (2011) Niche separation of methanotrophic archaea (ANME-1 and -2) in methane-seep sediments of the Eastern Japan Sea Offshore Joetsu. Geomicrobiol J 28(2):118–129. https://doi.org/10.1080/01490451003709334

Yang J, Jiang H, Wu G, Hou W, Sun Y, Lai Z, Dong H (2012) Co-occurrence of nitrite-dependent anaerobic methane oxidizing and anaerobic ammonia oxidizing bacteria in two Qinghai-Tibetan saline lakes. Front Earth Sci 6(4):383–391. https://doi.org/10.1007/s11707-012-0336-9

Zhao W, Wang Y, Liu S, Pan M, Yang J, Chen S (2013) Denitrification activities and N_2O production under salt stress with varying COD/N ratios and terminal electron acceptors. Chem Eng J 215–216:252–260. https://doi.org/10.1016/j.cej.2012.10.084

Zhou Z, Han P, Gu JD (2014) New PCR primers based on mcrA gene for retrieving more anaerobic methanotrophic archaea from coastal reedbed sediments. Appl Microbiol Biotechnol 98 (10):4663–4670. https://doi.org/10.1007/s00253-014-5599-5

Zhu B, van Dijk G, Fritz C, Smolders AJ, Pol A, Jetten MS, Ettwig KF (2012) Anaerobic oxidization of methane in a minerotrophic peatland: enrichment of nitrite-dependent methane-oxidizing bacteria. Appl Environ Microbiol 78(24):8657–8665. https://doi.org/10.1128/AEM.02102-12

The Biochemistry of Methane Monooxygenases

Sunney I. Chan and Seung Jae Lee

Contents

Abstract Selective methane (CH_4) oxidation is extremely difficult chemistry. Methane monooxygenases (MMOs) facilitate the biological conversion of CH_4

S. I. Chan
Noyes Laboratory, California Institute of Technology, Pasadena, CA, USA

Institute of Chemistry, Academia Sinica, Nankang, Taipei, Taiwan

S. J. Lee (✉)
Department of Chemistry and Institute for Molecular Biology and Genetics, Chonbuk National University, Jeonju, Republic of Korea
e-mail: slee026@jbnu.ac.kr

E. Y. Lee (ed.), *Methanotrophs*, Microbiology Monographs 32,
https://doi.org/10.1007/978-3-030-23261-0_3

into methanol in methanotrophs under ambient temperatures and pressures. Methanotrophs typically express the membrane-bound particulate form of MMO (pMMO), but the soluble form of MMO (sMMO) is often expressed under copper-limiting conditions. Numerous biochemical and biophysical approaches have been explored to elucidate the mechanisms of pMMO and sMMO over the past four decades, especially to examine the structures and the functional roles of the active sites. In this chapter, the biochemistry, including structural and functional features of MMOs, will be described. We first summarize the biochemical/biophysical studies that have led to the discovery of a unique tricopper cluster as the catalytic site in pMMO and culminated in the successful development of a biomimetic catalyst capable of mediating efficient CH_4 oxidation at room temperature. We then review the spectroscopic, kinetic, and structural studies that have contributed to clarification of the catalytic mechanisms near the non-heme diiron active sites in the hydroxylase of sMMO, as well as the roles played by the regulatory protein and the reductase in this chemistry.

1 Introduction

Methane (CH_4) is the final product in the metabolism of methanogenic bacteria. It is considered one of the major greenhouse gases (Hanson and Hanson 1996). Although smaller amounts of CH_4 gas are produced on planet earth compared with carbon dioxide (CO_2), CH_4 has a higher heat capacity relative to CO_2, and recent reports suggest that CH_4 is 33 times more detrimental molecule for molecule toward global warming (Shindell et al. 2009). Fortunately, CH_4 is regulated in ecological systems on earth through methanotrophic bacteria, or methanotrophs, which convert substantial amounts of the CH_4 produced into methanol (CH_3OH) in the presence of molecular oxygen (O_2). This oxidation process is an essential step in the metabolism of methanotrophs, as formaldehyde, an oxidation product of CH_3OH, is the key intermediate in the catabolism and anabolism during the oxidative assimilation of organic carbon into biomass in these bacteria (Hanson and Hanson 1996; Trotsenko and Khmelenina 2002). The biochemistry of CH_4 oxidation has also garnered attention among physical scientists because the chemistry involves the activation of one of the most stable C–H bonds in hydrocarbons (bond energy 104.9 kcal/mol), and methanotrophs mediate this chemistry with high efficiency under ambient conditions of temperature and pressure (Blanksby and Ellison 2003). To carry out this difficult process, most methanotrophs express the membrane-bound particulate form of MMO (pMMO), and specific type II and X methanotrophs also express the soluble form of MMO (sMMO) under limiting copper/biomass conditions (Murrell et al. 2000; Chidambaram-Padmavathy et al. 2015).

pMMO is a membrane-bound protein comprising three subunits (PmoA, PmoB, and PmoC) and many copper cofactors. As a membrane protein, this enzyme is unstable outside of the lipid bilayer, and accordingly it has been difficult to purify the pMMO to homogeneity for biochemical/biophysical studies. Even so, a form of the enzyme has been crystallized for X-ray structural analysis (Lieberman and Rosenzweig 2005; Hakemian et al. 2008; Smith et al. 2011; Sirajuddin et al. 2014). Strategies have also been developed to obtain membranes highly enriched in the active form of pMMO for in-depth biochemical/biophysical characterization (Yu et al. 2003a; Chan et al. 2011). These studies have led to the discovery of a tricopper cluster as the catalytic site of the enzyme and the successful development of a biomimetic catalyst capable of mediating efficient hydroxylation of CH_4 under ambient conditions (Chan et al. 2013; Chen et al. 2014; Nagababu et al. 2014; Wang et al. 2017).

Soluble MMO has been extensively studied owing to its solubility and relative ease of purification compared to pMMO. Essential proteins for CH_4 hydroxylation include the sMMO hydroxylase (MMOH), the regulatory protein (MMOB) and reductase (MMOR), which form the bacterial multicomponent monooxygenase (BMM) superfamily (Merkx et al. 2001; Murray and Lippard 2007; Wallar and Lipscomb 1996). These are the necessary components to complete the catalytic cycle and achieve the biological conversion from CH_4 into CH_3OH. The non-heme diiron active site is positioned in the MMOH α-subunit for generation of the O_2-activated intermediates, but apparently only one of these intermediates (*vide infra*) can hydroxylate CH_4 through C–H activation (Elango et al. 1997; Liu et al. 1995a; Rosenzweig et al. 1993; Sazinsky and Lippard 2006; Shu et al. 1997). Structural studies of sMMO provide crucial information toward understanding CH_4 hydroxylation, and the recent complex structure between hydroxylase and the regulatory protein (MMOH–MMOB) demonstrate how substrates such as molecular oxygen, electrons, and protons are regulated for CH_4 oxidation (Lee et al. 2013).

Both sMMO and pMMO are metalloenzymes. pMMO is a copper enzyme (Nguyen et al. 1994, 1996, 1998), whereas sMMO is a diiron protein, as mentioned earlier. Interestingly, the pMMO is a single protein of three subunits (PmoA, PmoB, and PmoC) (Nguyen et al. 1998; Lieberman and Rosenzweig 2005), whereas sMMO is a protein complex consisting of three distinct proteins: the hydroxylase, reductase, and a regulating protein (Lipscomb 1994). Clearly, both pMMO and sMMO have evolved during the early days of planet earth when CH_4 was abundant in the atmosphere of the planet and when O_2 was beginning to emerge in significant quantities from photosynthesis. However, sMMO was likely to have evolved later than the pMMO, given the greater structural complexity and substrate diversity of the sMMO.

As expected, the K_M for O_2 is low, ca. 100 μM, for both sMMO and pMMO (Semrau et al. 2010). The catalytic site needs to be activated by O_2 before it is in the position to mediate the hydroxylation of CH_4. The K_M for CH_4 are 19 μM and 92 μM for pMMO and sMMO at 45 °C, respectively; and the corresponding values of V_{max} are 450 and 363 nM/min/mg of protein (Trotsenko and Murrell 2008). We surmise that pMMO was designed specifically for selective CH_4 oxidation by methanotrophic organisms at the outset, especially in a habitat like soil where the CH_4 and O_2 concentrations are both low (Hanson and Hanson 1996). Regardless, it is the most efficient CH_4 oxidizer known under ambient conditions.

2 Particulate Methane Monooxygenase (pMMO)

We begin with pMMO. Despite its complexity as a membrane protein, pMMO is a relatively simple molecular machine, and the function of this enzyme can be related to its three-dimensional structure according to concepts familiar to most chemists and biochemists.

2.1 The Limited Substrate Range of pMMO and Its Remarkable Regio-specificity and Stereo-selectivity in Alkane Hydroxylation and Alkene Epoxidation

pMMO is found in all methanotrophs (Anthony 1982). Its function is to convert CH_4 into CH_3OH, which is then oxidized further along the electron transport chain in the metabolism of the organism. Besides CH_4, the pMMO enzyme exhibits activity toward a range of other substrates, although the substrate range is significantly narrower compared to sMMO. Only straight-chain hydrocarbons C1–C5 are converted into their alcohols, and a small number of linear alkenes (ethylene, propene, 1-butene, 2-butenes) are epoxidized (Elliott et al. 1997). In addition to this unique chemical specificity towards substrates, pMMO also shows remarkable regio-specificity and stereo-selectivity in the alkane hydroxylation. Aside from CH_3OH and ethanol, only the secondary alcohols are produced with the other alkanes. The hydroxylation occurs with a total retention of configuration at the carbon atom oxidized (Wilkinson et al. 1996; Yu et al. 2003b). These observations have been taken as evidence for a well-defined mechanism in the chemistry of hydroxylation of the C–H bonds in these alkanes (Chan and Yu 2008). Similarly, pMMO epoxidizes small alkenes via concerted electrophilic *syn* addition (Ng et al. 2008).

2.2 pMMO Genes, Protein Sequences, and Expression

The genomes of many methanotrophs have now been deciphered and the pMMO genes sequenced (Ward et al. 2004; Stein et al. 2010; see also http://www.methanotroph.org/wiki/genomes/). While the basic structure of the pMMO genes is the expected *pmoCAB* gene cluster encoding the subunits of pMMO (PmoABC), the overall gene structure can be quite complex. For example, the γ-proteobacterial methanotroph *Methylococcus capsulatus* (Bath) contains two complete copies of *pmoCAB*, as well as a third copy of *pmoC* (Stolyar et al. 1999). The two *pmoCAB* regions are almost identical at the nucleotide sequence level, differing in only 13 positions in 3183 base pairs; and at the amino acid level, each translated gene product contains only one different residue in each copy. On the other hand, the *pmoC3* sequence is more divergent from *pmoC1* and *pmoC2* at both the far *N*- and

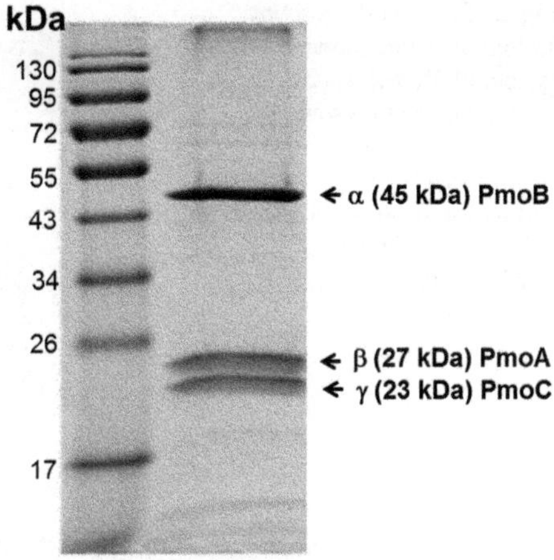

Fig. 1 Protein subunits of pMMO purified from membranes of *Methylococcus capsulatus* (Bath). Molecular weight markers are shown on the *left* panels. Adapted from Pham et al. (2015). Copyright 2015 Elsevier, Amsterdam, Netherlands

C-termini. Based on chromosomal insertion null mutants of pmoC1, pmoC2, pmoA1, pmoA2, pmoB1 and pmoB2, both gene copies were shown to be functional and essential for wild type pMMO activity.

The primary sequences of PmoA, PmoB and PmoC in pMMO *from M. capsulatus* (Bath) have been deduced from gene sequencing (Stolyar et al. 1999) and verified by peptide sequencing of the pMMO subunits by "bottom-up" proteomics using LC-MS/MS (Pham et al. 2013, 2015). The functional enzyme consists of one copy each of the three subunits PmoA, PmoB, and PmoC (Nguyen et al. 1998; Lieberman and Rosenzweig 2005). On denaturing sodium dodecyl sulfate polyacrylamide gel electrophoresis (SDS-PAGE) gels, these subunits appear as α, β, and γ (in the order of decreasing molecular weights) (Fig. 1), corresponding to PmoB (45 kDa), PmoA (27 kDa), and PmoC (23 kDa), respectively (Zahn and DiSpirito 1996; Nguyen et al. 1998; Basu et al. 2003; Yu et al. 2003a; Choi et al. 2003; Pham et al. 2015). According to the most recent high resolution matrix-assisted laser desorption/ionization time-of flight mass spectrometry (MALDI-TOF MS) study, the exact molecular masses of PmoA, PmoB, and PmoC in the pMMO isolated from *M. capsulatus* (Bath) are 28,302.25, 42,664.66, and 29,690.34 amu, respectively (Pham et al. 2013, 2015). The pMMO assembles into an $(\alpha\beta\gamma)_3$ trimer in the membrane according to electron microscopy (Myronova et al. 2006).

Most methanotrophs are capable of expressing the pMMO. Only a few strains produce both sMMO and pMMO, and even for these methanotrophs, the sMMO is expressed only under conditions of low copper availability (<5 μM) (Kao et al. 2004). Copper appears to be a transcriptional activator for the *pmmo* genes and a suppressor of the *smmo* genes. At copper/biomass ratios >2 μM, the overexpression of pMMO occurs, and the level of pMMO expression increases with the concentration of copper (Kao et al. 2004). The copper ions required during the growth of these methanotrophs not only control the differential expression of pMMO *versus* sMMO but also provide the main metal cofactors of the pMMO (Nguyen et al. 1994; Yu et al. 2003a).

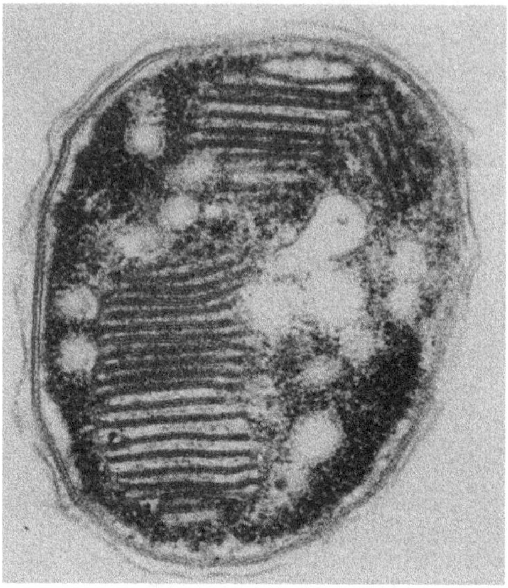

Fig. 2 Electron microscopy section of the membrane system of *Methylococcus capsulatus* (Bath) grown at high copper/biomass ratio. Reproduced with permission from Anthony (1982). Copyright 1982 Academic Press, Inc. (London)

Quantitative proteomic studies of *M. capsulatus* (Bath) indicate that expression of the pMMO is accompanied by enhanced expression of the genes downstream of the *pmmo* gene cluster that are involved in the metabolism of the cell, such as those that regulate the biosynthesis of membrane lipids (Kao et al. 2004). The co-expression of these metabolic genes concomitant with the expression of the *pmoCAB* gene cluster ensures that the subunits of the overproduced pMMO are properly assembled in the milieu of a lipid membrane to accomplish the catalytic function with the specific activity expected. When the expression of the MMO genes in *M. capsulatus* (Bath) switches from sMMO to pMMO, the membrane lipids are produced in such great abundance that the overproduction of the pMMO is accompanied by the formation of an extensive network of vesicular disk shaped bundles of intracytoplasmic membranes (Fig. 2).

2.3 Protein Structure

The X-ray crystal structure of pMMO from *M. capsulatus* (Bath) at 2.8-Å resolution was reported in 2005 (Lieberman and Rosenzweig 2005). Subsequently, the crystal structures of pMMO from *Methylosinus trichosporium* OB3b (Hakemian et al. 2008), *Methylocystis species* strain M (Smith et al. 2011), and *Methylocystis species* strain Rockwell (Sirajuddin et al. 2014) were also determined. All crystal structures show the trimeric $(\alpha\beta\gamma)_3$ ultrastructure obtained by electron microscopy of pMMO from *M. capsulatus* (Bath) at 23-Å resolution. The three-dimensional structure of the pMMO protomer from *M. capsulatus* (Bath) is depicted in Fig. 3. Within each protomer, the PmoA and PmoC subunits are mainly composed of membrane-spanning α-helices, whereas the PmoB subunit consists of two cupredoxin-like β-barrel

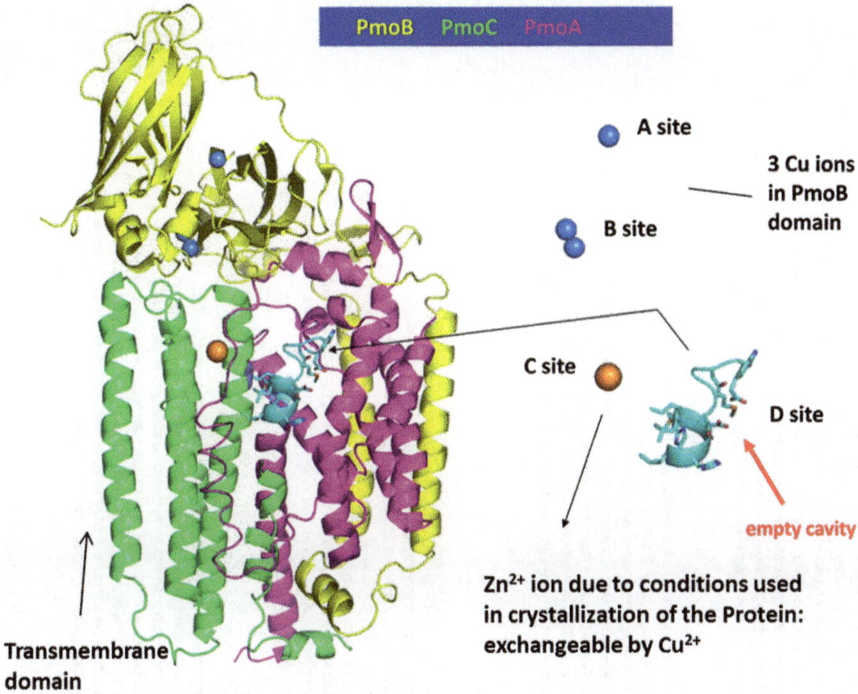

Fig. 3 X-ray crystal structure of the (Cu_3Zn)–pMMO from *Methylococcus capsulatus* (Bath) reported by Lieberman and Rosenzweig (2005). Copper and zinc atoms are presented in *blue* and *orange*, respectively. Reproduced with permission from Chan and Yu (2008). Copyright 2008 American Chemical Society

domains linked by two transmembrane α-helices resting at the membrane–aqueous interface on top of the cytoplasmic surface of the transmembrane helix bundle formed by the PmoA and PmoC subunits.

Three copper ions and one zinc ion are associated with the protein structure in each PmoABC protomer. The coppers are seen in the water-soluble domains of the PmoB subunit. One copper ion is coordinated to His48 and His72 in the *N*-terminal domain and linked to Gln404 in the *C*-terminal domain of the subunit (Fig. 4, *top panel*). This mononuclear copper site has been referred to as the **A** site. One of the histidine ligands is not conserved in the pMMOs from other methanotrophs. In pMMO from *Methylosinus trichosporium* OB3b, *Methylocystis species* strain M, and *Methylocystis species* strain Rockwell, His48 is replaced by an Asp residue instead. No copper ions were found at the **A** sites in the crystal structures of the PmoB subunits in these strains. Thus, the copper ion at the **A** site in the X-ray structure of pMMO in *M. capsulatus* (Bath) might not be critical for the function of the enzyme.

The remaining two copper ions form a dicopper center referred to as the **B** site: one of these copper atoms is ligated to two histidine ligands from the *N*-terminal

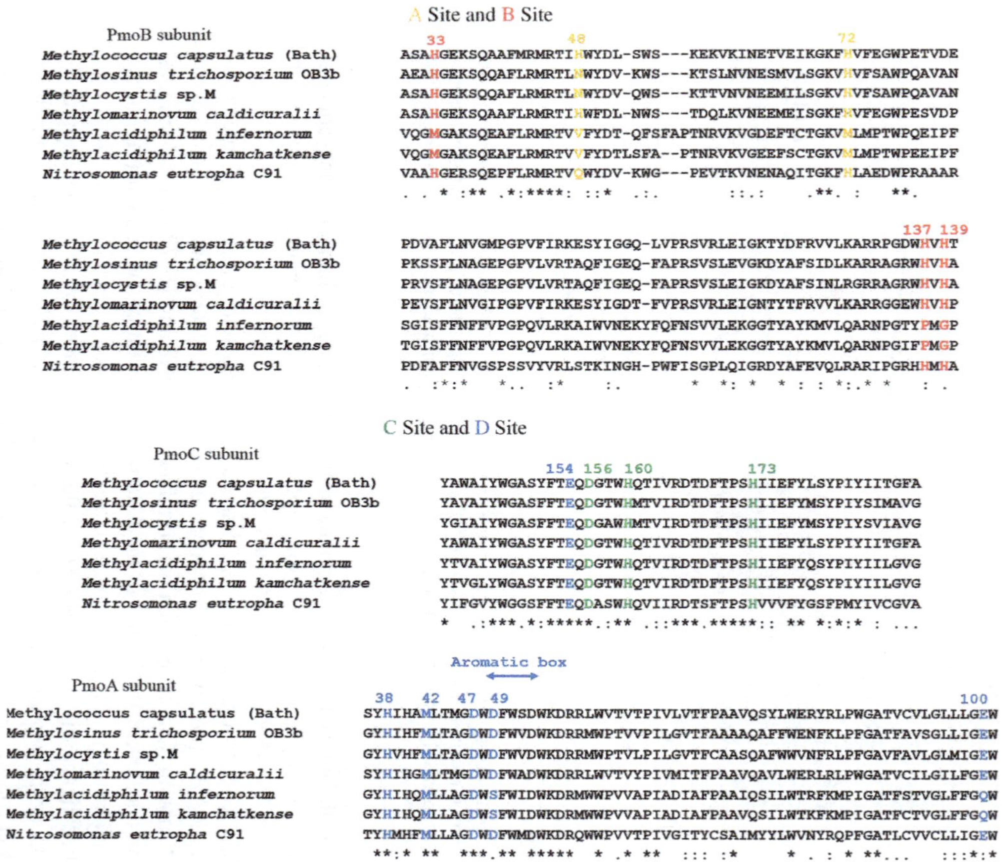

Fig. 4 Sequence alignments of pMMO from methanotrophic species and an AMO from *Nitrosomonas eutropha* C91. The amino acids near the different active sites proposed by different research groups are shown, and the proposed amino acids binding to the metals in the active sites are highlighted. Reproduced with permission from Wang et al. (2017). Copyright 2017 American Chemical Society

domain (His137 and His139), whereas the other copper is coordinated by the side-chain imidazole-δ nitrogen and the peptide nitrogen from the same His residue (His33) to form a histidine brace (Fig. 4, *second* panel from *top*). However, based on quantum refinement of the site, the electron density in this region has recently been reinterpreted without the histidine braced copper atom so that it might also be a mononuclear copper site as well (Cao et al. 2018). In any case, these three coordinating histidine ligands are not highly conserved among methanotrophs.

The zinc atom is located within the transmembrane domain associated with PmoC and PmoA and is referred to as the **C** site. The amino acids coordinating the Zn atom are Asp156, His160, and His173 from PmoC and Glu195 from PmoA, which are highly conserved (Fig. 4, *third* panel from *top*). However, it turns out the zinc ion observed in the X-ray crystal is an artifact of growing the crystals from concentrated $ZnSO_4$. A zinc containing buffer is required for the crystallization of pMMO from *M. capsulatus* (Bath) as well as pMMO from *Methylocystis species* strain M. In the X-ray crystal structure of pMMO from *Methylocystis* species strain Rockwell, a single copper atom is seen at the **C** site, but this copper ion is replaced by a zinc ion after soaking the crystals in a $ZnSO_4$ buffer without displacing the coppers at the **B** site (Sirajuddin et al. 2014). Evidently, the **C** site in pMMO is a copper-binding site as well.

A peculiar "empty" hydrophilic cavity has been identified in the X-ray structure of pMMO (Lieberman and Rosenzweig 2005). This region, which has been referred as the **D** site, consists of the amino acids His38, Met42, Asp47, Asp49, and Glu100 from the PmoA subunit and Glu154 from PmoC (Fig. 4, *third* panel from *top*). Interestingly, His38, Met42, Asp47, and Asp49 (highlighted in *red* below) are part of an alpha helix comprising

$$\mathbf{Y^{38}HIHA^{42}MLTMG^{47}DW^{49}DFWSDWKDR}$$

and this structural construct is totally conserved in all PmoA sequences that have been reported to-date (Fig. 4, *bottom* panel).

2.4 *Growth of* M. capsulatus *(Bath) and Isolation/ Purification of the pMMO*

In-depth studies of a protein are only possible when one has access to non-limiting quantities of the functional protein purified to homogeneity. Needless to say, the protein isolation and purification must be carried out with adequate controls to ensure the structural integrity for biochemical and biophysical measurements. The purified protein must have all the subunits intact and structurally folded into the physiological functioning conformation. It must also contain all the essential metal cofactors. These requirements present significant challenges to the protein chemist for a membrane-bound metalloenzyme like pMMO containing up to ca. 15 copper cofactors (*vide infra*). Moreover, as a membrane protein, it is difficult to isolate and purify the pMMO from the intracytoplasmic membranes of the cells because of its instability in detergents used to solubilize the protein as well as the potential loss of the metal cofactors during purification by column chromatography (*vide infra*).

As noted earlier, the expression of the *pmmo* gene in *M. capsulatus* (Bath) increases with increasing Cu^{2+} concentration in the habitat of the microorganism concomitant with the production of intracytoplasmic membranes to accommodate the pMMO produced. Accordingly, it is possible to produce the pMMO in large quantities in *M. capsulatus* (Bath) cells by growing and harvesting the microorganism at high concentrations of Cu^{2+} in the growth medium (Nguyen et al. 1994). For example, when the initial Cu^{2+} in the culture medium is ca. 30 μM, the pMMO produced can account for ca. 80% of the total proteins in the membranes. However, it was found that all the coppers in the fermenter were taken up by the cells and the pMMO in the membrane was heterogeneous in terms of the copper contents, with some protein molecules containing the full complement of copper ions and others with fewer copper ions. Some of the pMMO subunits produced were assembled without the full complement of copper ions as there were insufficient amounts of copper ions in the growth medium (Yu et al. 2003a). As expected, the specific activity of the protein isolated from a preparation of these membranes reflected this heterogeneity in the copper content (Table 1). This is a common problem with the cloning and overexpression of metalloenzymes.

To obtain pMMO-enriched membranes from *M. capsulatus* (Bath) with high activity and in high yields, a method was subsequently devised to process cell growth in a fermenter adapted with a hollow-fiber bioreactor that allowed easy control and quantitative adjustment of the copper ion concentration in the culture medium over the time course of the cell growth (Yu et al. 2003a; Chan et al. 2011). A schematic diagram of this flow reactor is shown in Fig. 5. This technical development allows the *M. capsulatus* (Bath) cells to be grown in the presence of well-

Table 1 The effects of copper concentration in the culture medium on the pMMO content in the pMMO-enriched membranes, the average copper content of the pMMO proteins, and the specific activity of the pMMO-enriched membranes using NADH as the reductant

Cu (μ*M*) in growth medium	pMMO/total protein (%)	Specific activity (nmol/min/mg of protein)	Copper/pMMO[a]
<0.3	19	19.6	9.4
5	27	30.5	10.5
10	44	42.6	10.3
20	68	59.6	10.8
30	80	65.9	13.4
40	73	88.9	13.8
60	65	63.5	15.3

Taken from Yu et al. (2003a). Copyright © American Society for Microbiology

[a]mol copper/mol of protein, determined assuming that the monomeric pMMO hydroxylase has a subunit composition of αβγ with a molecular mass of 99 kDa

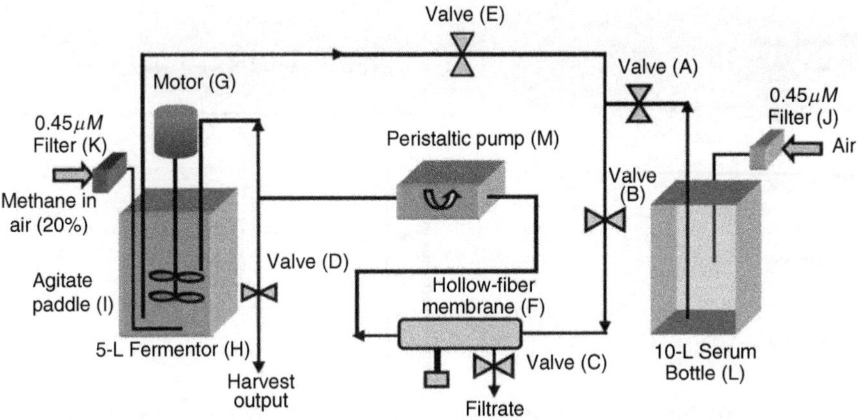

Fig. 5 Schematic of a fermenter adapted with a hollow fiber membrane bioreactor to culture *M. capsulatus* (Bath) for production of high-quality pMMO in high yields. Reproduced from Yu et al. (2003a). Copyright 2003 American Society for Microbiology

defined copper concentrations (30–35 μM) so that the cells are not subjected to gross variations in the copper ion levels during the time course of the culturing. It was demonstrated that the cells grew optimally when the copper concentration was sustained at this level. At lower steady state copper concentrations, not all the pMMO was assembled with the full complement of copper ions. However, at copper concentrations in excess of 50 μM, the cells were not tolerant of the copper and did not grow well. As a Cu^{I} enzyme (*vide infra*), it has proven difficult to replenish the missing copper cofactors by simply adding additional $CuSO_4$ to the harvested cells or to the isolated membranes. It seems that copper transporters, e.g., methanobactin (Kim et al. 2004), are required to introduce some of the copper cofactors during assembly of the protein in the membranes.

Based on the above strategy, it was possible to obtain membranes of *M. capsulatus* (Bath) highly enriched in the pMMO with high specific activity in sufficient quantities for biochemical and biophysical studies (Yu et al. 2003a). The pMMO accounted for >80% of the total intracytoplasmic membrane proteins with a specific activity as high as 90 nmol of propylene oxide/min/mg of protein with NADH as the reductant. The copper stoichiometry was ca. 13 atoms per pMMO molecule. Analysis of other metal contents provided no evidence of zinc, and only traces of iron were present in the pMMO-enriched membranes.

Membranes enriched in the pMMO, with the three subunits of the protein properly assembled together with the full complement of ca. 15 copper cofactors, are required before the purified enzyme could be obtained from these membranes with high activity and in high yields. Two approaches were subsequently developed to isolate and purify pMMO from the pMMO-enriched membranes (Chan et al. 2011). Both methods began with the solubilization of the membranes by the detergent dodecyl β-D-maltoside. In one protocol, the solubilization was followed by fractionation of the protein–detergent complexes according to the molecular sizes

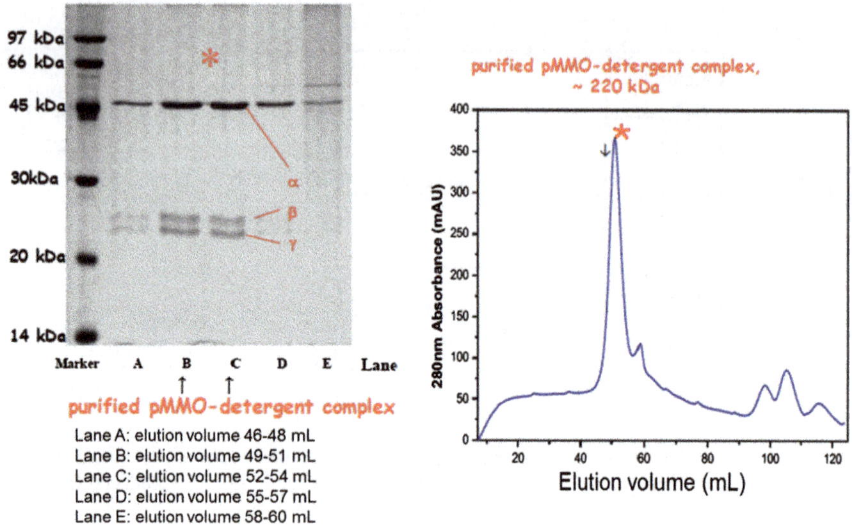

Fig. 6 *Right*: The elution profile of the detergent-solubilized pMMO-enriched membranes from a Sephacryl S-300HR gel filtration column. *Left*: SDS-PAGE of the eluant at different elution volumes. Adapted from Yu et al. (2003a). Copyright 2003 American Society for Microbiology

of the micelles by gel filtration chromatography (Yu et al. 2003a). In the other, the solubilized proteins were subjected to ion-exchange chromatography to separate membrane proteins of varying sizes and hydrophobicity. The gel filtration method is gentle and works well when the membrane protein of interest is present in abundance, which is the case when the pMMO-enriched membranes are obtained as described above. The second method is considerably harsher, but it has been shown to work, provided precautions are taken to ensure that the copper ions remain reduced, namely not oxidized, so that they do not dissociate from the protein during the fractionation and chromatography steps (Nguyen et al. 1998).

When pMMO accounts for 80–90% of the total proteins in the purified membranes, the bulk of the detergent–protein complexes will contain the competent pMMO when the membranes are solubilized in a non-denaturing detergent. If a sufficient amount of detergent is used to solubilize the membranes so that [detergent]/[protein] is ~1000, then the typical detergent–protein micelle particle would contain on the average only one protein molecule per micelle (Yu et al. 2003a). The mass distribution of the micelles should also be fairly narrow given that most of the micelles would contain the abundant protein. Figure 6 shows the elution profile of the proteins after a preparation of the membranes from cells cultured and harvested at a copper concentration of 30 mM is solubilized in 5% (w/v) dodecyl β-D-maltoside (or ~3 mg of detergent/mg of protein) and subjected to size exclusion column chromatography on a Sephacryl S-300HR column (Yu et al. 2003a). The elution buffer contains 50 mM PIPES, 150 mM NaCl, 50 mM imidazole, 5 mM ascorbate, and 0.05% (w/v) dodecyl β-D-maltoside (pH ~ 7.2). The strong band appearing

Table 2 The copper contents and specific activities of the pMMO in various preparations

Preparation	Cu content (atoms/ protein)	Reductant	Specific activities[a] (nmol/min/mg of protein)
pMMO-enriched membranes	12–15	NADH[b]	88.9
Solubilized membranes	13.3	NADH[b]	10.2
	13.3	Duroquinol[c]	93.5
pMMO–detergent complex			
Method 1[d]	14.3	NADH[b]	21.5
	14.3	Duroquinol[c]	15.6
Method 2a[e]	13.3	NADH	5.1
Method 2b[f]	14.1	NADH	2.6

[a]Conversion of propylene to propylene oxide
[b]The concentration of NADH used is 5 m*M*
[c]The concentration of duroquinol used is 1 m*M*
[d]The pMMO is purified by direct transfer into detergent micelles followed by gel filtration using a Sephacryl S-300HR column
[e]The pMMO is purified by ion-exchange chromatography using a L-lysine-agarose column
[f]The pMMO is purified by ion-exchange chromatography using a DEAE-sepharose column

around the elution volume of ~50 mL corresponds to the purified pMMO/detergent complex. SDS-PAGE gels of cuts derived in the vicinity of this peak show the bands associated with the 45-, 27-, and 23-kDa (or α, β, and γ) subunits of pMMO without contamination by other proteins. Thus, the gel filtration procedure outlined here leads to a homogeneous preparation of pMMO–detergent complexes. The solubilized pMMO–detergent complex was shown to be an αβγ protein monomer with ~240 detergent molecules for a total molecular mass of 220 kDa according to MALDI-TOF mass spectrometry of the solubilized pMMO–detergent complex.

The copper contents and specific activities of the pMMO in various preparations as described above are summarized in Table 2. ICP-MS revealed essentially the same copper content per protein molecule (13–14 atoms/protein) among the various preparations. Solubilization of the membranes into detergent micelles did not materially affect the specific activities of the pMMO, especially if quinols are available to turn over the enzyme within the detergent/membrane micelles. As discussed later, the catalytic site of pMMO resides in the transmembrane domain. When quinol in the membranes is limiting, the enzyme complex uses an auxiliary pathway driven by NADH that accounts for *ca.* 20% of the specific activity. Thus, it is gratifying that the NADH-driven activities are rather similar between the detergent-solubilized membranes and the purified protein–detergent complexes. However, reconstitution of the purified pMMO into lipid bilayer vesicles from solubilized-detergent micelles of the protein remains a challenge.

2.5 Biochemistry and Biophysical Studies: Summary of Key Findings

During the late 1990s and early 2000s, many studies were reported on the biochemistry and biophysics of the purified pMMO-enriched membranes isolated from *M. capsulatus* (Bath) as prepared above. These studies included, among others, MALDI-TOF mass spectrometry (Yu et al. 2003a; Pham et al. 2013, 2015); peptide mass fingerprinting (Yu et al. 2003a; Chan et al. 2004; Pham et al. 2015); quantitative proteomics (Kao et al. 2004); polarized attenuated total reflection Fourier transform infrared spectroscopy-(FTIR) (Vinchurkar et al. 2004); inductively coupled plasma mass spectrometry (ICP-MS) (Nguyen et al. 1998; Yu et al. 2003a); Cu *K*-edge X-ray absorption spectroscopy (Nguyen et al. 1996; Yu et al. 2003a); low temperature saturation magnetization and magnetic susceptibility measurements (Chan et al. 1993); electron paramagnetic resonance (EPR) spectroscopy (Nguyen et al. 1996; Yu et al. 2003a; Chen et al. 2004); and redox potentiometry EPR spectroscopy (Chan et al. 2007). Some of the key findings are summarized here.

(a) The molecular masses of the pMMO subunits separated on SDS-PAGE gels were determined by MALDI-TOF mass spectrometry, and the α, β, and γ subunits were identified as PmoB, PmoA, and PmoC, respectively, by proteomics analysis of the peptides derived from in-gel proteolytic digestion of the protein subunits (Yu et al. 2003a; Pham et al. 2015).
(b) Analysis of the secondary structure of the membrane-embedded domains of pMMO by attenuated total reflection FTIR revealed that the transmembrane domain comprises mostly α-helices (Vinchurkar et al. 2004).
(c) In situ trypsin digestion of the water-exposed domains of the membrane fragments and peptide mass fingerprinting showed that the peptides released originated from only the PmoB subunit (Fig. 7) (Chan et al. 2004). These results are consistent with protein hydropathy plots predicting some 15 transmembrane α-helices (seven for PmoA, two for PmoB, and six for PmoC), together with a hydrophilic domain contributed mainly by the 45 kDa PmoB subunit on the cytosolic face of the plasma membrane.
(d) According to analysis by ICP-MS, pMMO is a copper only enzyme with 13–14 copper ions per PmoABC (Nguyen et al. 1998; Yu et al. 2003a). A trace amount of iron (0.9 Fe per protein) is usually associated with the membrane preparation. This iron is adventitious, as metal analysis of the purified pMMO–detergent complex obtained by size exclusion column chromatography on a Sephacryl S-300HR column yielded a high copper-to-iron ratio of 80:1 (average copper and iron contents of 0.143 and ~0.002 mmol/mg of protein, respectively). The copper content of the pMMO/detergent complex prepared using DEAE-Sepharose ion-exchange chromatography is 0.141 mmol/mg of protein, which corresponds to 14.1 Cu atoms per protein, assuming a molecular mass of 100 kDa for the apo-protein. The residual level of Fe in the purified pMMO corresponds to a Cu/Fe of ~70–100/1. The purified pMMO/detergent complexes showed the same subunit composition in SDS-PAGE as the purified enriched membranes.

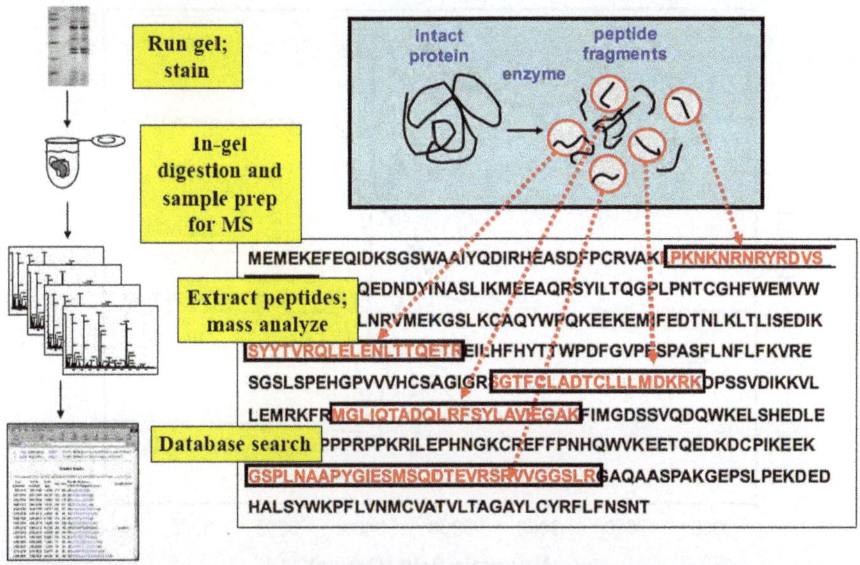

Fig. 7 In-gel digestion and peptide mass fingerprinting of pMMO-enriched membranes

(e) The Cu K_{α} X-ray absorption edge at 8984 eV indicated that ca. 60–70% of the copper ions in the purified pMMO-enriched membranes as isolated are Cu^{I} (Nguyen et al. 1996; Yu et al. 2003a). The same result was obtained for the purified pMMO/detergent complexes regardless of the method of preparation. These Cu^{I} atoms are resistant to air oxidation to Cu^{II}.

(f) 77 K EPR of the purified pMMO-enriched membranes revealed a type 2 Cu^{II} EPR signal with intensity corresponding to one Cu^{II} ion per protein molecule superimposed on an almost isotropic signal centered near 2.1 of similar intensity (Fig. 8) (Chen et al. 2004). The latter isotropic Cu^{II} EPR signal was subsequently assigned to a tricopper cluster consisting of a triad of three ferromagnetically coupled Cu^{II} ions ($S_T = 3/2$ ground state) on the basis of magnetization and magnetic susceptibility measurements (Chan et al. 1993). The purified pMMO/detergent complex exhibited the same spectral features as the pMMO-enriched membranes.

(g) Redox potentiometry EPR experiments poised at different electrochemical cell potentials allowed the type 2 Cu^{II} and tricopper cluster EPR signals to be observed separately at different redox potentials (Fig. 9, *left* panel) as well as the observation of additional type 2 Cu^{II} EPR signals corresponding to 10 Cu^{II} ions at significantly higher potentials (Fig. 9, *right* panel). Some of these high-potential copper ions were subsequently identified in the *C*-terminal domain of the PmoB subunit when the latter was cloned and overexpressed in *E. coli* (Yu et al. 2007). They were found to be inert toward oxidation by O_2 as well.

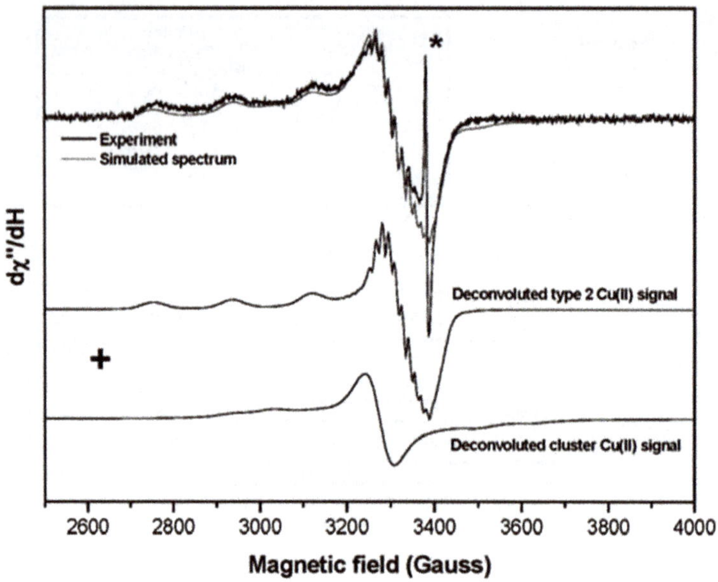

Fig. 8 77 K EPR spectrum of purified pMMO-enriched membranes as isolated. Taken from Chen et al. (2004). Copyright 2004 Taipei & Wiley-VCH Verlag GmbH & Co.

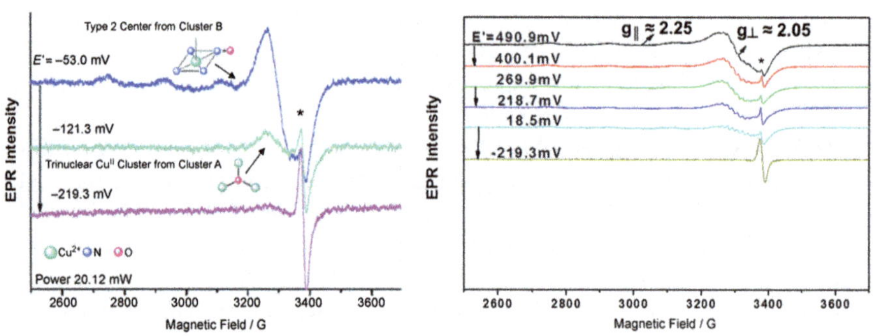

Fig. 9 Redox potentiometry EPR experiments on purified pMMO-enriched membranes. *Left*: Redox titrations at low cell potentials separate composite type 2 Cu^{II} and tricopper cluster EPR signals according to the redox potentials of the copper sites producing these spectroscopic signatures. *Right*: Redox titration at high potentials. Taken from Chan et al. (2007). Copyright 2007 John Wiley & Sons

It is evident, upon comparing these findings with the X-ray crystal structure, that there is general agreement over the structure of the protein fold and the overall topology of the protein in the membrane. There is also a consensus that pMMO is a copper only protein; and iron and zinc atoms are not part of the functioning enzyme. However, there is substantial disagreement on the metal cofactors, not only the number of copper atoms, the nuclearity of the copper centers, but also their locations within the protein structure. The most obvious source of the discrepancies is that there had been a loss of copper ions during purification of the protein for the crystallization of the pMMO. Indeed, when the solubilization and purification of the pMMO-enriched membranes was repeated according to the procedures used to obtain the protein preparation for the protein crystallization, loss of coppers was observed at every stage of the purification process, and the final purified pMMO contained only 2–3 copper atoms, in agreement with the X-ray structure (Chan and Yu 2008). Concomitant with the loss of copper cofactors, the enzyme was essentially inactive.

2.6 *Rebuilding the Copper Sites into the Protein Scaffold*

Over the past decade, considerable efforts have been made to rebuild the copper sites into the pMMO protein structure. These efforts are essential if the discrepancies between the biochemical/biophysical studies and the X-ray structure are to be resolved. Moreover, only then, will the structure and function of pMMO become fully understood.

Beginning with the empty hydrophilic cavity at the **D** site within the transmembrane domain, it was conjectured that this might be the site of the tricopper cluster implicated by the biophysical studies. Indeed, it was possible to model a tricopper cluster at this site (Chan et al. 2007). A model of the fully oxidized tricopper cluster with a capping oxo is depicted in Fig. 10. The ligands to the three Cu^{II} ions are:

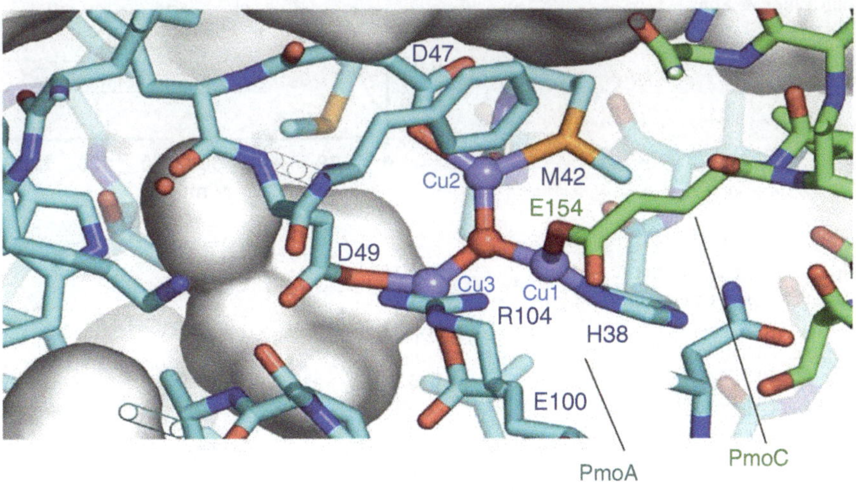

Fig. 10 The tricopper cluster modeled as a $Cu^{II}Cu^{II}Cu^{II}$ triad with a capping "oxo" at the **D** site of the X-ray crystal structure. Taken from Chan et al. (2007). Copyright 2007 John Wiley & Sons

Cu1: PmoA His38 and PmoC Glu154
Cu2: PmoA Met42 and Asp47
Cu3: PmoA Asp49 and Glu100.

Since, except for PmoC Glu154, all the ligands are associated with the highly conserved PmoA helix, ^{38}HIHAMLTMGDW^{49}D……..^{100}E, the validity of the model could be verified by peptide chemistry (Chan et al. 2013). Indeed, the peptide **H**I**H**A**M**LTMG**D**W**D** forms both 1:1 $Cu^{I}Cu^{I}Cu^{I}$ and $Cu^{II}Cu^{II}Cu^{II}$ complexes in the presence of excess acetate (also chloride, but not nitrate) with the residues highlighted in bold as ligands. The EXAFS of both complexes showed Cu···Cu backscattering expected for a copper triad. The EPR of the $Cu^{II}Cu^{II}Cu^{II}$–peptide complex showed the isotropic featureless signal at $g = 2.1$. More importantly, when the $Cu^{I}Cu^{I}Cu^{I}$–peptide complex was immobilized on mesoporous carbon and activated by O_2, the activated tricopper cluster complex mediated the facile hydroxylation of CH_4 to CH_3OH (Fig. 11) as well the epoxidation of propene to propylene oxide (Fig. 12), albeit a single turnover, as no provisions were built into the system to regenerate the $Cu^{I}Cu^{I}Cu^{I}$ complex for multiple turnovers.

Building on this chemistry, a biomimetic tricopper cluster complex has been constructed capable of efficient and selective conversion of CH_4 into CH_3OH under ambient conditions (Figs. 13 and 14) (Chan et al. 2013; Chen et al. 2014; Nagababu et al. 2014). It was possible to accomplish multiple turnovers by using H_2O_2 as a reductant to regenerate the catalyst, with a catalytic efficiency approaching 90%,

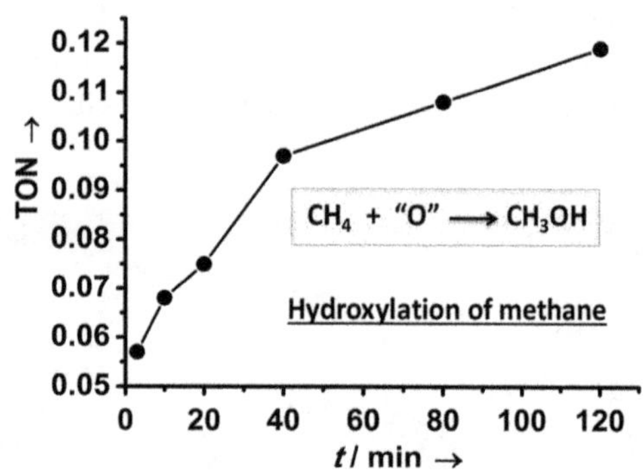

Fig. 11 Reactivity of the $Cu^{I}Cu^{I}Cu^{I}$–peptide complex toward hydroxylation of CH_4. Taken from Chan et al. (2013). Copyright 2013 John Wiley & Sons

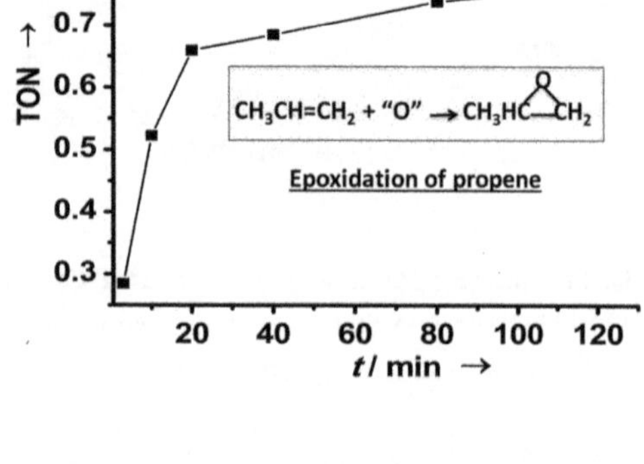

Fig. 12 Reactivity of the $Cu^{I}Cu^{I}Cu^{I}$–peptide complex toward epoxidation of propene. Taken from Chan et al. (2013). Copyright 2013 John Wiley & Sons

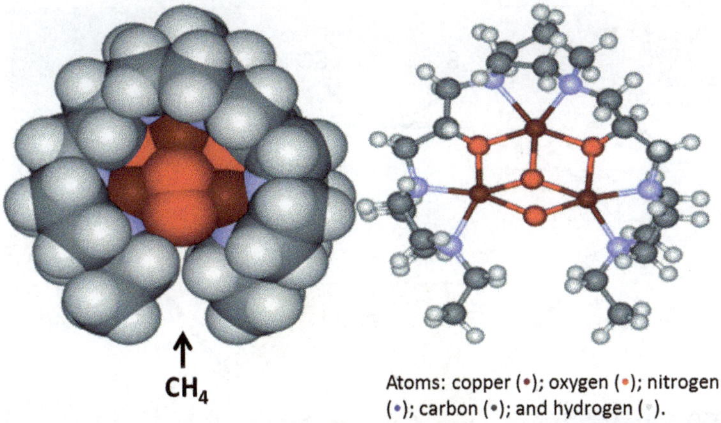

Fig. 13 A biomimetic tricopper complex that is capable of efficient conversion of CH_4 into CH_3OH at room temperature and normal pressures upon activation by O_2. Adapted from Chan et al. (2013). Copyright 2013 John Wiley & Sons

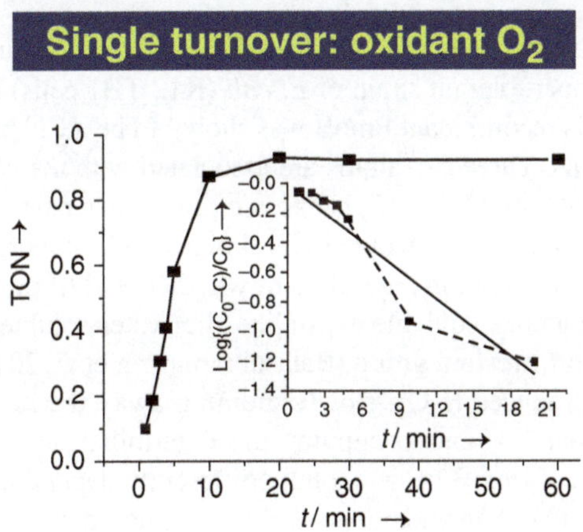

Fig. 14 Single turnover of biomimetic tricopper complex to convert CH_4 into CH_3OH in the presence of O_2. Taken from Chan et al. (2013). Copyright 2013 John Wiley & Sons

with almost one CH_3OH produced per H_2O_2 used to drive the catalytic turnover (Fig. 15) (Nagababu et al. 2014; Liu et al. 2016). Moreover, in the absence of hydrocarbon substrate, the activated tricopper complex was found to be rapidly reduced by one electron reduction to produce the fully oxidized $Cu^{II}Cu^{II}Cu^{II}$ tricopper complex with its characteristic isotropic EPR signature (Maji et al. 2012). This outcome is reminiscent of the chemistry that has been observed in the case of the pMMO enzyme where the putative activated tricopper cluster occupying the **D** site is deactivated by a reducing equivalent from the mononuclear Cu^{I} ion at the nearby **C** site, generating the isotropic EPR signal superimposed on a type 2 Cu^{II} EPR spectrum (Chan et al. 2004).

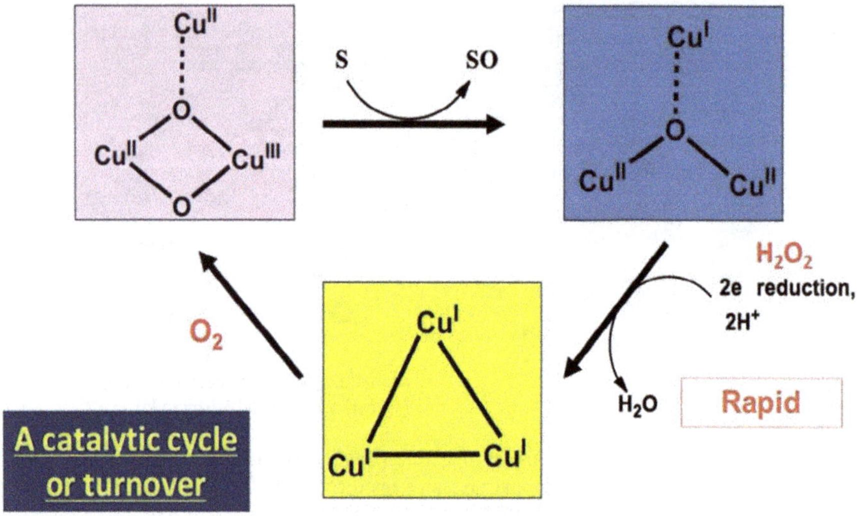

Fig. 15 A scheme to accomplish multiple turnovers during catalytic oxidation of a substrate (S) mediated by the tricopper cluster complex

Recently, the successful cloning and over-expression of the PmoB subunit in the membranes of a copper tolerant strain of *E. coli* (K12 TB1 cells) has been achieved (Lu et al. 2019). This recombinant PmoB was shown to be a Cu^I protein, with a total of 9–10 copper atoms. Three Cu^I atoms are associated with the *N*-terminal domain, and there is clear evidence of Cu···Cu backscattering from a pair of Cu^I atoms. Thus, the **B** site is without a doubt a dicopper site. Interestingly, this recombinant PmoB does not mediate CH_4 oxidation regardless of whether NADH or duroquinol is used as the source of reducing equivalents, unlike the water-soluble spmob construct developed by the Northwestern group (Balasubramanian et al. 2010). Instead, in the presence of O_2, it produces H_2O_2, slowly draining away the reducing equivalents from the six or seven Cu^I ions occupying the *C*-terminal domain. Evidently, the cupredoxin domains of PmoB behave totally differently depending on whether they are attached to the two transmembrane helices anchored into the membrane, or simply exist together as a soluble protein in solution. Based on these findings, it is evident that the functional pMMO is a Cu^I protein with a total of 13–14 copper atoms.

2.7 *Suicide Substrates as Probes of the Catalytic Site*

A suicide substrate is a substrate analogue that interacts with the active site of an enzyme, but during the normal catalysis reaction, reacts with the active site to form a reactive product that chemically modifies a residue in the protein rendering the enzyme no longer functional toward normal substrates. Acetylene and propyne

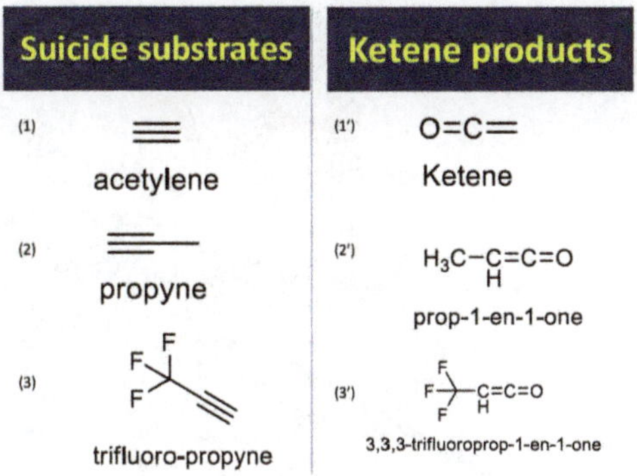

Fig. 16 Suicide substrates of pMMO and their ketene products

(also trifluoropropyne) are suicide substrates in the case of pMMO. These substrates are oxidized at the catalytic site by the addition of an "O" atom across the triple bond of the alkyne to form the highly reactive ketene intermediate (Fig. 16), which then migrates to a site within the protein where it undergoes a nucleophilic addition reaction and irreversibly inactivates the enzyme. In the case of acetylene, the highly reactive ketene chemically modifies a lysine (K196 PmoC) within the transmembrane domain of the enzyme according to proteomics analysis of peptides (Pham et al. 2015). Alkylation of K196 PmoC alone totally inhibits the activity of pMMO toward propene oxidation. This lysine residue resides within the transmembrane domain of the protein on the periplasmic face of the membrane away from the cytoplasmic surface, where the electron input takes place during turnover of the enzyme (see Fig. 19). It has been suggested that chemical modification of K196 PmoC blocks the substrate channel, or inhibits the formation of the substrate channel by suppressing the protein dynamics and conformational fluctuations, so that CH_4 and other small alkanes can no longer access and reach the catalytic site of the enzyme from the periplasm of the cell.

Ketenes are very reactive molecules, the details of their reactivity depending on the environment in which they are produced. The reactivity will differ dramatically whether the ketene is formed in an aqueous solution, or in the transmembrane region of a membrane protein. In case of pMMO, since the alkylation by the ketene takes place within the transmembrane domain of the enzyme, the reactive ketene must be formed within the membrane. Moreover, the pMMO and a synthetic model tricopper cluster complex exhibit the same ketene chemistry toward acetylene, whereas a similar formation of ketene is not observed with a model dicopper complex. Thus, the catalytic site or site of hydroxylation of CH_4 in pMMO must be the tricopper cluster site.

2.8 *The Catalytic Site: Substrate Binding Near the Site of Hydroxylation*

Identification of a hydrophobic binding pocket for CH_4 and other known alkane substrates has allowed us to home in on the site of hydroxylation in pMMO. The

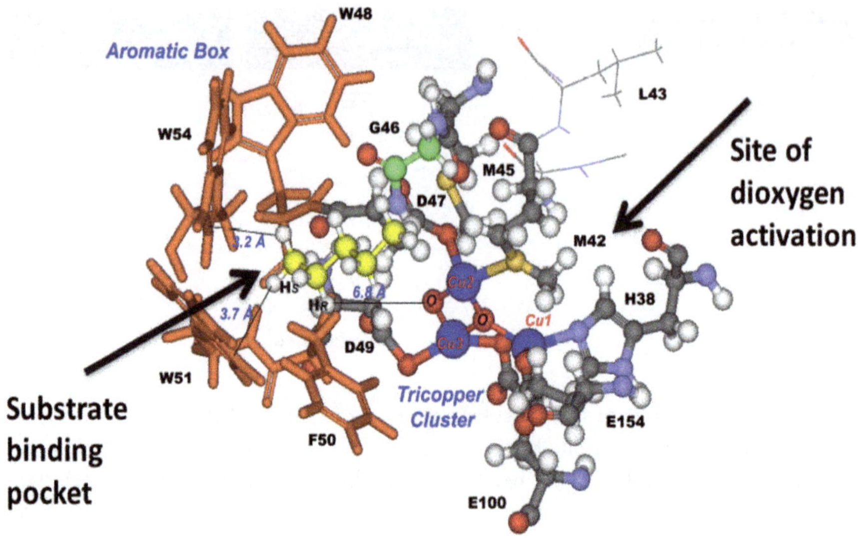

Fig. 17 The catalytic site of pMMO, showing both the activated tricopper cluster and a pentane molecule sequestered within the hydrophobic binding pocket (the "aromatic box")

catalytic site of the enzyme must include an alkane binding site in juxtaposition to the tricopper cluster generating the highly reactive "oxene" for O-atom transfer to the hydrocarbon substrate if the chemistry is indeed mediated there (Ng et al. 2008). Pentane was used as a probe to search for the hydrocarbon substrate binding site by DockLigand (LigandFit). Pentane was chosen because it is the substrate with the largest volume to surface area that could be oxidized by pMMO. One hydrophobic cavity was identified within 6 Å of the tricopper cluster at the **D** site (Fig. 17). With the alkane molecule bound here, the **B** and **C** sites are too far away to be directly involved in the hydroxylation chemistry (26 Å and 13 Å, respectively). This putative alkane binding site is composed of the aromatic residues Trp48, Phe50, Trp51 and Trp54 from PmoA. Gly46 is located at the open substrate entrance of the pocket. These residues are all absolutely conserved in methanotrophs (see Fig. 4). It has been suggested that van der Waals interactions of the C–H bonds of the alkane molecule with the π-electrons of the surrounding aromatic residues forming the pocket contribute to the binding affinity of the cavity for the hydrocarbon substrate. Thus, the binding pocket has been referred to as an "aromatic box".

The putative hydrophobic pocket is sufficiently long to bind only C1–C5 hydrocarbons, and it is wide enough to accommodate only straight-chain alkanes (Ng et al. 2008; Elliott et al. 1997). With pentane, the hydrocarbon fits snugly within the cavity, oriented to facilitate direct O-atom insertion across the C–H_R and C–H_s bonds of the secondary carbon (Fig. 18). Experimentally, only (2*R*)-pentan-2-ol and (2*S*)-pentan-2-ol were observed, with an *ee* = 80% (Elliott et al. 1997). From the stereo-selectivity observed, it is evident that there is preference for the alkane to be oriented with the *pro-R* face directed toward the activated tricopper cluster.

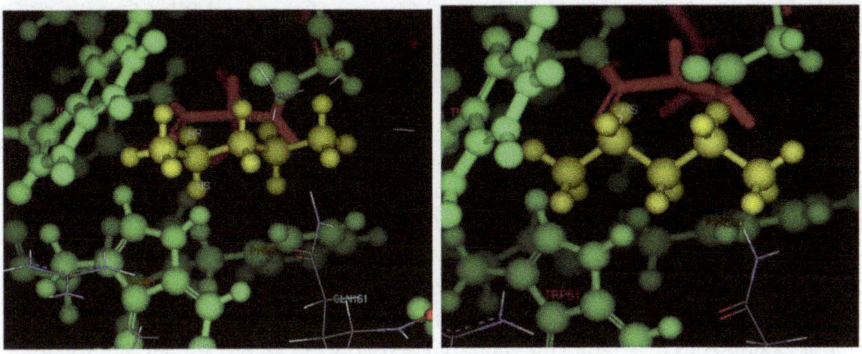

Fig. 18 A pentane molecule oriented within the hydrophobic pocket to facilitate direct O-atom insertion across the C–H_R (*left* panel) and C–H_S (*right* panel) bonds of the secondary carbon (viewed from side)

The above picture accounts for the regio-selectivity and stereo-selectivity observed for both butane and pentane. As noted earlier, the O-atom transfer occurs with total retention of configuration at the carbon-center oxidized (Wilkinson et al. 1996; Yu et al. 2003b). So the *ee* ratios observed upon the hydroxylation of these alkanes by pMMO can be accounted for only by different presentations of the *R*- and *S*-faces of the hydrocarbon to the activated tricopper cluster (Elliott et al. 1997). For normal alkanes, there is a stronger preference for exposing the *pro-R* C–H bond to the tricopper cluster relative to the *pro-S* C–H bond. Analysis of results obtained from stereo-chemical experiments on *n*-butane, and the (2*R*,3*R*)-[2-^{2}H,3-^{2}H] and 2*S*,3*S*)-[2-^{2}H,3-^{2}H] chiral butanes, yielded an equilibrium constant of ca. 0.2 for the *pro-S* facing orientation relative to the *pro-R* facing orientation, and a kinetic isotope effect (k_H/k_D) of ca. 5 for the insertion of the O-atom (Yu et al. 2003b). This kinetic isotope effect is the same as the value of 5.2–5.5 reported from experiments on cryptically chiral ethanes (Wilkinson et al. 1996). A k_H/k_D of this magnitude is consistent with an early transition state in the O-atom transfer chemistry, namely, a concerted O-atom insertion across the C–H bond of the alkane oxidized.

Similarly, pMMO also epoxidizes small alkenes via concerted electrophilic *syn* addition (Ng et al. 2008). Oxo-transfer to vinylic hydrocarbons gives rise to poor chiral selectivity (propene: *ee* = 18%; 1-butene: *ee* = 37%) because the O-atom can approach the *re* and *si* faces with almost equal probability. As expected, pMMO epoxidation of *cis*- and *trans*-2-butene gave exclusively only the *meso*-2,3-dimethyl oxirane and the *d,l*-2,3-dimethyloxiranes, respectively, without any evidence of epimerization.

2.9 *Rationale for the Number of Copper Ions and Their Functions*

The functioning of the pMMO requires as many as 13–14 copper ions. There seems now to be compelling evidence that the catalytic site is a fully reduced tricopper cluster ($Cu^ICu^ICu^I$) that becomes functional upon activation by O_2. However, as a molecular machine, the enzyme is expected to contain additional cofactors for the input and storage of electrons as well as to gate or control the flow of reducing equivalents to the catalytic site to mitigate futile cycles. It seems that some of these additional Cu^I ions perform these functions.

For example, to regenerate the functioning tricopper catalyst for another round of catalytic turnover, it is necessary to provide two additional reducing equivalents to reduce the "spent" catalyst that is produced after O-atom transfer from the activated tricopper cluster to the alkane substrate. This is the role played by the Cu^ICu^I cluster occupying the dicopper site (the **B** site), which can react with O_2 to produce the H_2O_2 reductant, with the assistance of two protons (Wang et al. 2017; Lu et al. 2019). The reducing equivalents could be derived from the high potential Cu^I ions located in the *C*-terminal domain of the PmoB subunit. These copper ions can tune their redox potentials via allosteric linkage with the O_2 chemistry at the dicopper site (the **B** site) to facilitate transfer of reducing equivalents to the **B** site for the H_2O_2 production. Thus, the 6–7 Cu^I ions residing in the *C*-terminal domain of the PmoB subunit serve as a reservoir of reducing equivalents to turn over the enzyme. These Cu^I ions sequestered within the *C*-terminal domain of the PmoB subunit are in close contact with the cytosol and accordingly, they should be in redox equilibrium with NADH and other sources of reductants, given their high intrinsic redox potentials. Of course, since the catalytic site (the **D** site) is embedded in the transmembrane domain, the reducing equivalents needed to regenerate the tricopper cluster catalyst during turnover could also come from quinols residing within the intracytoplasmic membrane. The relative contribution of the two pathways should depend on the availability of quinols within the quinol pool, which in turn can vary with the level of over-expression of the pMMO in the methanotroph.

Finally, given the close proximity of the Cu^I at the **C** site to the tricopper cluster occupying the **D** site (13 Å), and the collective behavior manifested by this mononuclear copper site and the tricopper site when the enzyme turns over in the absence of CH_4, very likely this mononuclear Cu^I provides the reducing equivalent to abort the activated tricopper cluster when it is formed in the absence of a hydrocarbon substrate. In other words, the enzyme has a built-in protective mechanism to mitigate it against undesirable oxidations by the highly reactive oxene produced. Instead, the O_2 activating the tricopper cluster is converted into water to create a futile cycle under these conditions.

2.10 A Scheme for the Overall Operations of the pMMO Enzyme

When we try to integrate all the above findings into a conceptual framework on how pMMO operates, the overall scheme summarized in Fig. 19 emerges. First, pMMO is a Cu^{I} enzyme with about 13–14 copper ions. The catalytic site consists of a tricopper cluster sequestered by PmoA and PmoC subunits within the transmembrane domain. Upon activation by O_2, a highly reactive oxene is generated at this site. When a molecule of CH_4 is available, this oxene is efficiently transferred across one of the C–H bonds of the alkane. In the absence of CH_4 or an alkane substrate, the hot oxene is aborted by a reducing equivalent from the mononuclear Cu^{I} to produce a water molecule with the assistance of two protons.

The CH_4 oxidation is selective and takes place with total retention of configuration of the carbon center oxidized. The product CH_3OH is ejected into the cytoplasmic space of the cell, where it is further metabolized to formaldehyde by methanol dehydrogenase as the next step in the electron transport or metabolic chain. There is a substrate channel within the transmembrane domain to allow the substrate to traverse from the periplasmic space of the methanotroph to the tricopper site where it becomes eventually oxidized. Suicide substrates like acetylene and propyne

Fig. 19 How pMMO operates

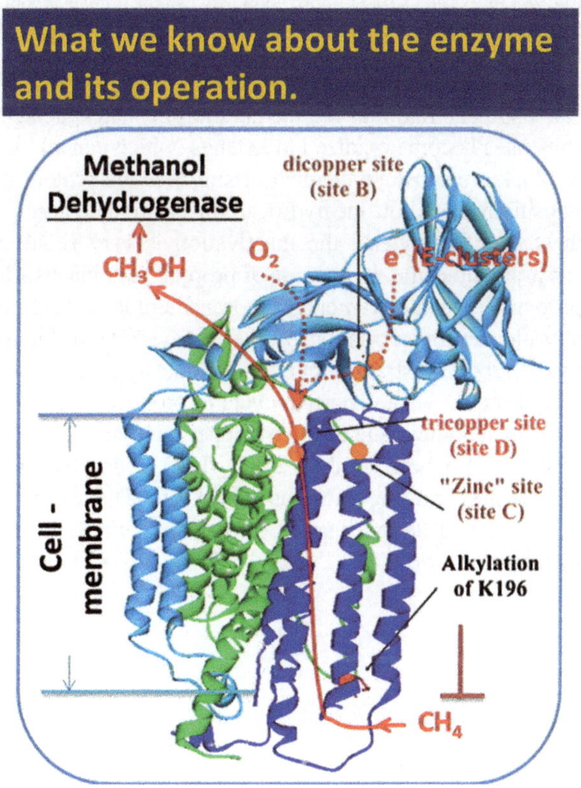

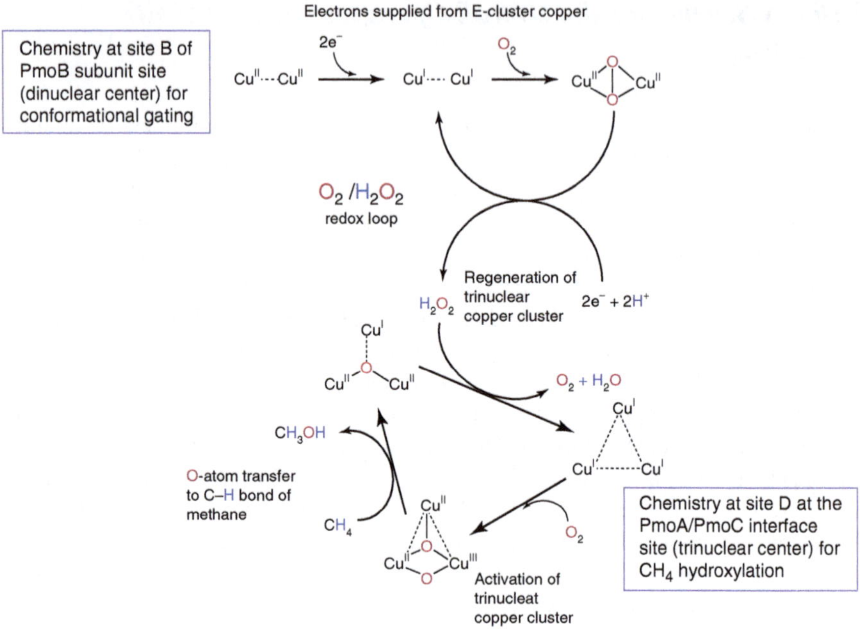

Fig. 20 Proposed O_2/H_2O_2 redox loop between the dinuclear copper site (**B** site) and the trinuclear copper site (**D** site). Adapted from Wang et al. (2017). Copyright 2017 American Chemical Society

can also penetrate into the membrane in exactly the same manner. However, these substrates become oxidized to ketenes, which can alkylate an amino acid side chain to block the substrate channel or suppress the protein dynamics or conformational transitions that allow the hydrocarbon substrate to access the tricopper cluster site. In the case of acetylene, the inactivation is very rapid, requiring only one suicide substrate molecule. In the case of propyne, the inactivation is slower, and additional propyne molecules can enter the transmembrane fold before the channel is eventually blocked by alkylation by the methyl ketene. Interestingly, with access of the hydrocarbon substrate blocked and alkane hydroxylation inactivated, the pMMO can still turn over via an abortive cycle to produce water.

The catalytic cycle at the tricopper catalytic site is illustrated in Fig. 20 (Wang et al. 2017), where we show the O_2/H_2O_2 redox loop between the dinuclear copper site (**B** site) and the trinuclear copper site (**D** site). As mentioned earlier, the spent catalyst could be regenerated directly by quinols from the quinol pool within the membrane as well.

3 Soluble Methane Monooxygenase (sMMO)

We now turn to sMMO. As a multicomponent protein complex requiring three distinct proteins working together in synergy to accomplish its capability of diverse substrate specificities, it is evident that sMMO must be a biochemical machinery of considerable complexity.

3.1 Large and Multi-component Enzymes for Small Substrate Activation

Advanced spectroscopic methods support the mechanistic elucidation of the oxidation of organic substrates by sMMO (Merkx et al. 2001; Wallar and Lipscomb 1996; Wang et al. 2017). The oxidation of CH_4 requires at least three essential components, including the large hydroxylase (MMOH, 251 kDa), the small regulatory protein (MMOB, 15.9 kDa), and the reductase (MMOR, 37.8 kDa) to achieve catalytic turnover. Hydroxylation of CH_4 requires elaborate orchestration of these complexes. Artificially varying the molar ratios of MMOB and MMOR to MMOH affects the catalytic activity (Gassner and Lippard 1999; Liu et al. 1994, 1995a). For other organic substrates, different product distributions are also obtained. Complex formation between MMOH and MMOB exhibits dissociation constants at two binding sites with K_{d1} and K_{d2} of 0.3 and 0.7 μM at 25 °C for the system from *M. capsulatus* (Bath) (Gassner and Lippard 1999). For optimal activity, 2 equiv. of MMOB to MMOH is required. Excess amounts of MMOB retard the activity of MMOH, suggesting that association of additional MMOB might interfere with the binding of MMOR to MMOH. The corresponding K_ds between MMOR–MMOH complexes are 0.9 (K_{d1}) and 2.6 μM at 25 °C.

Structural and spectroscopic information have confirmed the involvement of the different sMMO components, and catalytic intermediates have been proposed from studies on the enzymes purified from both *M. capsulatus* (Bath) and *M. trichosporium* OB3b. While recent studies have provided details about the geometry of the intermediate Q and the complex formation between MMOH and MMOB/MMOR (Banerjee et al. 2015; Lee et al. 2013; Wang et al. 2014), the interactions between MMOH and other components of sMMO still need to be investigated further before the mechanism could be fully understood. Moreover, the interpretation of product distributions and catalytic activity data is complex when experiments are performed on heterogeneous mixtures of MMOB–MMOH and MMOR–MMOH complexes in solution.

The largest component in sMMO, MMOH, is a homodimer ($\alpha_2\beta_2\gamma_2$), and each protomer consists of an α- (60.6 kDa), β- (40.5 kDa), and γ-subunit (19.8 kDa), which are expressed from *mmoX*, *mmoY*, and *mmoZ* genes, respectively, as shown in Fig. 21 (Cardy et al. 1991; Stainthorpe et al. 1990). MMOB is expressed from *mmoB* located between *mmoY* and *mmoZ*. This specific gene map is conserved in all sMMO-expressing type II and type X methanotrophs. The genetic specificity is

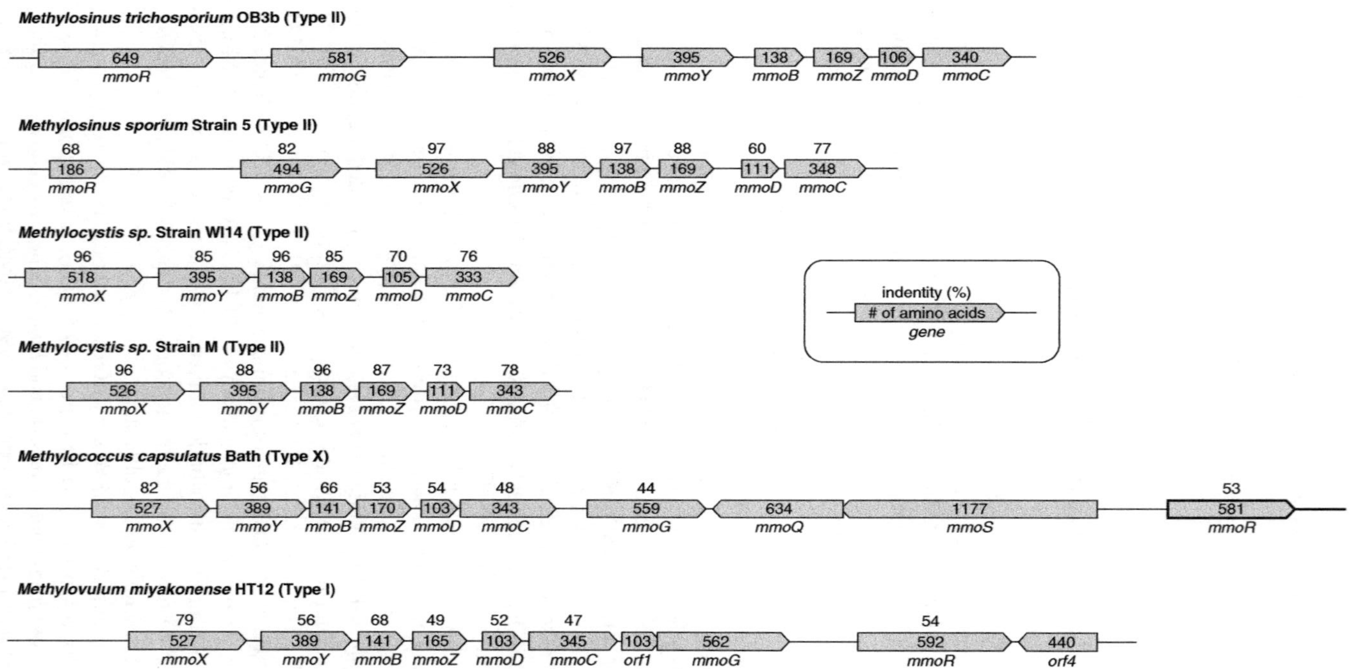

Fig. 21 Gene maps of sMMO from organisms and their encoded proteins. Physical maps of sMMO from type I, type II, and type X methanotrophs. Numbers above box represent sequence identity (%) of amino acids to *Methylosinus trichosporium* OB3b and numbers in the side box are residue numbers of amino acids in each protein. The order of the structural core genes *mmoXYBZDC* is conserved in most organisms

not yet fully explained, but it holds the key for the expression of MMOH in heterologous expression systems, such as *E. coli*. It has been proposed that dimer formation and/or folding of MMOH can be regulated by chaperone proteins to generate soluble MMOH. MMOG (*mmoG*) has been suggested as the putative GroEL homolog chaperone for complex generation among α-, β-, and γ-subunits (Stafford et al. 2003), although its function still remains under investigation. MMOD (*orfY*), the putative inhibitory component of sMMO, is located in the 5′-region of MMOR (*mmoC*) (Merkx and Lippard 2002). The functions of other genes are still unclear, including *mmoQ* and *mmoR* located in the 5′- or 3′-region, although they are considered as regulators of sMMO expression. The solubility of sMMO has made this enzyme system easier to study its function compared to that of pMMO, but the involvement of so many proteins in this multi-component system and the complex interactions among them render the task of understanding the structure and function of sMMO extremely challenging.

It has been proposed that interactions between MMOH and other components, including MMOB and MMOR, affect O_2 activation, electron transfer, and regio-selectivity of substrate oxidations (Davydov et al. 1999a, b; Pulver et al. 1993, 1994; Tinberg and Lippard 2011; Wallar and Lipscomb 2001). The redox potential of the diiron active sites is changed upon MMOB addition, but the origin of this redox regulation by the cofactorless and small protein was not fully understood prior to the structural reports on the MMOH–MMOB complex. It was evident that the diiron active sites are affected upon MMOB binding; spectroscopic studies, including ENDOR, ESEEM, and EXAFS, confirmed the changing environments near the diiron center (Derose et al. 1993; Dewitt et al. 1991; Fox et al. 1991; Rosenzweig et al. 1997; Shu et al. 1997; Thomann et al. 1993). X-ray crystallography shows that MMOB binds near the diiron active site, and it regulates the pathways of substrate delivery and the coordination of the substrate to the active site through conformational changes within the four-helix bundle (Lee et al. 2013). The unstructured *N*-terminal region (1–35) of MMOB seen in solution forms a specific circular structure with helices H and 4 of MMOH through hydrogen bonds and hydrophobic interactions. The well-folded core region of MMOB is positioned above the diiron active sites around helices E and F and mediates allosteric effects through opening and closing of the pore and cavity. The detailed component interactions of the MMOH–MMOB complex will be discussed in Sect. 3.3.

Reduction of the diiron active sites is required to initiate O_2 activation and achieve the catalytic cycle in the sMMO system (Liu et al. 1997; Lund et al. 1985). Reports demonstrated that two domains of MMOR, including the [2Fe–2S] cluster containing the ferredoxin domain (MMOR–Fd) and FAD/NADH-binding domain (MMOR–FAD), shuttle electrons from NADH to the diiron active sites of MMOH (Chatwood et al. 2004; Muller et al. 2002). The FAD cofactor accepts a hydride (H^-) from NADH, and it transfers an electron to the [2Fe–2S] cluster. MMOR–Fd in the *N*-terminus shows high sequence homology with ferredoxins from archaebacteria, plants, and cyanobacteria. The MMOR–FAD domain is positioned in the *C*-terminal region of MMOR and tightly holds the FAD cofactor as a distal form. Although NMR studies have provided structural information of each domain, the full-length MMOR structure has yet to be reported. The electron-transfer pathways of MMOR (Sect. 3.4) have been deciphered by time-resolved spectroscopic techniques, which characterize and identify intermediates with NADH-dependent complexes (Blazyk et al. 2005; Gassner and Lippard 1999; Kopp et al. 2001). An advanced EPR study proposed that MMOB shares binding sites with MMOR in the canyon region of MMOH; however, limited structural results are available to understand the MMOH–MMOR complex (Wang et al. 2014).

3.2 *Tuning of the Hydroxylase Structure*

The O_2 reaction and C–H activation occur in the diiron active site of the MMOH α-subunit with support from other components, including MMOB and MMOR. In 1993, the first X-ray crystallographic structure of MMOH was obtained from *M. capsulatus* Bath, and it provided overall structures and active site geometry at atomic levels (Rosenzweig et al. 1993; Whittington and Lippard 2001). Subsequently, the X-ray structure of MMOH from *M. trichosporium* OB3b was also reported (Elango et al. 1997), and structures of the MMOB, MMOR–Fd, and MMOR–FAD domains were characterized by nuclear magnetic resonance (NMR) methods (Fig. 22) (Chang et al. 1999; Chatwood et al. 2004; Muller et al. 2002; Walters et al. 1999).

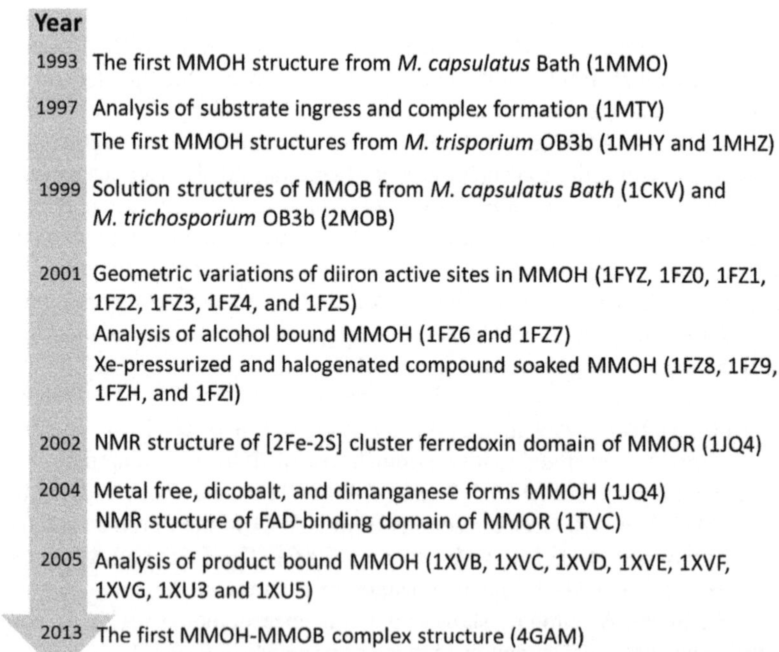

Fig. 22 Discovering history of sMMO structures, including MMOH, MMOB, [2Fe–2S] cluster ferredoxin domain, and FAD/NADH-binding domain. Parentheses represent PDB accession

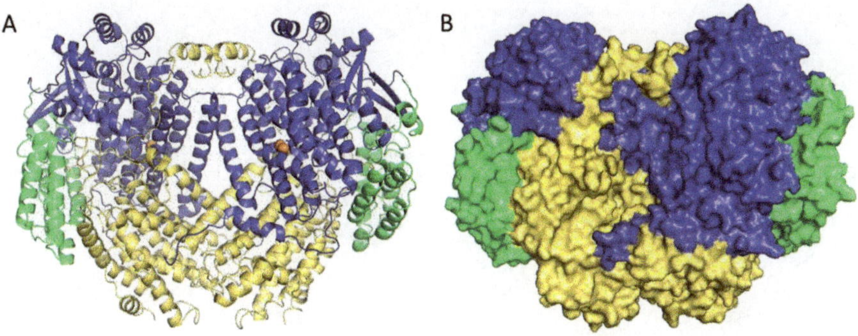

Fig. 23 Structure of hydroxylase (MMOH) from *M. capsulatus* Bath (PDB: 1MMO). (**a**) Ribbon diagram and (**b**) Surface model. α-, β-, and γ-subunits represented as *blue*, *yellow*, and *green*. Orange balls indicate diiron

MMOH consists primarily of the major secondary structure α-helices, and there is high structural similarity between the α- and β-subunits (Fig. 23). Both subunits have identical numbers of long α-helices with more than 25 residues and four short-helices. Most interactions between the αβγ protomers occur through the β-subunits in the dimer. This dimerization generates the canyon region, which provides the essential space for MMOB interaction to regulate access of the substrates to the active sites.

The diiron (Fe1–Fe2) is coordinated by four glutamate and two histidine residues housed in the four-helix bundle from helices B, C, E, and F. These residues (Glu114, Glu144, His147, Glu209, Glu243, and His246) show high sequence identity to the non-heme diiron coordinating amino acids among sMMO-expressing methanotrophs (Rosenzweig et al. 1993, 1995, 1997; Whittington and Lippard 2001). The resting state of diiron, Fe^{III}–Fe^{III}, must be reduced to the diferrous state, Fe^{II}–Fe^{II}, to activate molecular O_2, and structural studies have shown that the coordination environments are different from oxidation states. One of the most significant changes observed is the shift of the side chain from Glu243 positioned in helix F. Glu243 is coordinated to Fe2 in a monodentate manner in the resting state, but it is coordinated in the chelating mode in the reduced state (Fig. 24). The reported MMOH–MMOB complex shows that Glu243 is coordinated with Fe2 in a chelating manner, but the bond distances between the oxygen atoms of carboxylate moieties to the iron atoms are changed. This observation suggests that O_2 activation for the CH_4 hydroxylation is regulated elaborately in the diiron active site of the four-helix bundle. As described, the overall secondary structure of sMMO involves α-helices. Interestingly, the π-turn occurs around Thr213 in helix E. The π-turn is usually seen near active sites in enzymes, and these residues transfer or block protons or electrons from the surface of the protein to the active site positioned 12 Å below the surface. MMOB binding also changes the positions of the side chains of Thr213, Asp214, and Glu240 near the diiron active sites; these residues regulate cavity and pore formation during the catalytic cycle (Lee et al. 2013). MMOH has three hydrophobic cavities between the protein surface and

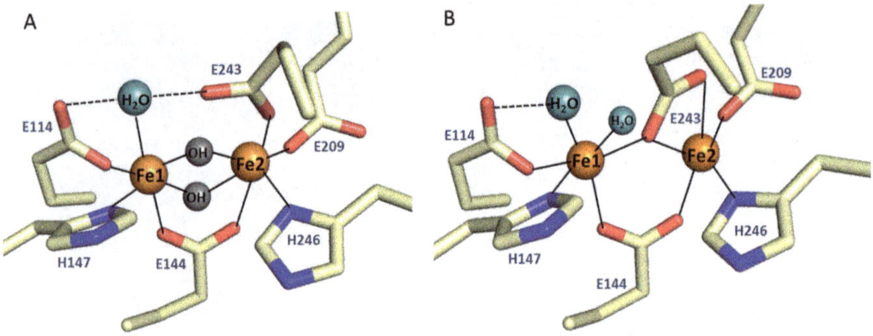

Fig. 24 Geometries of the diiron active sites from oxidized (**a**, PDB: 1MTY) and reduced (**b**, PDB: 1FYZ) states of MMOH

the diiron active sites. Cavities 2 and 1 are disconnected by Phe188, but MMOB binding changes the location of this residue to allow for substrate tunneling between the two cavities.

In MMOH, the major function of the β-subunit is to mediate dimerization of the two protomers through hydrogen and ionic interactions, but the overall structure as α-helices of the β-subunit has high structural similarity without sequence identity to the α-subunit. The *N*-terminal MMOH β-subunit has one short α-helix that interacts mostly with helices A, B, and C in the α-subunit. This interaction may affect MMOH activity, because helices B and C have Fe1 coordinating residues, such as Glu114, Glu144, and His 147. Other possible functions of this region might include regulation of the four-helix bundle, which controls the cavities for CH_4 and O_2. The smallest subunit of MMOH (γ-subunit), which consists of eight helices, does not interact with each other, and it generates a complex through the α- and β-subunits. The difficulties of sMMO expression in the heterologous system for enzymatic studies were caused by the solubility of α- and β-subunits, but the γ-subunit showed reasonable solubility during expression in the *E. coli* system.

3.3 *Regulation of Methane Oxidation by MMOB*

As noted earlier, the regulatory component, MMOB, is essential for achieving maximal activities of sMMO. Spectroscopic studies demonstrate that MMOB binding affects the geometry of the non-heme diiron active site in MMOH (Davydov et al. 1997; Dewitt et al. 1991; Fox et al. 1991; Froland et al. 1992; Liu et al. 1995a). Structural information on this protein from *M. capsulatus* Bath and *M. trichosporium* OB3b was unveiled in 1999 through NMR spectroscopies (Chang et al. 1999; Walters et al. 1999). MMOB has three regions, including unstructured *N*-terminus (1–35), core region (36–128), and *C*-terminus (129–138), based on its folding states. Solution structures of MMOB show a well-folded core

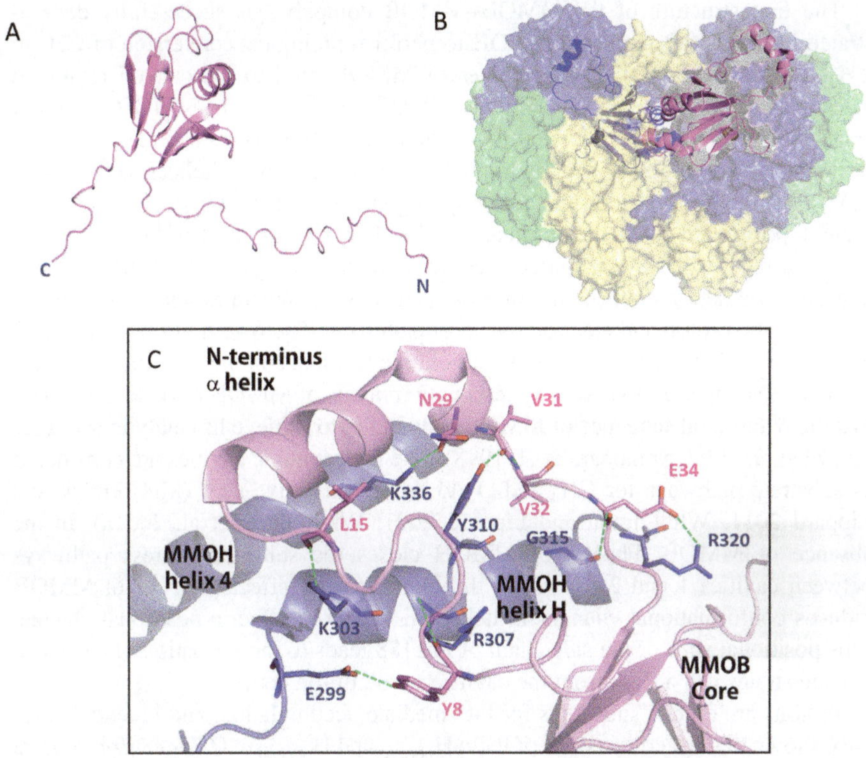

Fig. 25 MMOB regulates substrates through allosteric effects in the MMOH. (**a**) Unstructured *N*-terminus and well-folded core region of MMOB in the absence of MMOH (PDB: 1CKV). (**b**) Two molecules of MMOB (front: *magenta*; rear: *blue*) bind to MMOH. (**c**) Structure of the MMOH–MMOB complex with the *N*-terminus generating specific hydrogen and hydrophobic interactions

region with three α-helices and seven β-strands (*M. capsulatus* Bath) or three α-helices and six β-strands (*M. trichospoiurm* OB3b) forming two *anti*-parallel β-sheets (Fig. 25a). Based on features of these organized secondary structures, it was proposed that the core region of MMOB will dock on to helices E and F, located on the surface of MMOH, to control the activity of the diiron center in the MMOH α-subunit (Walters et al. 1999). Different molar ratios of MMOB/MMOH changes the position of the product hydroxylation and reactivity (Froland et al. 1992; Gassner and Lippard 1999; Liu et al. 1995a). Kinetic studies have shown that 2 molar equiv. of MMOB yields the highest reaction rate, but excess MMOB retards oxygenation. It has been reported that MMOB regulates the redox potentials of the diiron active site and the regio-selectivities of the products (Fox et al. 1989; Liu et al. 1995a; Pulver et al. 1993, 1994). Therefore, these results suggest that specific interactions of the MMOB with the diiron center will change the structures and environments for CH_4 hydroxylation.

The first structure of the MMOH–MMOB complex has successfully demonstrated how MMOB regulates MMOH to perform biological conversion of CH_4 to CH_3OH (Lee et al. 2013). Two copies of MMOB bind to the canyon region of MMOH, with most interactions with the MMOB occurring in the MMOH α-subunit (Fig. 25b). The unstructured *N*-terminal region of MMOB forms a circular structure through hydrogen bonds and hydrophobic interactions with helices H and 4 in MMOH (Fig. 25c). Further, the core region generates interactions between helices E and F positioned above the diiron active site. The rigid structure of the *N*-terminus and core region of MMOB induces conformational changes in MMOH, and this allosteric regulation controls the ingress pathways of substrates and pore regions, including the proton and electron transport pathways. The *N*-terminus of MMOB is conserved in sMMO, and this region has unique features common to other members of the BMM superfamily. Activity tests with mutants of MMOB have demonstrated that the *N*-terminal sequence of MMOB is essential to achieve its catalytic activities (Lloyd et al. 1997; Shinohara et al. 1998). The hydrophobic cavities are considered as substrate pathways for CH_4 and O_2 to the diiron active sites (McCormick and Lippard 2011; Whittington and Lippard 2001; Whittington et al. 2001a). In the absence of MMOB, Phe188 of MMOH closes the substrate ingress pathways between cavities 1 and 2. However, through allosteric effects, Ser111 of MMOB induces conformational changes in helices E and D to effect a positional change. This positional shift of the side chain of Phe188 leads to the formation of canals so that substrates can access from the cavities to the diiron center.

Protons are crucial substrates for intermediate accumulation, and kinetic studies have shown that the conversions of P∗ to H_{peroxo} and H_{peroxo} to Q (*vide infra*) require proton transfers (Tinberg and Lippard 2009, 2010, 2011). The MMOB core blocks the pore region of MMOH, which shows conformational changes upon MMOB binding. This process may transfer protons to Glu243, a putative final proton acceptor, which generates side chain shift near Fe2 (Lee et al. 2013; Wang et al. 2015). The pore region is also considered to be part of the product egress pathway, because hydrophilic residues export CH_3OH to solvent-accessible area. MMOB binding also induces allosteric effects through helices E and F, with coordinating residues to Fe2 on the surface of MMOH. Glu243 on helix F shows conformational changes upon MMOB interaction, and the side chains of Thr213 and Asn214 in helix E are rotated sequentially. These events reorient the position of the phenyl moiety of Phe188 in helix D, which is not part of the four-helix bundle (Fig. 26). These structural results support the conclusion that rapid O_2 activation and the formation of transient intermediates at the diiron active site of MMOH are accelerated in the presence of MMOB.

3.4 *Electron Transfer to the Diiron Active Sites by MMOR*

MMOH requires electron transfers from MMOR for the catalytic cycle. $MMOH_{red}$ then reacts with O_2 to generate the intermediates to hydroxylate one of the C–H

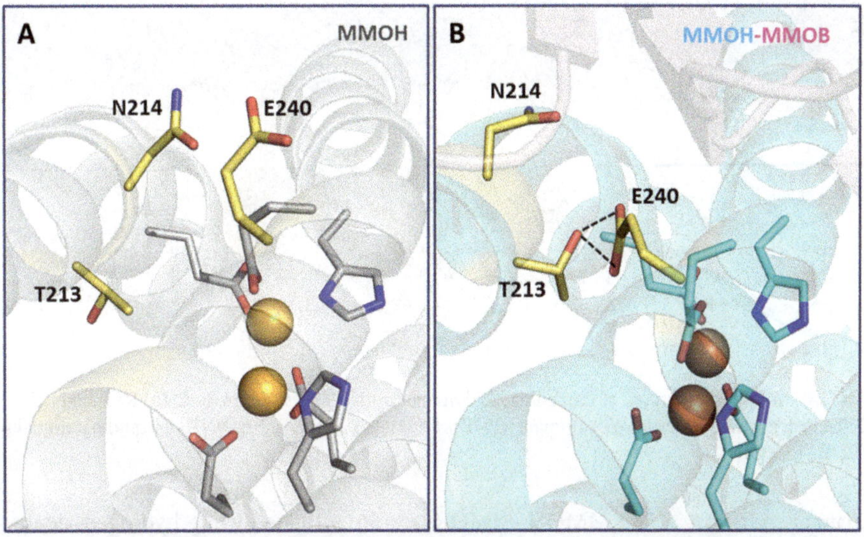

Fig. 26 Diiron active site is precisely regulated through MMOB interactions. Conformational changes near the diiron active site of MMOH, including Glu240, Thr213, and Asn214, show that the pore region is controlled by MMOB. Figures represent MMOB (**a**) and MMOH-MMOB (**b**).

bonds of CH_4. MMOR contains two prosthetic groups: the flavin adenine dinucleotide (FAD) and the [2Fe–2S] cluster; and these cofactors shuttle electrons from NADH to MMOH (Chatwood et al. 2004; Colby et al. 1977; Green and Dalton 1989; Lund et al. 1985; Muller et al. 2002). The structure of the full-sequence MMOR has not yet been reported, but solution structures of the ferredoxin domain (MMOR–Fd) containing the [2Fe–2S] cluster, and the NADH/FAD-binding domain (MMOR–FAD), provide detailed information to delineate the electron transfer pathways in sMMO (Fig. 27). MMOR–Fd is positioned in the *N*-terminus (1–98) of MMOR, which consists of six β-strands and three α-helices. The spectroscopic data for MMOR–Fd are similar to other ferredoxins from plants (Fig. 27a). Four cysteine residues, including Cys42, 47, 50, and 82, coordinate with the [2Fe–2S] cluster. Further, these residues show sequence identity to reductases from other methanotrophs. The third α-helix (Pro71–Gly77) of MMOR–Fd has five consecutive glutamic acids (Glu72–76) with high sequence similarity with other ferredoxin domains. These negatively charged carboxylic groups may aid in interactions with the positively charged region of MMOH. A docking model study proposed that MMOR–Fd could fit into the canyon region of MMOH, similar to MMOB (Bailey et al. 2008; Sazinsky et al. 2006). This result supports the notion that MMOB and MMOR compete with each other to bind to the MMOH canyon region, which is consistent with the proposed catalytic cycle (Sect. 3.5). Oxidized MMOH docks with MMOR to receive electrons, and the $MMOH_{red}$ interacts with MMOB for O_2 and C–H bond activation. The linkage of this sequence of events is controlled in a timely

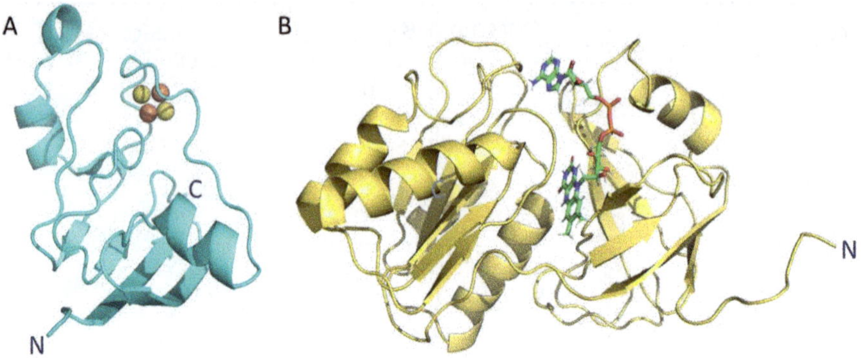

Fig. 27 Solution structures of the reductase domains from *M. capsulatus* Bath. (**a**) Reduced [2Fe–2S] cluster MMOR ferredoxin domain (PDB: 1JQ4) (**b**) MMOR FAD/NADH-binding domain (PDB: 1TVC)

manner. A report of the MMOH–MMOR was investigated using hydrogen–deuterium exchange coupled mass-spectrometry (HDX-MS), and results demonstrated that the [2Fe–2S] cluster of MMOR–Fd binds to the canyon region of MMOH ((Wang et al. 2014), see Sect. 3.6).

The MMOR–FAD domain, positioned towards the *C*-terminus (99–348), shows high structural similarity to other FAD-containing oxidoreductases, including benzoate-1,2-dioxygenase reductase from *Acinetobacter sp.* strain ADP1 (Blazyk and Lippard 2004; Chatwood et al. 2004). This region includes a FAD cofactor-containing domain (Ser108–Phe207) and a NADH-binding domain (Leu209–Ala308) like other flavoproteins. FAD cofactors are tightly bound, and the residues interacting with the FAD and NADH are highly conserved. Adenosine is distal from the isoalloxazine ring in an extended conformation in both of these two domains. In MMOR–FAD, the FAD-binding domain has a β-barrel with six anti-parallel β-strands and one α-helix, and the NADH-binding domains consist of five parallel β-strands with the classical fold for nucleotide binding (Fig. 27b). The first 10 residues (Cys99–Gly110) of MMOR–FAD are unstructured, and it is considered as a linker between the Fd- and FAD-binding domains. The FAD-binding and NADH-binding domains are also connected by 15 amino acids (Gly202–Pro216).

MMOR was cloned and expressed in *E. coli* to achieve high yields of the purified enzyme. EPR spectra of the fully reduced MMOR reveal a rhombic signal from the $[2Fe–2S]^+$ cluster with almost identical g-values (2.047, 1.958, and 1.871), similar to the native MMOR (Blazyk and Lippard 2002; Gassner and Lippard 1999; Kopp et al. 2001). Extinction coefficients from UV-visible spectra of MMOR–Fd (332, 418, and 467 nm) and MMOR–FAD (394 and 458 nm) show identical values with those of the full sequence MMOR. Optical spectra of the MMOR–Fd are not influenced by pH, but the MMOR–FAD showed significant alterations to the chromophores (FAD_{ox}, FAD_{sq}, and FAD_{hq}) with pH. The pK_a values of FAD_{ox} and FAD_{hq} are 6.5 and 7.1, respectively, and the blue natural FAD_{sq} is deprotonated to form the red anionic semiquinone at pH greater than 8.5.

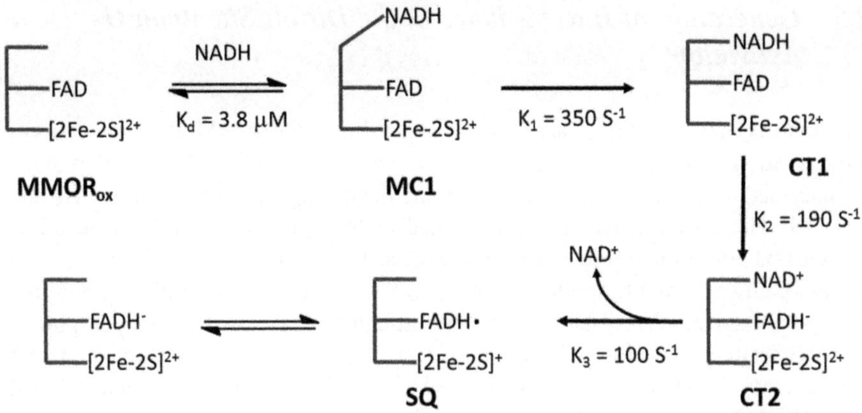

Fig. 28 Intermediates formed during the electron transfer reactions from NADH to the full-length $MMOR_{ox}$

NADH is the sole electron source to reduce MMOR, and the dissociation constants (K_d) between MMOR–FAD and NAD^+ is 43 ± 2 μM (Blazyk and Lippard 2002). Stopped-flow UV-visible spectroscopy reveals four intermediates during the electron transfer process from NADH to MMOR–FAD. The first intermediate is generated between MMOR–FAD and NADH, which induces a charge transfer band (CT1, $\lambda_{max} \approx 570$ nm) between the nicotine amide ring from NADH and isoalloxazine in the FAD cofactor. The rate of formation of CT1 varies with the NADH concentration and the curve is hyperbolic. These data imply that a Michaelis complex (MC1) is generated before CT1 formation. The full-length MMOR has a stronger binding affinity for NADH ($K_d = 3.8$ μM) as compared to the MMOR–Fd owing to different binding environments. Hydride transfer to the FAD_{ox} produces the second intermediate (CT2), and this charge transfer band ($\lambda_{max} > 700$ nm) is induced by NAD^+ and reduced flavin (FAD_{hq}). CT1 and CT2 are formed in less than 7 and 10 ms, and the intermediate CT2 decays in greater than 10 ms to become FAD_{hq}. The intermediates formed during electron transfer from NADH to the full-length $MMOR_{ox}$ are described in Fig. 28. The formation of intermediates in the case of the full-length MMOR, such as MC1, CT1, and CT2, are similar, and partial electron transfer from $FADH^-$ to $[2Fe–2S]_{ox}$ occurs concomitantly to yield FADH· and $[2Fe–2S]^+$ (Blazyk et al. 2005; Blazyk and Lippard 2004). In the presence of $MMOH_{ox}$, the rate of NADH binding and hydride transfer is almost similar, but the subsequent electron-transfer steps to $MMOH_{ox}$, including the intermolecular electron transfer, are slower. However, one equiv. of MMOB/MMOH enhances the transfer rate from the MMOR.

3.5 Generation of Intermediates at the Diiron Site upon O_2 Activation

CH_4 hydroxylation requires O_2 activation of the diiron site in MMOH. The details of the activation of the O_2 by the non-heme diiron center, including identification of the O_2 intermediates is key toward understanding the biological conversion of CH_4 into CH_3OH. The diiron active site, coordinated to four glutamates and two histidine residues, is housed in a four-helix bundle consisting of helices B, C, E, and F. The ingress-pathways of O_2 to the diiron active site have been confirmed through Xe-pressurized X-ray crystallography (Whittington et al. 2001a). A hydrophobic pathway from the exterior to the interior is formed, and xenon atoms, similar van der Waals radius with O_2 and CH_4, are detected along this hydrophobic route developed in most BMMs (McCormick and Lippard 2011; Murray and Lippard 2007; Wang et al. 2015). Single-turnover reactions of $MMOH_{red}$ with 2 equiv. of MMOB and with O_2 reveal the formation of a number of intermediates, including P*, H_{peroxo}, Q, and Q* in the absence of substrates (Lee and Lipscomb 1999; Tinberg and Lippard 2009; Wallar and Lipscomb 2001). Protons, an essential substrate of sMMO, are required for the catalytic cycle, and the MMOH–MMOB structure proposes that MMOB may bring protons from a solvent-accessible area to the interior of MMOH via Glu240. In the absence of MMOB, MMOH has a well-developed pore region near the diiron center (Fig. 26). MMOB blocks this pore region by shifting Glu240 in MMOH. During this carboxylate shift of Glu240, protons or electrons might be brought to the diiron active site. This proton transfer facilitates formation of H_{peroxo} and Q as monitored in the kinetic studies.

Single-turnover stopped-flow and rapid freeze-quench (SF-RFQ) methods have identified and characterized intermediates at the non-heme diiron center, and these results have been confirmed by X-ray crystallography, Mössbauser, EPR, MCD, EXAFS, and UV-visible spectroscopies. Detailed results are extensively well summarized elsewhere (Jasniewski and Que 2018; Merkx et al. 2001; Wallar and Lipscomb 1996). Slightly different transient intermediates have been proposed for sMMO from *M. capsulatus* Bath and *M. trichosporium* OB3b. Six intermediates are proposed from *M. capsulatus* Bath by mixing $MMOH_{red}$ with 2 equiv. of MMOB and O_2 (Tinberg and Lippard 2010; Wang et al. 2015; Fig. 29). $MMOH_{red}$ decays very quickly in 24 s^{-1} at pH 7.0 and 4 °C, accumulating the P* intermediate, namely, the peroxodiiron(III) species, with an estimated 370 ms of maximal accumulation time (t_{max}). The optical coefficients at 420 and 720 nm of P* are similar with H_{peroxo} ($t_{max} = 2.1$ s), and these observations are suggestive of the peroxo-to-iron(III) charge-transfer band like other peroxodiiron(III) intermediates from carboxylate-bridged diiron enzymes (Tinberg and Lippard 2009, 2011). The optical similarities between P* and H_{peroxo} indicate that they have very similar electronic and geometric environments, but pH-dependent experiments indicate that this conversion requires one proton. Glu243 may be the acceptor of this proton from the electron donor (Glu240), but this needs to be confirmed by mutant studies. In any case, evidence for this conformational change of Glu240 has been obtained from structural comparison

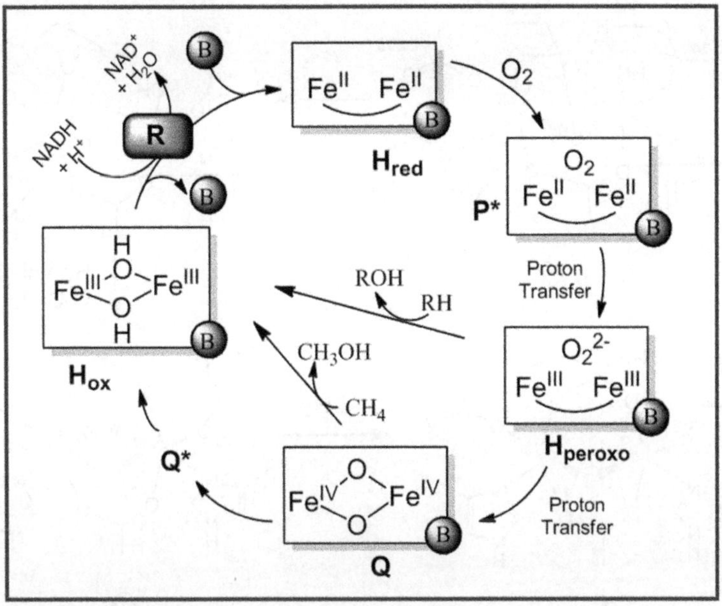

Fig. 29 Proposed catalytic cycle for sMMO from *M. capsulatus* (Bath)

between MMOH and the MMOH–MMOB complex. Double-mixing SF studies demonstrate that H_{peroxo} can activate the C–H bond from electron-rich substrates, such as ethers and alcohols, but this intermediate is not a potent oxidant (unlike Q) for CH_4 hydroxylation (Tinberg and Lippard 2010).

H_{peroxo} rapidly decays to the Q intermediate in the absence of substrates, which is one of the most potent oxidants for hydroxylation of CH_4 and other substrates. From RFQ-EXAFS data, it has been proposed that this transient intermediate has the di (μ-oxo)diiron(IV) with a short Fe–Fe distance of 2.46 Å and Fe-μ-O bonds with short (≈1.77 Å) and long (≈2.05 Å) distances (Shu et al. 1997). The antiferromagnetically coupled diamond core shows optical bands at 350 nm ($\varepsilon = 3600\ M^{-1}\ cm^{-1}$), 420 nm ($\varepsilon = 7200\ M^{-1}\ cm^{-1}$), and 720 nm ($\varepsilon = 1000\ M^{-1}\ cm^{-1}$) and Mössbauer parameters ($\delta = 0.17$ mm/s and $\Delta E_Q = 0.53$ mm/s) for the MMOH from *M. trichosporium* OB3b. A recent report has confirmed the geometry of the Q intermediate by time-resolved Raman spectroscopy and proposed the bis-μ-oxo diamond core with both oxygen atoms of O_2 bridging the diiron (Banerjee et al. 2015). This result supports that O–O bond breaking occurs through homolytic cleavage. The authors also presented evidence for intermediate T, which holds one of the oxygen atoms from O_2 and the other oxygen atom incorporated into the product formed. The maximal accumulation time of intermediate Q from *M. trichosporium* OB3b is 3 s. This intermediate has a specific vibrational mode at 690 cm^{-1}. Spectroscopic assays comparing isotopes (^{16}O and ^{18}O) confirmed that Q was formed from oxygen atoms after O–O cleavage

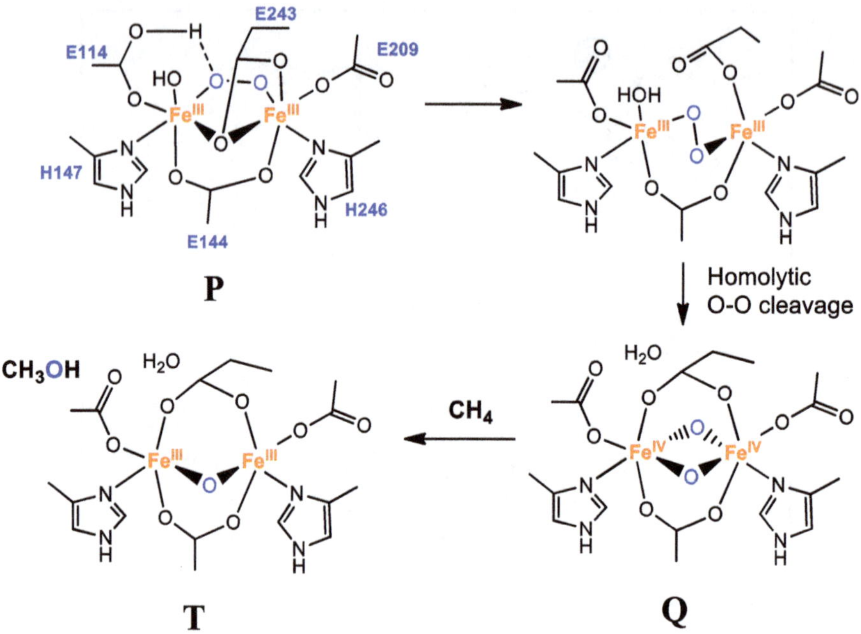

Fig. 30 SF-optical and RFQ-rR spectrocopies using isotopes (^{16}O and ^{18}O) demonstrate that the oxo-bridges of the Q intermediate originate from O_2. *Blue* oxygen represents the sMMO substrate O_2

of O_2. Decay of the Q intermediate causes formation of the T, mono-μ-oxo core, by transferring one oxygen atom to the product (Fig. 30).

Alternatively, a recent high-energy–resolution fluorescence detected X-ray absorption spectroscopy (HERFD-XAS) study of the Q intermediate and related model compounds revealed that the open-core structure with the terminal $Fe^{IV}=O$ moiety cannot be ruled out for the Q intermediate, implying that additional experimental definition may be required to reach a consensus on the structure of the Q intermediate (Castillo et al. 2017). This proposed Q intermediate can be applied to achieve more detailed mechanisms of sMMO and structural information from the complex provides substrate ingress and egress route to explain where the intermediate is isolated.

The decay rate of the Q intermediate is slow in the absence of substrates, and this intermediate requires two protons and two electrons to reach the resting state (Tinberg and Lippard 2009). This unknown process was discovered in kinetic studies through the identification of Q* (t_{max} = 186 s), which shows a unique shoulder at 455 nm. The intermediate Q* does not react with CH_4, and pH changes in the buffer do not affect its formation.

3.6 Regulation of Substrates Through Harmony of Component Interactions

Hydroxylation of CH_4 requires four substrates, including CH_4, O_2, electrons, and protons, with three essential enzymes as described. Current knowledge indicates that complex formation between MMOH–MMOB and MMOH–MMOR precisely controls the catalytic cycle for substrate regulation in sMMO, although structural information has yet to be reported on the MMOH–MMOR complex. The rates of hydroxylation and epoxidation change with the molar ratio of MMOB/MMOH. Increasing the concentration of MMOB to 2.0 equiv. shows maximal rate, but additional MMOB reduces the rates of hydroxylation (Fox et al. 1991; Gassner and Lippard 1999; Liu et al. 1995b). From these results, it was proposed that one MMOB molecule regulates activity of a MMOH protomer. In addition, excess MMOB interferes with MMOR interactions to generate $MMOH_{red}$, although $MMOH_{ox}$ requires the presence of MMOB for reduction by the reductase. The regio-selectivity of substrates has also been studied in the presence of MMOB and MMOR. These experiments yielded different product distributions (Fox et al. 1988; Froland et al. 1992). Kinetic studies suggest that O_2 activation at the diiron active site of $MMOH_{red}$ requires proton transfer before C–H activation of various hydrocarbons. Reports have extensively provided indirect evidence of these component interactions obtained by various methods during the early days of the field owing to the absence of structural information on these complexes (Deeth and Dalton 1998; Feig and Lippard 1994; Wallar and Lipscomb 1996, 2001; Wilkinson et al. 1996).

The first complex structure, MMOH–MMOB, provided direct evidence of the mechanism of MMOB regulation for O_2 and CH_4 ingress to the diiron active site as well as the putative proton and electron transfer routes (Lee et al. 2013). The cavities developed inside of MMOH have been considered as O_2 and CH_4 transport routes from the exterior to the diiron center in cavity 1. The distance of cavity 3 to the diiron center via the hydrophobic residues is 36–40 Å. Xe-pressurized MMOH crystal has been applied to understand O_2 pathways since the two molecules have similar sizes and hydrophobicity (Whittington et al. 2001a, b). MMOH crystals has also been soaked with halogenated substrates, such as CH_2Br_2 and CH_3CH_2I, as hydrocarbon analogues to take advantage of their high electron density (Whittington et al. 2001a). These model compounds were detected around cavities 3 and 2, but electron density maps of substrates were not detected in cavity 1. Gatekeepers, including Phe188 and Leu110, block cavities 1 and 2. Upon MMOB docking of the canyon region of MMOH, Tyr8 from *N*-terminal region, and Ser111 from the core region induce a conformational shift of helices H and E. This allosteric effect causes a shift of the Phe188 side chains, and cavities 1 and 2 become connected to transfer substrates. The complex formation of MMOH–MMOB based on oxidation states has also been investigated by double electron–electron resonance (DEER) spectroscopy (Wang and Lippard 2014). The MMOB *N*-terminal region is flexible in the $MMOH_{ox}$ state, but the relaxed MMOB *N*-terminus acquires an ordered conformation upon MMOH reduction. Structure of the MMOH–MMOB complex suggests that Tyr8 in the *N*-

terminus regulates residues to open the cavities for O_2 access to the reduced diiron (Fe^{II}–Fe^{II}) active site.

Electron transfer to the diiron center initiates the catalytic cycle. A recent HDX-MS study demonstrates a putative MMOR binding site on MMOH (Wang et al. 2014). MMOR covers the pore region, which indicates that it shares a binding site with MMOB. MMOR transfers electrons from the [2Fe–2S] cluster to the diiron center. This pore is considered as the shortest route to reach the diiron active site, and side chain shifts of Glu240, Asn214, and Thr213 have been observed in MMOH-MMOB complex. These conformational changes generate a hydrogen-bonding network for electron transfer. The gate residues of the pore include Asn214 and Glu240, which are bound to the [2Fe–2S] cluster, as in toluene o-xylene monooxygenase (ToMO, (Wang et al. 2015)). As part of this process, the carboxylate side chain of Glu240 is shifted inward, facilitating proton coupled electron transfers to the diiron center. Proton is an essential substrate of the O_2 activation, and MMOB blocks the pore to avoid undesired products such as water or hydrogen peroxide. The sharing of a binding site between MMOB and MMOR can regulate O_2 activation and electron transfer more precisely and in a timely manner. Electrons from MMOR generates the Fe^{II}–Fe^{II} active site, and MMOB binding permits O_2 activation without additional electron transfer(s) to the diiron center to avoid quenching of the intermediate formed.

3.7 Proposed Mechanisms of Hydrocarbon Hydroxylation

The geometries of the transient intermediates were proposed based on spectroscopic data and structural information of $MMOH_{red}$ and $MMOH_{ox}$. Mechanistic elucidations of O_2 intermediates, especially the Q, reactions with CH_4 and other hydrocarbons, are a major interest. However, no intermediates have been observed when Q reacted with substrates. To understand the C–H activation mechanisms, various substrates have been studied and different product distributions are seen, but CH_4 is the only physiological substrate in sMMO. Double-mixing stopped-flow studies indicate that the Q intermediate is the only transient intermediate for CH_4 oxygenation, although H_{peroxo} could oxidize electron-rich substrates, such as propylene (Tinberg and Lippard 2010). No k_H/k_D kinetic isotope effect (KIE) is seen during the formation of Q, but this intermediate rapidly decays in the presence of substrates. The Q decay-rate constant, however, exhibits a large KIE (50–100) for CH_4 and CD_4 (Lipscomb et al. 1994; Nesheim and Lipscomb 1996), decreasing linearly in the series of isotope homologs ($CH_4 > CH_3D > CH_2D_2 > CHD_3 > CD_4$). This result supports a primary KIE associated with CH_4 hydroxylation. For larger substrates (ethane and propane), a secondary KIE is observed for deuterated substrates. Here, it is assumed that substrate binding is the rate-determining step rather than C–H bond cleavage. Product analysis can also be used to determine the k_H/k_D KIE in single-turnover experiments of MMOH with CH_4 and its deuterated homologs. Intermolecular KIE of 19.3 was observed from a 1:1 mixture of CH_4 and CD_4, and a smaller intramolecular KIE of 3.9 was detected for CH_3D (Lipscomb et al. 1994; Nesheim and Lipscomb 1996). These KIEs demonstrate that C–H activation occurs in transition state, but the KIE values are different from those for the Q decay. These different values of KIE indicate that the process of product formation could be more complicated than a simple reaction between Q and the CH_4 substrate.

The presence of the regulatory protein and reductase also affect product distribution. Hydroxylation of substrate is performed at the active sites, MMOR transfers electrons to the non-heme diiron center for the O_2 activation, but MMOB regulates these electron transfers and the specifics of the oxidation chemistry for a given substrate. If this assumption is accepted, MMOB addition can not only change the reaction rate, but also produce different products with different products distributions. Binding of the regulatory protein may change the substrate access route and the geometry of the diiron center in MMOH through allosteric effects. sMMO exhibits a broad range of substrates, including alkanes, alkenes, aromatics, chlorinated compounds, and heterocyclic compounds. The diverse substrate specificity is, in fact, a unique feature of sMMO in the BMM superfamily. Presumably, different substrates exhibit different access routes to the diiron active site, different rates of formation of the kinetic intermediates, and even the mechanism of substrate oxygenation. Formation of H_{peroxo} and Q intermediates depend on the ratio of MMOB/MMOH, but, more significantly, on the substrate as well. Thus, interactions of the substrate with the protein scaffold near the active site must also contribute to the effects of MMOB and MMOR on the enzymology.

4 Epilogue

Over the past several decades, substantial progress has been made toward understanding how methanotrophs mediate the conversion of CH_4 into CH_3OH. These microbes oxidize CH_4 generated from anaerobic methanogenic metabolism to CO_2 as their sole source of carbon and energy in the first step of the metabolic chain with O_2 as the oxidant for the process. Two enzymes are known to perform this chemistry efficiently and selectively under ambient conditions depending on cellular conditions: pMMO and sMMO. The overall architectures of these two enzymes are now known and the roles played by various metal cofactors are more or less established. In fact, based on our knowledge of the pMMO, a biomimetic catalyst has been recently developed capable of mediating efficient and selective CH_4 oxidation under ambient conditions (Chan et al. 2013; Chen et al. 2014; Nagababu et al. 2014; Liu et al. 2016, 2018). This catalyst works amazingly well and just like the enzyme. In retrospect, pMMO is the simpler enzyme system relative to sMMO, even though it is membrane bound. It seems that the protein scaffold of this three-subunit membrane protein was designed primarily, or at least, in part, to anchor the 13 Cu^{I} cofactors, especially the unique tricopper cluster, to carry out the CH_4 oxidation in the simplest and most efficient way. It is a relatively primitive molecular machine. The molecular machinery of sMMO, on the other hand, is much more intricate and sophisticated.

sMMO utilizes a system of three enzymes, and protein–protein interactions as well as protein dynamics, to orchestrate the delivery of electrons, O_2, protons, and the CH_4 to the catalytic site, so that the various substrates are delivered in an orderly fashion, in the proper sequence and with precise timing, for kinetic control and optimal performance of the machinery. As sMMO is designed to oxidize a diverse range of organic substrates besides CH_4, we can expect an individual organic substrate to feedback on to the enzyme near the catalytic site in its unique way, in order to fine tune the protein–protein interactions and protein dynamics, and even to alter certain specifics of the catalytic mechanism, to accomplish the desired molecular outcome. Thus, sMMO is a sophisticated molecular machine indeed.

The goal of the research in this field has been to understand the mechanisms of the chemistry behind the MMOs so that we can learn from the microbes on how to design efficient and robust catalysts that might eventually be adapted for CH_3OH production from natural gas on an industrial scale. Considerable efforts have been expended over the past couple of decades toward this end. Accordingly, it is gratifying that some of this work has culminated in the successful development of a platform technology capable of liquefaction of CH_4 into CH_3OH (Liu et al. 2016), and even natural gas into their component alcohols and ketones (Liu et al. 2018). However, the scaling up of this technology, or the development of related technology platforms based on the basic chemical principles embodied in the tricopper cluster, remains a considerable, if not formidable, challenge. We will end this chapter with anticipation, if not a certain degree of guided optimism.

References

Anthony C (1982) The biochemistry of methylotrophs. Academic, London

Bailey LJ, McCoy JG, Phillips GN, Fox BG (2008) Structural consequences of effector protein complex formation in a diiron hydroxylase. Proc Natl Acad Sci U S A 105:19194–19198

Balasubramanian R, Smith SM, Rawat S, Yatsunyk LA, Stemmler TL, Rosenzweig AC (2010) Oxidation of methane by a biological dicopper centre. Nature 465:115–119

Banerjee R, Proshlyakov Y, Lipscomb JD, Proshlyakov DA (2015) Structure of the key species in the enzymatic oxidation of methane to methanol. Nature 518:431–434

Basu P, Katterle B, Andersson KK, Dalton H (2003) The membrane-associated form of methane mono-oxygenase from *Methylococcus capsulatus* (Bath) is a copper/iron protein. Biochem J 369:417–427

Blanksby SJ, Ellison GB (2003) Bond dissociation energies of organic molecules. Acc Chem Res 36:255–263

Blazyk JL, Gassner GT, Lippard SJ (2005) Intermolecular electron-transfer reactions in soluble methane monooxygenase: a role for hysteresis in protein function. J Am Chem Soc 127:17364–17376

Blazyk JL, Lippard SJ (2002) Expression and characterization of ferredoxin and flavin adenine dinucleotide binding domains of the reductase component of soluble methane monooxygenase from *Methylococcus capsulatus* (Bath). Biochemistry 41:15780–15794

Blazyk JL, Lippard SJ (2004) Domain engineering of the reductase component of soluble methane monooxygenase from *Methylococcus capsulatus* (Bath). J Biol Chem 279:5630–5640

Cao L, Caldararu O, Rosenzweig AC, Ryde U (2018) Quantum refinement does not support dinuclear copper sites in crystal structures of particulate methane monooxygenase. Angew Chem Int Ed 57:162–166

Cardy DL, Laidler V, Salmond GP, Murrell JC (1991) The methane monooxygenase gene cluster of *Methylosinus trichosporium*: cloning and sequencing of the *mmoC* gene. Arch Microbiol 156:477–483

Castillo RG, Banerjee R, Allpress CJ, Rohde GT, Bill E, Que L, Lipscomb JD, DeBeer S (2017) High-energy-resolution fluorescence-detected X-ray absorption of the Q intermediate of soluble methane monooxygenase. J Am Chem Soc 139:18024–18033

Chan SI, Nguyen HHT, Shiemke AK, Lidstrom ME (1993) The copper ions in the membrane-associated methane monooxygenase. In: Karlin KD, Tyeklar Z (eds) Bioinorganic chemistry of copper. Chapman and Hall, New York

Chan SI, Chen KHC, Yu SSF, Chen CL, Kuo SSJ (2004) Toward delineating the structure and function of the particulate methane monooxygenase from methanotrophic bacteria. Biochemistry 43:4421–4430

Chan SI, Wang VCC, Lai JCH, Yu SSF, Chen PPY, Chen KHC, Chen CL, Chan MK (2007) Redox potentiometry studies of particulate methane monooxygenase: support for a trinuclear copper cluster active site. Angew Chem Int Ed 46:1992–1994

Chan SI, Yu SSF (2008) Controlled oxidation of hydrocarbons by the membrane-bound methane monooxygenase: the case for a tricopper cluster. Acc Chem Res 41:969–979

Chan SI, Nguyen HHT, Chen KHC, Yu SSF (2011) Overexpression and purification of the particulate methane monooxygenase from *Methylococcus capsulatus* (Bath). In: Rosenzweig AC, Ragsdale SW (eds) Methods in methane metabolism, methods in enzymology, vol 495. Academic, Burlington, MA, pp 177–193

Chan SI, Lu YJ, Nagababu P, Maji S, Hung MC, Lee MM, Hsu IJ, Minh PD, Lai JCH, Ng KY, Ramalingam S, Yu SSF, Chan MK (2013) Efficient oxidation of methane to methanol by dioxygen mediated by tricopper clusters. Angew Chem Int Ed 52:3731–3735

Chang SL, Wallar BJ, Lipscomb JD, Mayo KH (1999) Solution structure of component B from methane monooxygenase derived through heteronuclear NMR and molecular modeling. Biochemistry 38:5799–5812

Chatwood LL, Muller J, Gross JD, Wagner G, Lippard SJ (2004) NMR structure of the flavin domain from soluble methane monooxygenase reductase from *Methylococcus capsulatus* (Bath). Biochemistry 43:11983–11991

Chen KHC, Chen CL, Tseng CF, Yu SSF, Ke SC, Lee JF, Nguyen HT, Elliott SJ, Alben JO, Chan SI (2004) The copper clusters in the particulate methane monooxygenase (pMMO) from *Methylococcus capsulatus* (Bath). J Chin Chem Soc 51:1081–1098

Chen PPY, Nagababu P, Yu SSF, Chan SI (2014) Development of the tricopper cluster as a catalyst for the efficient conversion of methane into MeOH. ChemCatChem 6:429–437

Chidambaram-Padmavathy K, Oblulisamy PK, Heimann K (2015) Role of copper and iron in methane oxidation and bacterial biopolymer accumulation. Eng Life Sci 15:387–399

Choi DW, Kunz RC, Boyd ES, Semrau JD, Antholine WE, Han JI, Zahn JA, Boyd JM, de la Mora AM, DiSpirito AA (2003) The membrane-associated methane monooxygenase (pMMO) and pMMO-NADH: quinone oxidoreductase complex from *Methylococcus capsulatus* (Bath). J Bacteriol 185:5755–5764

Colby J, Stirling DI, Dalton H (1977) Soluble methane mono-oxygenase of *Methylococcus capsulatus* (Bath). Its ability to oxygenate n-alkanes, n-alkenes, ethers, and alicyclic, aromatic and heterocyclic-compounds. Biochem J 165:395–402

Davydov A, Davydov R, Graslund A, Lipscomb JD, Andersson KK (1997) Radiolytic reduction of methane monooxygenase dinuclear iron cluster at 77 K - EPR evidence for conformational change upon reduction or binding of component B to the diferric state. J Biol Chem 272:7022–7026

Davydov R, Hoffman BM, Valentine AM, Lippard SJ, Sligar SG, Ikeda-Saito M (1999a) EPR and ENDOR studies on cryoreduced metalloproteins. J Inorg Biochem 74:110

Davydov R, Valentine AM, Komar-Panicucci S, Hoffman BM, Lippard SJ (1999b) An EPR study of the dinuclear iron site in the soluble methane monooxygenase from *Methylococcus capsulatus* (Bath) reduced by one electron at 77 K: the effects of component interactions and the binding of small molecules to the diiron(III) center. Biochemistry 38:4188–4197

Deeth RJ, Dalton H (1998) Methane activation by methane monooxygenase: free radicals, Fe-C bonding, substrate dependent pathways and the role of the regulatory protein. J Biol Inorg Chem 3:302–306

DeRose VJ, Liu KE, Kurtz DM, Hoffman BM, Lippard SJ (1993) Proton ENDOR identification of bridging hydroxide ligands in mixed-valent diiron centers of proteins: methane monooxygenase and semimet azidohemerythrin. J Am Chem Soc 115:6440–6441

Dewitt JG, Bentsen JG, Rosenzweig AC, Hedman B, Green J, Pilkington S, Papaefthymiou GC, Dalton H, Hodgson KO, Lippard SJ (1991) X-ray absorption, moessbauer, and EPR studies of the dinuclear iron center in the hydroxylase component of methane monooxygenase. J Am Chem Soc 113:9219–9235

Elango N, Radhakrishnan R, Froland WA, Wallar BJ, Earhart CA, Lipscomb JD, Ohlendorf DH (1997) Crystal structure of the hydroxylase component of methane monooxygenase from *Methylosinus trichosporium* OB3b. Protein Sci 6:556–568

Elliott SJ, Zhu M, Tso L, Nguyen HHT, Yip JHK, Chan SI (1997) Regio- and stereoselectivity of particulate methane monooxygenase from *Methylococcus capsulatus* (Bath). J Am Chem Soc 119:9949–9955

Feig AL, Lippard SJ (1994) Reactions of non-heme iron(II) centers with dioxygen in biology and chemistry. Chem Rev 94:759–805

Fox BG, Froland WA, Dege JE, Lipscomb JD (1989) Methane monooxygenase from *Methylosinus trichosporium* OB3b - purification and properties of a three-component system with high specific activity from a type II methanotroph. J Biol Chem 264:10023–10033

Fox BG, Liu Y, Dege JE, Lipscomb JD (1991) Complex formation between the protein components of methane monooxygenase from *Methylosinus trichosporium* OB3b - identification of sites of component interaction. J Biol Chem 266:540–550

Fox BG, Surerus KK, Munck E, Lipscomb JD (1988) Evidence for a μ-oxo-bridged binuclear iron cluster in the hydroxylase component of methane monooxygenase - Mössbauer and EPR studies. J Biol Chem 263:10553–10556

Froland WA, Andersson KK, Lee SK, Liu Y, Lipscomb JD (1992) Methane monooxygenase component B and reductase alter the regioselectivity of the hydroxylase component-catalyzed reactions - a novel role for protein-protein interactions in an oxygenase mechanism. J Biol Chem 267:17588–17597

Gassner GT, Lippard SJ (1999) Component interactions in the soluble methane monooxygenase system from *Methylococcus capsulatus* (Bath). Biochemistry 38:12768–12785

Green J, Dalton H (1989) A stopped-flow kinetic study of soluble methane mono-oxygenase from *Methylococcus capsulatus* (Bath). Biochem J 259:167–172

Hakemian AS, Kondapalli KC, Telser J, Hoffman BM, Stemmler TL, Rosenzweig AC (2008) The metal centers of particulate methane monooxygenase from *Methylosinus trichosporium* OB3b. Biochemistry 47:6793–6801

Hanson RS, Hanson TE (1996) Methanotrophic bacteria. Microbiol Rev 60:439–471

Jasniewski AJ, Que L (2018) Dioxygen activation by nonheme diiron enzymes: diverse dioxygen adducts, high-valent intermediates, and related model complexes. Chem Rev 118:2554–2592

Kao WC, Chen YR, Yi EC, Lee H, Tian Q, Wu KM, Tsai SF, Yu SSF, Chen YJ, Aebersold R, Chan SI (2004) Quantitative proteomic analysis of metabolic regulation by copper ions in *Methylococcus capsulatus* (Bath). J Biol Chem 279:51554–51560

Kim HJ, Graham DW, DiSpirito AA, Alterman MA, Galeva N, Larive CK, Asunskis D, Sherwood PMA (2004) Methanobactin, a copper-acquisition compound from methane-oxidizing bacteria. Science 305:1612–1615

Kopp DA, Gassner GT, Blazyk JL, Lippard SJ (2001) Electron-transfer reactions of the reductase component of soluble methane monooxygenase from *Methylococcus capsulatus* (Bath). Biochemistry 40:14932–14941

Lee SJ, McCormick MS, Lippard SJ, Cho US (2013) Control of substrate access to the active site in methane monooxygenase. Nature 494:380–384

Lee SK, Lipscomb JD (1999) Oxygen activation catalyzed by methane monooxygenase hydroxylase component: proton delivery during the O–O bond cleavage steps. Biochemistry 38:4423–4432

Lieberman RL, Rosenzweig AC (2005) Crystal structure of a membrane-bound metalloenzyme that catalyses the biological oxidation of methane. Nature 434:177–182

Lipscomb JD (1994) Biochemistry of the soluble methane monooxygenase. Annu Rev Microbiol 48:371–399

Lipscomb JD, Lee SK, Nesheim JC, Froland WA, Fox BG, Munck E (1994) Reactive intermediates in the methane monooxygenase catalyzed oxidation of methane and other hydrocarbons. Abstr Pap Am Chem Soc 207:273

Liu CC, Mou CY, Yu SSF, Chan SI (2016) Heterogeneous formulation of the tricopper complex for efficient catalytic conversion of methane into methanol at ambient temperature and pressure. Energy Environ Sci 9:1361–1374

Liu CC, Janmanchi D, Wen DR, Oung JN, Mou CY, Yu SSF, Chan SI (2018) Catalytic oxidation of light alkanes mediated at room temperature by a tricopper cluster complex immobilized in mesoporous silica nanoparticles. ACS Sustain Chem Eng 6(4):5431–5440

Liu KE, Valentine AM, Wang DL, Huynh BH, Edmondson DE, Salifoglou A, Lippard SJ (1995a) Kinetic and spectroscopic characterization of intermediates and component interactions in reactions of methane monooxygenase from *Methylococcus capsulatus* (Bath). J Am Chem Soc 117:10174–10185

Liu KE, Wang DL, Huynh BH, Edmondson DE, Salifoglou A, Lippard SJ (1994) Spectroscopic detection of intermediates in the Reaction of dioxygen with the reduced methane monooxygenase/hydroxylase from *Methylococcus capsulatus* (Bath). J Am Chem Soc 116:7465–7466

Liu Y, Nesheim JC, Lee SK, Lipscomb JD (1995b) Gating effects of component B on oxygen activation by the methane monooxygenase hydroxylase component. J Biol Chem 270:24662–24665

Liu Y, Nesheim JC, Paulsen KE, Stankovich MT, Lipscomb JD (1997) Roles of the methane monooxygenase reductase component in the regulation of catalysis. Biochemistry 36:5223–5233

Lloyd JS, Bhambra A, Murrell JC, Dalton H (1997) Inactivation of the regulatory protein B of soluble methane monooxygenase from *Methylococcus capsulatus* (Bath) by proteolysis can be overcome by a Gly to Gln modification. Eur J Biochem 248:72–79

Lu YJ, Hung MC, Chang BTA, Lee TL, Lin ZH, Tsai IK, Chen YS, Chang CS, Tsai YF, Chen KHC, Chan SI, Yu SSF (2019) The PmoB subunit of particulate methane monooxygenase (pMMO) in *Methylococcus capsulatus* (Bath): the Cu^{I} sponge and its function. J Inorg Biochem 196:110691

Lund J, Woodland MP, Dalton H (1985) Electron transfer reactions in the soluble methane monooxygenase of *Methylococcus capsulatus* (Bath). Eur J Biochem 147:297–305

Maji S, Lee JCM, Lu YJ, Chen CL, Hung MC, Chen PPY, Yu SSF, Chan SI (2012) Dioxygen activation of a trinuclear $Cu^{I}Cu^{I}Cu^{I}$ cluster capable of mediating facile oxidation of organic substrates: competition between O-atom transfer and abortive intercomplex reduction. Chem Eur J 18:3955–3968

McCormick MS, Lippard SJ (2011) Analysis of substrate access to active sites in bacterial multicomponent monooxygenase hydroxylases: X-ray crystal structure of xenon-pressurized phenol hydroxylase from *Pseudomonas sp.* OX1. Biochemistry 50:11058–11069

Merkx M, Kopp DA, Sazinsky MH, Blazyk JL, Muller J, Lippard SJ (2001) Dioxygen activation and methane hydroxylation by soluble methane monooxygenase: a tale of two irons and three proteins. Angew Chem Int Ed 40:2782–2807

Merkx M, Lippard SJ (2002) Why OrfY? Characterization of MMOD, a long overlooked component of the soluble methane monooxygenase from *Methylococcus capsulatus* (Bath). J Biol Chem 277:5858–5865

Muller J, Lugovskoy AA, Wagner G, Lippard SJ (2002) NMR structure of the [2Fe-2S] ferredoxin domain from soluble methane monooxygenase reductase and interaction with its hydroxylase. Biochemistry 41:42–51

Murray LJ, Lippard SJ (2007) Substrate trafficking and dioxygen activation in bacterial multicomponent monooxygenases. Acc Chem Res 40:466–474

Murrell JC, McDonald IR, Gilbert B (2000) Regulation of expression of methane monooxygenases by copper ions. Trends Microbiol 8:221–225

Myronova N, Kitmitto A, Collins RF, Miyaji A, Dalton H (2006) Three-dimensional structure determination of a protein supercomplex that oxidizes methane to formaldehyde in *Methylococcus capsulatus* (Bath). Biochemistry 45:11905–11914

Nagababu P, Yu SSF, Maji S, Ramu R, Chan SI (2014) Developing an efficient catalyst for controlled oxidation of small alkanes under ambient conditions. Cat Sci Technol 4:930–935

Nesheim JC, Lipscomb JD (1996) Large kinetic isotope effects in methane oxidation catalyzed by ethane monooxygenase: evidence for C–H bond cleavage in a reaction cycle intermediate. Biochemistry 35:10240–10247

Ng KY, Tu LC, Wang YS, Chan SI, Yu SSF (2008) Probing the hydrophobic pocket of the active site in the particulate methane monooxygenase (pMMO) from *Methylococcus capsulatus* (Bath) by variable stereoselective alkane hydroxylation and olefin epoxidation. Chembiochem 9:1116–1123

Nguyen HHT, Shiemke AK, Jacobs SJ, Hales BJ, Lidstrom ME, Chan SI (1994) The nature of the copper ions in the membranes containing the particulate methane monooxygenase from *Methylococcus capsulatus* (Bath). J Biol Chem 69:14995–15005

Nguyen HHT, Nakagawa KH, Hedman B, Elliott SJ, Lidstrom ME, Hodgson KO, Chan SI (1996) X-ray absorption and EPR studies on the copper ions associated with the particulate methane monooxygenase from *Methylococcus capsulatus* (Bath). Cu(I) ions and their implications. J Am Chem Soc 118:12766–12776

Nguyen HHT, Elliott SJ, Yip JHK, Chan SI (1998) The particulate methane monooxygenase from *Methylococcus capsulatus* (Bath) is a novel copper-containing three-subunit enzyme—isolation and characterization. J Biol Chem 273:7957–7966

Pham MD, Yu SSF, Han CC, Chan SI (2013) Improved mass spectrometric analysis of membrane proteins based on rapid and versatile sample preparation on nanodiamond particles. Anal Chem 85:6748–6755

Pham MD, Lin YP, Vuong QV, Nagababu P, Chang BTA, Ng KY, Chen CH, Han CC, Chen CH, Li MS, Yu SSF, Chan SI (2015) Inactivation of the particulate methane monooxygenase (pMMO) in *Methylococcus capsulatus* (Bath) by acetylene. Biochim Biophys Acta Proteins Proteomics 1854:1842–1852

Pulver S, Froland WA, Fox BG, Lipscomb JD, Solomon EI (1993) Spectroscopic studies of the coupled binuclear non-heme iron active site in the fully reduced hydroxylase component of methane monooxygenase: comparison to deoxy and deoxy-azide hemerythrin. J Am Chem Soc 115:12409–12422

Pulver S, Froland WA, Fox BG, Lipscomb JD, Solomon EI (1994) Spectroscopic studies of the coupled binuclear non-heme iron active site in the fully reduced hydroxylase component of methane monooxygenase: comparison to deoxy and deoxy-azide hemerythrin. J Am Chem Soc 116:4529–4529

Rosenzweig AC, Brandstetter H, Whittington DA, Nordlund P, Lippard SJ, Frederick CA (1997) Crystal structures of the methane monooxygenase hydroxylase from *Methylococcus capsulatus* (Bath): implications for substrate gating and component interactions. Proteins Struct Funct Genet 29:141–152

Rosenzweig AC, Frederick CA, Lippard SJ, Nordlund P (1993) Crystal structure of a bacterial non-haem iron hydroxylase that catalyses the biological oxidation of methane. Nature 366:537–543

Rosenzweig AC, Nordlund P, Takahara PM, Frederick CA, Lippard SJ (1995) Geometry of the soluble methane monooxygenase catalytic diiron center in two oxidation states. Chem Biol 2:409–418

Sazinsky MH, Dunten PW, McCormick MS, DiDonato A, Lippard SJ (2006) X-ray structure of a hydroxylase-regulatory protein complex from a hydrocarbon-oxidizing multicomponent monooxygenase, *Pseudomonas sp.* OX1 phenol hydroxylase. Biochemistry 45:15392–15404

Sazinsky MH, Lippard SJ (2006) Correlating structure with function in bacterial multicomponent monooxygenases and related diiron proteins. Acc Chem Res 39:558–566

Semrau JD, DiSpirito AA, Yoon S (2010) Methanotrophs and copper. FEMS Microbiol Rev 34:496–531

Shindell DT, Faluvegi G, Koch DM, Schmidt GA, Unger N, Bauer SE (2009) Improved attribution of climate forcing to emissions. Science 326:716–718

Shinohara Y, Uchiyama H, Yagi O, Kusakabe I (1998) Purification and characterization of component B of a soluble methane monooxygenase from *Methylocystis sp.* M. J Ferment Bioeng 85:37–42

Shu LJ, Nesheim JC, Kauffmann K, Munck E, Lipscomb JD, Que L (1997) An $Fe_2^{IV}O_2$ diamond core structure for the key intermediate Q of methane monooxygenase. Science 275:515–518

Sirajuddin S, Barupala D, Helling S, Marcus K, Stemmler TL, Rosenzweig AC (2014) Effects of zinc on particulate methane monooxygenase activity and structure. J Biol Chem 289:21782–21794

Smith SM, Rawat S, Telser J, Hoffman BM, Stemmler TL, Rosenzweig AC (2011) Crystal structure and characterization of particulate methane monooxygenase from *Methylocystis species* Strain M. Biochemistry 50:10231–10240

Stafford GP, Scanlan J, McDonald IR, Murrell JC (2003) *rpoN, mmoR* and *mmoG*, genes involved in regulating the expression of soluble methane monooxygenase in *Methylosinus trichosporium* OB3b. Microbiology 149:1771–1784

Stainthorpe AC, Lees V, Salmond GPC, Dalton H, Murrell JC (1990) The methane monooxygenase gene cluster of *Methylococcus capsulatus* (Bath). Gene 91:27–34

Stein LY, Yoon S, Semrau JD, DiSpirito AA, Crombie A, Murrell JC, Vuilleumier S, Kalyuzhnaya MG, den Camp HJMO, Bringel F, Bruce D, Cheng JF, Copeland A, Goodwin L, Han SS, Hauser L, Jetten MSM, Lajus A, Land ML, Lapidus A, Lucas S, Médigue C, Pitluck S, Woyke T, Zeytun A, Klotz MG (2010) Genome sequence of the obligate methanotroph *Methylosinus trichosporium* Strain OB3b. J Bacteriol 192:6497–6498

Stolyar S, Costello AM, Peeples TL, Lidstrom ME (1999) Role of multiple gene copies in particulate methane monooxygenase activity in the methane-oxidizing bacterium *Methylococcus capsulatus* Bath. Microbiology 145:1235–1244

Thomann H, Bernardo M, McCormick JM, Pulver S, Andersson KK, Lipscomb JD, Solomon EI (1993) Pulsed EPR studies of mixed valent [Fe(II)Fe(III)] forms of hemerythrin and methane monooxygenase: evidence for a hydroxide bridge. J Am Chem Soc 115:8881–8882

Tinberg CE, Lippard SJ (2009) Revisiting the mechanism of dioxygen activation in soluble methane monooxygenase from *M. capsulatus* (Bath): evidence for a multi-step, proton-dependent reaction pathway. Biochemistry 48:12145–12158

Tinberg CE, Lippard SJ (2010) Oxidation reactions performed by soluble methane monooxygenase hydroxylase intermediates Hperoxo and Q proceed by distinct mechanisms. Biochemistry 49:7902–7912

Tinberg CE, Lippard SJ (2011) Dioxygen activation in soluble methane monooxygenase. Acc Chem Res 44:280–288

Trotsenko YA, Khmelenina VN (2002) Biology of extremophilic and extremotolerant methanotrophs. Arch Microbiol 177:123–131

Trotsenko YA, Murrell JC (2008) Metabolic aspects of aerobic obligate methanotrophy. Adv Appl Microbiol 63:183–229

Vinchurkar MS, KHC C, SSF Y, Kuo SJ, Chiu HC, Chien SH, Chan SI (2004) Polarized ATR-FTIR spectroscopy of the membrane-embedded domains of the particulate methane monooxygenase. Biochemistry 43:13283–13292

Wallar BJ, Lipscomb JD (1996) Dioxygen activation by enzymes containing binuclear non-heme iron clusters. Chem Rev 96:2625–2657

Wallar BJ, Lipscomb JD (2001) Methane monooxygenase component B mutants alter the kinetics of steps throughout the catalytic cycle. Biochemistry 40:2220–2233

Walters KJ, Gassner GT, Lippard SJ, Wagner G (1999) Structure of the soluble methane monooxygenase regulatory protein B. Proc Natl Acad Sci U S A 96:7877–7882

Wang VCC, Maji S, Chen PPY, Lee HK, Yu SSF, Chan SI (2017) Alkane oxidation: methane monooxygenases, related enzymes, and their biomimetics. Chem Rev 117:8574–8621

Wang WX, Iacob RE, Luoh RP, Engen JR, Lippard SJ (2014) Electron transfer control in soluble methane monooxygenase. J Am Chem Soc 136:9754–9762

Wang WX, Liang AD, Lippard SJ (2015) Coupling oxygen consumption with hydrocarbon oxidation in bacterial multicomponent monooxygenases. Acc Chem Res 48:2632–2639

Wang WX, Lippard SJ (2014) Diiron oxidation state control of substrate access to the active site of soluble methane monooxygenase mediated by the regulatory component. J Am Chem Soc 136:2244–2247

Ward N, Larsen O, Sakwa J, Bruseth L, Khouri H, Durkin AS, Dimitrov G, Jiang LX, Scanlan D, Kang KH, Lewis M, Nelson KE, Methé B, Wu M, Heidelberg JF, Paulsen IA, Fouts D, Ravel J, Tettlin H, Ren Q, Read T, DeBoy RT, Seshadri R, Salzberg SL, Jensen HB, Birkeland NK, Nelson WC, Dodson RJ, Grindhaug SH, Holt I, Eidhammer I, Jonasen I, Vanaken S, Utterback T, Feldblyum TV, Fraser CM, Lillehaug JR, Eisen JA (2004) Genomic insights into methanotrophy: the complete genome sequence of *Methylococcus capsulatus* (Bath). PLoS Biol 2:1616–1628

Whittington DA, Lippard SJ (2001) Crystal structures of the soluble methane monooxygenase hydroxylase from *Methylococcus capsulatus* (Bath) demonstrating geometrical variability at the dinuclear iron active site. J Am Chem Soc 123:827–838

Whittington DA, Rosenzweig AC, Frederick CA, Lippard SJ (2001a) Xenon and halogenated alkanes track putative substrate binding cavities in the soluble methane monooxygenase hydroxylase. Biochemistry 40:3476–3482

Whittington DA, Sazinsky MH, Lippard SJ (2001b) X-ray crystal structure of alcohol products bound at the active site of soluble methane monooxygenase hydroxylase. J Am Chem Soc 123:1794–1795

Wilkinson B, Zhu M, Priestley ND, Nguyen HHT, Morimoto H, Williams PG, Chan SI, Floss HG (1996) A concerted mechanism for ethane hydroxylation by the particulate methane monooxygenase from *Methylococcus capsulatus* (Bath). J Am Chem Soc 118:921–922

Yu SSF, Chen KHC, Tseng MYH, Wang YS, Tseng CF, Chen YJ, Huang DS, Chan SI (2003a) Production of high-quality particulate methane monooxygenase in high yields from *Methylococcus capsulatus* (Bath) with a hollow-fiber membrane bioreactor. J Bacteriol 185:5915–5924

Yu SSF, Wu LY, Chen KHC, Luo WI, Huang DS, Chan SI (2003b) The stereospecific hydroxylation of [2, 2-^{2}H2]butane and chiral dideuteriobutanes by the particulate methane monooxygenase from *Methylococcus capsulatus* (Bath). J Biol Chem 278:40658–40669

Yu SSF, Ji CZ, Wu YP, Lee TL, Lai CH, Lin SC, Yang ZL, Wang VCC, Chen KHC, Chan SI (2007) The C-terminal aqueous-exposed domain of the 45 kDa subunit of the particulate methane monooxygenase in *Methylococcus capsulatus* (Bath) is a Cu(I) sponge. Biochemistry 46:13762–13774

Zahn JA, DiSpirito AA (1996) Membrane-associated methane monooxygenase from *Methylococcus capsulatus* (Bath). J Bacteriol 178:1018–1029

Multi-omics Understanding of Methanotrophs

Yue Zheng and Ludmila Chistoserdova

Contents

Abstract Over the past two decades, application of high-throughput technologies for molecular analysis, collectively referred to as omics, have revolutionized the field of biology. These technologies enable collecting large amounts of data on DNA and RNA sequences, protein and metabolite identities, which can then be incorporated into metabolic models that can simulate metabolisms. Omics have been applied to studying methanotrophy for a long time, and progress in this area has been summarised previously, including recent reviews. In this chapter, we only highlight the very latest novel insights into methanotrophy through omics. Some of the highlights are the newly uncovered environmental dominance of the *Methylobacter* species, the discovery of novel methanotroph taxa through metagenome-assembled genome reconstruction, further insights into the role of lanthanides in methanotrophy, and further details on detection of the presence and activities of

Y. Zheng
Department of Chemical Engineering, University of Washington, Seattle, WA, USA

CAS Key Laboratory of Urban Pollutant Conversion, Institute of Urban Environment, Chinese Academy of Sciences, Xiamen, China

University of Chinese Academy of Sciences, Beijing, China

L. Chistoserdova (✉)
Department of Chemical Engineering, University of Washington, Seattle, WA, USA
e-mail: milachis@u.washington.edu

E. Y. Lee (ed.), *Methanotrophs*, Microbiology Monographs 32,
https://doi.org/10.1007/978-3-030-23261-0_4

different guilds of methanotrophs across geochemical gradients. We also touch briefly on the novel developments in understanding of the communal function in methanotrophy, and on the role of model organisms in further advancements of our knowledge.

1 Introduction and Overview

The methane cycle on Earth is an important part of the global carbon cycle. A major portion of methane is produced biogenically, in anaerobic environments, as part of anaerobic degradation of organic matter (Conrad 2009). The microbes ultimately responsible for methane production are known as methanogenic archaea, belonging to several disparate phylogenetic groups (Evans et al. 2019). Some of the methane is also emitted from abiogenic sources (Sherwood Lollar et al. 2006), and some is formed in aerobic environments (Keppler et al. 2006; Karl et al. 2008; Fig. 1). Methane is a valuable source of both carbon and energy for organisms that evolved to possess specific biochemistries to enable utilization of methane, these organisms known as the methanotrophs. Methanotrophy, as a metabolic phenomenon, has originally been characterized in aerobic bacteria, and methane oxidation in these species has been demonstrated to be dependent on molecular oxygen (Trotsenko and Murrell 2008). The specific enzymes/pathways for methane oxidation (Fig. 2) and assimilation have been extensively characterized over years of focused studies (Chistoserdova 2011; Chistoserdova and Lidstrom 2013). After many years of speculations on the possibility of oxygen-independent methane oxidation, based on the geochemical data (reviewed by Valentine and Reeburgh 2000), the process has been confirmed through molecular evidence, identifying consortia of methane-oxidizing archaea (the ANME types) and sulfate-reducing bacteria (reviewed in

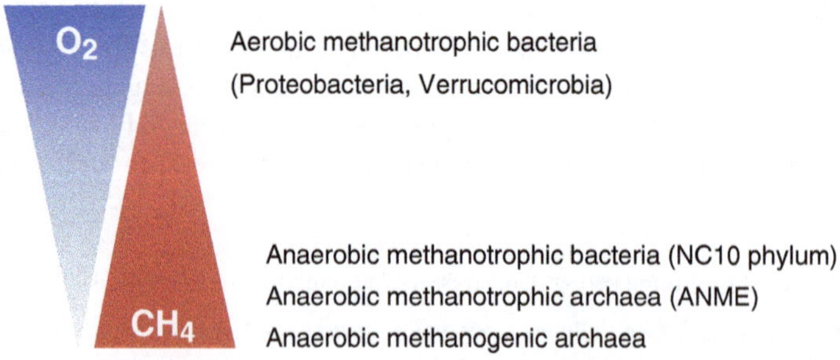

Fig. 1 In many environments, methane and dioxygen form counter gradients. The methanogenic archaea are typically found in anoxic niches. Anaerobic methanotrophic archaea and the NC10 bacteria are also typically found in anoxic niches, while aerobic methanotrophic bacteria thrive in niches where oxygen is present. However, recent data suggest that aerobic methanotrophs may be present and active in hypoxic and even anoxic environments, and that the anaerobic methanotrophs and even methanogens may be present in oxygenated environments

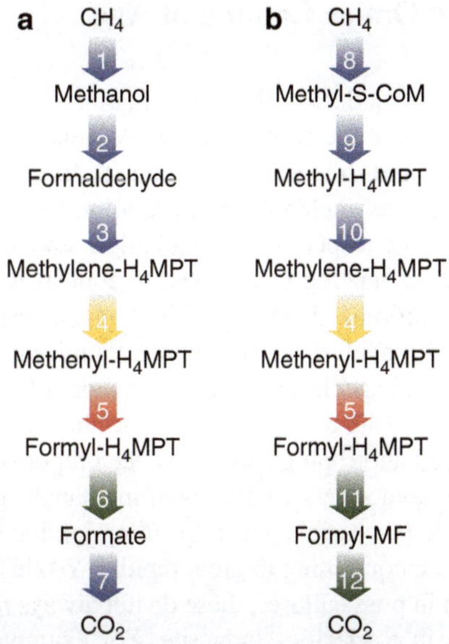

Fig. 2 The common and the distinct reactions in aerobic (**a**) and anaerobic (**b**) methane oxidation pathways. (1) Methane monooxygenase; (2) methanol dehydrogenase; (3) formaldehyde activating enzyme; (4) methylene-tetrahydromethanopterin (H_4MPT) dehydrogenase; (5) methenyl-H_4MPT cyclohydrolase; (6) formyltransferase hydrolase complex; (7) formate dehydrogenase; (8) methyl-coenzyme M reductase; (9) methyl-H_4MPT coenzyme M methyltransferase; (10) methylene-H_4MPT reductase (not present in all species, see Evans et al. 2019); (11) formyl-methanofuran H_4MPT formyltransferase; (12) formyl-methanofuran dehydrogenase. Grey arrows, distinct reactions catalyzed by distinct enzymes; red arrows, same reactions catalyzed by homologous enzymes; yellow arrows, same reactions catalyzed by non-homologous enzymes; green arrows, similar but not the same reactions catalyzed by homologous proteins

Knittel and Boetius 2009). The ANME archaea utilize the reverse methanogenesis pathway for methane oxidation (Knittel and Boetius 2009; Fig. 2). Subsequently, alternative electron acceptors for anaerobic methane oxidation have been uncovered, such as nitrate, iron, manganese, as well as humic substances (reviewed in Martinez-Cruz et al. 2017; Evans et al. 2019). Alternative players in anaerobic methane oxidation have also been identified, namely members of the NC10 phylum within bacteria. These organisms utilize the traditional methanotrophy pathway for their metabolism, while relying on nitrite as an electron acceptor, and these organisms have been proposed to generate dioxygen necessary for methane oxidation intracellularly (Ettwig et al. 2010). Thus, the contemporary understanding of microbes with methane-oxidizing potential includes (1) aerobic methanotrophic bacteria phylogenetically positioned within Proteobacteria and Verrucomicrobia; (2) anaerobic methanotrophic bacteria of the NC10 phylum, and methanotrophic archaea, closely related to the methanogenic archaea (Fig. 1).

2 Methanotrophy Omics Coming of Age

Omics collectively refer to modern high-throughput technologies that allow for fast and efficient measurements of the contents of DNA (genomics and metagenomics), RNA (transcriptomics and metatranscriptomics), proteins (proteomics and metaproteomics), as well as metabolites (metabolomics) and metabolite fluxes (fluxomics). Access to the respective technologies, and especially to the next generation nucleic acid sequencing techniques, in concert with the dedicated software packages have revolutionized the field of biology and especially microbiology, providing opportunities for analysis of environmental microbes in their entirety and not limited to the cultivated species. Culture-independent techniques for molecular detection of methylotrophs (the early omics) date to the early 1990s (Murrell et al. 1993), while the first methanotroph genome has been reported in 2004 (Ward et al. 2004). As of today, we have access to dozens of methanotroph genomes generated from pure culture DNA (accessible via https://img.jgi.doe.gov/cgi-bin/mer/main.cgi), and the databases are continuing to grow rapidly. While (some) methanotrophs are rather easy to obtain in pure cultures, these do not always reflect the nature of the major functional types in a specific study site. For example, *Methylomonas* and *Methylocystis* have been easily isolated from soils and lake sediments (e.g. Auman et al. 2000). However, metagenomic studies have revealed that they may not be the major active species, and species that are harder to cultivate in pure cultures, such as *Methylobacter*, may be the dominant players, based on rapid accumulation by these species of carbon from isotopically labeled methane (Kalyuzhnaya et al. 2008, 2015; Beck et al. 2013). Functional metagenomics following the fate of the labeled methane have also uncovered label transfer to non-methanotrophic members of the community, primarily *Methylophilaceae*, suggestive of the existence of a food chain involved in metabolism of methane (Kalyuzhnaya et al. 2008). In this early metagenomic study, genomic binning was accomplished for the major metabolic guilds involved in methane metabolism, allowing for reconstruction of metabolisms of major actors in methane consumption, the *Methylobacter* species and the partner *Methylotenera* species (Kalyuzhnaya et al. 2008).

When it comes to the major actors in anaerobic methane oxidation, meta-omics are the main tools to obtain insights into their metabolisms, as none of the ANME archaea or the NC10 bacteria, are available in pure cultures. Thus, omics approaches are of major significance for uncovering the behavior and the functional roles of the major players in methane oxidation in natural methane-oxidizing communities. As of today, methanotrophy omics are coming of age, with almost three decades of research reliant on molecular tools, in addition to analysis of single genomes and the classic physiology experiments (Chistoserdova 2017; Chistoserdova and Kalyuzhnaya 2018). Generation of metagenome assembled genomes (MAGs) is now a standard approach in analysing the metagenomic data (see below). As single genome information is prerequisite for the follow up physiological analyses of model organisms in pure cultures, analysis of MAGs is prerequisite for metabolic reconstruction of the carbon flow through methane-consuming communities, as well as for focused efforts toward cultivation of members of the major functional guilds active in specific ecological niches. The flowchart of a current state of the art omics protocol for characterization of un-enriched environmental samples as well as enrichments selecting for specific methanotroph guilds is presented in Fig. 3.

3 Environmental Dominance of *Methylobacter*

While *Methylobacter* species that are available in pure cultures do not perform in laboratory as robustly as other methanotrophs, such as *Methylomonas* or *Methylocystis* (Yu et al. 2016), it has been noted that they appear to dominate in methane-consuming communities not only in Arctic and sub-Arctic environments, but in most freshwater and terrestrial environments in the Northern hemisphere. The evidence comes from the stable isotope probing (DNA-SIP) experiments, microcosm incubations, as well as from environmental detection of DNA and RNA molecules (reviewed in Chistoserdova 2015, 2017). Several recent studies support prior findings. The recent omics trends, given the current access to relatively inexpensive sequencing of both DNA and RNA molecules from environmental samples, extend beyond analysis of fragments of nucleic acids. The new trend in metagenomic analysis is assembly of single genomes from metagenomic sequence data, known as MAGs. These, when sufficient sequence data are collected, may approximate the quality of 'draft genomes' produced from single organism DNA. However, the quality of MAGs depends on the nature of the communities analyzed via metagenomics, such as the heterogeneity of the populations in question, which can result in MAGs consisting of very fragmented DNA sequences, while rare but clonal genomes could produce very high quality MAGs, truly approximating single organism genomic sequences. Thus, the quality of MAGs is not necessarily a reflection of abundance of a specific organism. Nevertheless, MAGs are the new metagenomics outputs that provide very important insights into key players in biogeochemical processes taking place in specific environments, especially when supported by data from other omics approaches.

As part of one such multi-omics approach, the significance of *Methylobacter* in methane oxidation in oxic/hypoxic/anoxic freshwater wetland soils has been recently demonstrated (Smith et al. 2018). First, via 16S rRNA gene profiling, it has been determined that only a handful of operational taxonomic units (OTUs) were representing the fifths most abundant taxonomic order, the Methylococcales, across sampling seasons, soil depths, and land covers, suggesting that they were the active methane oxidizers. Second, several MAGs were reconstructed, and methanotrophy and denitrification pathways were delineated through metabolic reconstruction. Given the fact that *Methylobacter* sequences were found in hypoxic and anoxic layers, the authors also compiled an inventory of genes encoding functions that may enable metabolism at low dioxygen levels. Next, they matched transcript data with the MAGs, and determined that the *Methylobacter* species were some of the most active species in the surface sediments, the genes for particulate methane monooxygenase

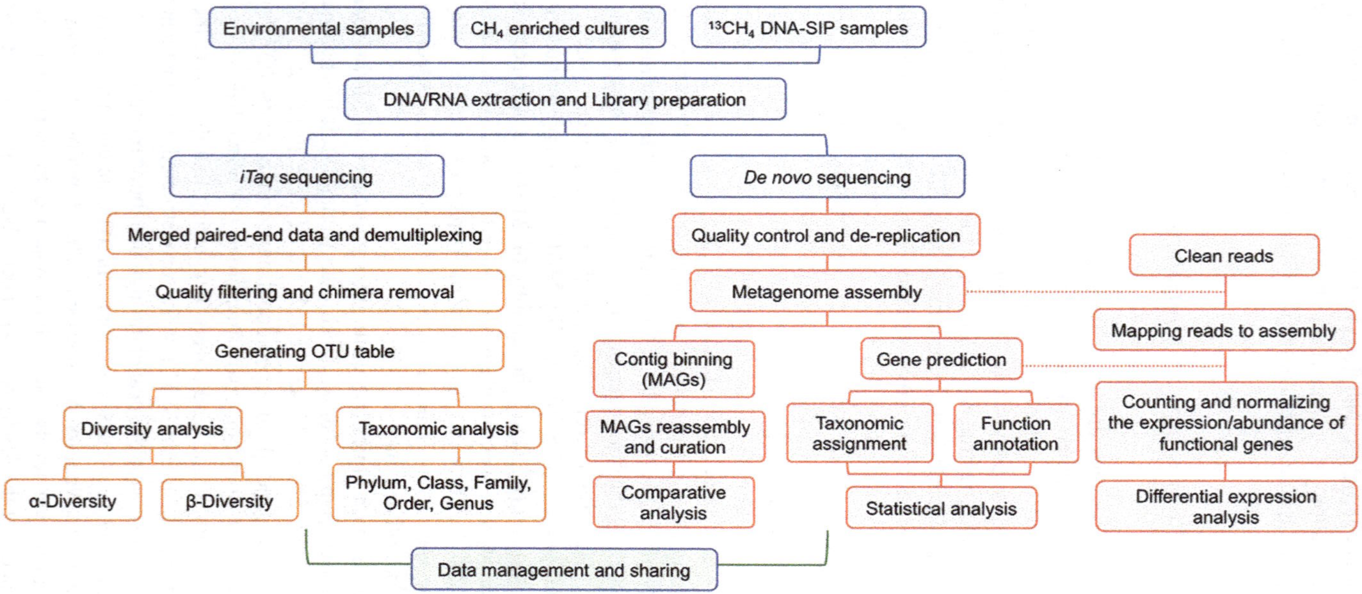

Fig. 3 A typical flowchart for analysis of key omics data. DNA and RNA can be isolated from un-manipulated environmental samples or from samples enriched through incubation or through stable isotope probing (DNA-SIP). iTAG sequencing is frequently used to determine community structure, and metagenomic sequencing, including assembly and MAG reconstruction are used to obtain insights into metabolic capabilities of communities and single organisms. RNA sequences can be assembled or matched to MAGs and assembled DNA contigs

(pMMO) being among the top 3% of the most highly transcribed genes. Other methylotrophy genes were also transcribed, while the denitrification gene transcripts or transcripts of other genes that could be involved in low dioxygen response were not detectable (Smith et al. 2018). Finally, the distribution of the specific lineage of *Methylobacter* identified in this study was tested via mining the metagenomic and metatranscriptomic public databases, uncovering the presence of the lineage in other methane-emitting hydric soils and sediments across North America, Europe, Russia, and Asia, further supporting the notion that *Methylobacter* species are important cosmopolitan methanotrophs active across ecosystems (Smith et al. 2018).

MAG reconstruction and respective 16S rRNA gene sequence dynamics have been employed recently in testing a model predicting major biogeochemical processes in a model freshwater lake ecosystem (Arora-Williams et al. 2018). In this work, among other MAGs, MAGs for *Methylobacter tundripaludum* and *Methylotenera versatilis*, whose populations are frequently found to correlate with the populations of *Methylobacter*, were reconstructed, and, through a special algorithm, 16S rRNA genes were tied to the MAGs. Population dynamics then were followed via 16S rRNA gene profiling, identifying some deviations with the original model, among the poorly predicted processes being methane oxidation. MAG-based metabolic reconstruction identified the denitrification capability for *Methylobacter*, and incorporation of nitrate-linked methane oxidation into the biogeochemical model improved its predictive power. However, the transcriptomic component was missing in this study, leaving this proposal at the hypothesis level (Arora-Williams et al. 2018).

Methylobacter MAGs have also been reconstructed from a permafrost thaw gradient (Singleton et al. 2018), and these organisms were found to be most abundant in the fully thawed fen, where methane fluxes were the highest. While alphaproteobacterial methanotrophs were found more abundant in the partially thawed bog samples and in the intact permafrost, metatranscriptimics revealed that these organisms showed very low activity, while the *Methylobacter* species, where present, were highly expressing their pMMO and other methylotrophy genes, in both oxic and hypoxic samples (Singleton et al. 2018).

4 Identifying Novel Methanotrophs Through Metagenomics

One exciting aspect of MAGs is their potential for allowing for genomic analysis of novel microbes possessing specific metabolic capabilities, such as methanotrophy, under varying conditions of methane and oxygen availability. Through MAG reconstruction, Singleton et al. (2018) were able to predict a potential for methanotrophy in representatives of *Hyphomicrobiaceae*, the *Rhodomicrobium* species, expanding the phylogenetic space of aerobic methanotrophy. These organisms appear to encode both particulate (pMMO) and soluble (sMMO) methane monooxygenases, and both enzymes appear to be divergent from the known enzymes, while still clustering with alphaproteobacterial sequences (Singleton et al. 2018). Further metabolic reconstruction identified the pathways necessary for the assimilation and dissimilation of carbon from methane, suggesting an autotrophic life style for these species. Singleton et al. (2018) were also able to reconstruct, through MAG analysis, the metabolism of the elusive, uncultivated upland soil cluster alphaproteobacterial methanotrophs (USCα), implicated in oxidation of methane at atmospheric concentrations. These were placed with the *Beijerinckiaceae* family and shown to possess the traditional pathways for methanotrophy. However, these organisms did not appear to encode the pMMO2 enzyme that has been previously proposed to be responsible for the high-affinity methanotrophy, a somewhat controversial result (Singleton et al. 2018).

A MAG representing the uncultivated upland soil cluster gammaproteobacterial methanotrophs (USCγ), also implicated in atmospheric methane oxidation, has been reconstructed from a metagenome of incubated mineral cryosols from Antarctica (Edwards et al. 2017). The 16S rRNA gene that was part of the genomic bin was phylogenetically placed within the order Chromatiales, related to Methylococcales. The metabolic pathways were reconstructed, identifying genes for pMMO, methanol dehydrogenase and formaldehyde and formate oxidation. In terms of carbon assimilation, nearly complete serine cycle was predicted, typically operational in alphaproteobacterial methanotrophs, but not the ribulose monophosphate cycle, typically operational in Methylococcales (Edwards et al. 2017).

Members of the *Crenothrix* genus are another group of uncultivated methanotrophs, characterized by the large size of cells arranged into multicellular filaments. MAGs representing *Crenothrix* species from lacustrine and sand filter samples have been reconstructed from respective metagenomes, providing first glimpses into these long known but mysterious microbes (Oswald et al. 2017). The metabolic pathway reconstruction revealed the presence of genes for the traditional methanotrophy pathways. Remarkably, the long-standing believe that *Crenothrix* species encode a divergent pMMO has been dismissed, by identifying genes for a traditional, gammaproteobacterial pMMO in the reconstructed *Crenothrix* MAGs, while assigning the divergent genes to other members of the communities (Oswald et al. 2017). The denitrification pathways have also been delineated in the genomic bins, and incorporation of nitrogen from nitrate into cell material has been proven via nanoSIMS (Oswald et al. 2017).

MAGs representing genomes of novel marine methanotrophs have also been reconstructed. Padilla et al. (2017) reported a nearly complete genome of a representative of the OPU3 clade, so far lacking in any cultivated species, from an anoxic oxygen minimum zone. These organisms are now characterized as divergent members of the Methylococcales, possessing small genomes, as previously reported for other pelagic methylotrophs (reviewed in Chistoserdova 2017). Transcriptomic analysis indicated that OPU3 clade microbes express the traditional methylotrophy pathways as well as the partial denitrification pathway, suggestive of the metabolic linkage between methanotrophy and denitrification (Padilla et al. 2017).

MAG reconstruction from a metagenome representing a community in another oxygen-limited niche, a hydrothermal vent, provided an outlook at the metabolic peculiarities of a *Methylothermaceae* species, so far represented by only a single cultivated strain (Skennerton et al. 2015; even if the validity of the nomenclature is contested, Tindall 2019). Analysis of this assembled genome identified the traditional methanotrophy pathways as well as the denitrification pathway, suggestive, once again, of the potential interdependence of these pathways, given the environmental circumstance of low oxygen availability (Skennerton et al. 2015).

5 Role of Lanthanides in Methanotrophy

Recently, rare earth elements (REE) lanthanides emerged as important players in methanotrophy (Chistoserdova 2016; Yu et al. 2019). A novel methanol dehydrogenase (MDH), designated as XoxF, has been demonstrated to require REEs in its active centre, and, in addition, REEs were demonstrated to regulate expression of alternative MDH enzymes, by activating transcription of the genes for XoxF and repressing transcription of genes for the alternative MDH, designated as MxaFI and requiring calcium for activity (Chistoserdova 2019). Based on such a regulatory pattern, it has been proposed that XoxF must be the main MDH in nature, as long as REEs are bioavailable, and MxaFI has been proposed to be a secondary enzyme (Chu and Lidstrom 2016; Ochsner et al. 2019). Indeed, organisms have been reported that only encode XoxF and not MxaF, representing disparate phylogenetic backgrounds (Beck et al. 2014; Vekeman et al. 2016; Versantvoort et al. 2018), supporting this hypothesis. Analysis of some high-quality MAGs from natural environments further supports the wider environmental occurrence of REE-dependent MDH enzymes, compared to their calcium-dependent counterparts. Smith et al. (2018) reconstructed MAGs that are highly similar to the available single species genomes of *Methylobacter*. However, these only encoded XoxF and not MxaFI enzymes, the MAG-based metabolic reconstruction further supported by the metatranscriptomic data. While all the so far cultivated *Methylococcaceae* species do encode MxaFI (Kalyuzhnaya et al. 2015), higher frequencies of *xoxF* genes compared to *mxaF* genes have been previously inferred from lake sediment metagenomic data (Beck et al. 2013). The *Methylophilaceae* species isolated from the same study site show variable contents of *xoxF* genes and variable presence of *mxaFI* (Beck et al. 2014; McTaggart et al. 2015).

In the high-quality MAGs representing the *Methylocystis* species, the upper soil cluster alphaproteobacterial methanotrophs or the novel *Hyphomicrobiaceae* methanotrophs, Singleton et al. (2018) could only identify *xoxF* and not *mxaFI* genes. Similarly, only *xoxF* could be identified in the pelagic gammaproteobacterial methanotroph MAG reconstructed from the oxygen minimum zone (Padilla et al. 2017). XoxF-based methylotrophy has also been predicted for the candidate phylum bacteria Rokubacteria and for novel members of Gemmatimonadetes, through proteogenomics and MAG and metabolic reconstruction, from soil metagenomes (Butterfield et al. 2016). These data support the prevalence of REE-dependent methanol oxidation in a variety of environmental niches and suggest that successful enrichment and isolation of many methylotrophs may require REE supplements.

6 Aerobic Versus Anaerobic Methanotrophy

It has been noted that, while methanotrophic bacteria require oxygen for activating the molecule of methane, they have also been detected in many hypoxic and even anoxic environments, and in many cases their activity was supported by RNA detection and/or methane consumption measurements (reviewed in Chistoserdova 2015, 2017; Chistoserdova and Kalyuzhnaya 2018). Additional experimental evidence became available more recently. When anoxic sediment core samples collected from a thermokarst sub-Arctic lake were incubated in laboratory with methane in the absence of dioxygen, biological oxidation of methane was observed, based on measurements of ^{13}C-methane consumption and $^{13}C\text{-}CO_2$ production (Martinez-Cruz et al. 2017). ^{13}C-labeled phospholipid fatty acid profiling and metagenomic sequencing of ^{13}C-labeled DNA revealed that the main species consuming methane were members of the Methylococcales, and the dominant species were the *Methylobacter* species. Nitrate, nitrite or other potential electron acceptors were ruled out in this case, based on pore water analysis, as was the seepage of dioxygen into the incubation vials. No sequences for anaerobic methane oxidizers, such as ANME type archaea or NC10 phylum bacteria were detected in the isotopically labeled DNA. Thus, the source of oxygen for methane activation by *Methylobacter*, or the nature of alternative electron acceptors potentially involved remained poorly understood (Martinez-Cruz et al. 2017).

Similar incubation experiments have been expanded to include samples of lake sediments spanning global climatic and trophic gradients, to estimate the potential for anaerobic methanotrophy in freshwater environments (Martinez-Cruz et al. 2018). Again, in most samples, strong methane oxidation activity was observed, over the background of methane production. Tests for potential electron acceptors supporting this activity were not necessarily conclusive, as nitrogen species were limited in most cases, and the potential for sulfate reduction-linked ANME archaea activities have not been tested (Martinez-Cruz et al. 2018). However, it has been observed that the methane-oxidizing activity, once again, was correlated with the presence of the aerobic methanotrophs, while the NC10 phylum anaerobic methanotrophs were present at significantly lesser abundances (Martinez-Cruz et al. 2018).

It is now clearly realized that the potentials for both aerobic and anaerobic methanotrophy as well as methanogenesis, once considered to be incompatible processes, can be present in the very same environmental niches. For example, recent metagenomic and metatranscriptomic analyses of a 2-year anaerobic enrichment in the presence of methane and nitrate revealed that, while the main active species was the archaeon '*Ca.* Methanoperedens nitroreducens', a small fraction of the aerobic methanotrophs was present, along with other bacterial species (Guerrero-Cruz et al. 2018). Only 24-h exposure to oxygen resulted in rapid shift in the fraction of transcripts belonging to the aerobic methanotrophs, especially the *Methylomonas* species, suggesting that these species remained alive and active over years of anaerobic incubation.

Metagenomic analysis of a dynamic development of a deep-sea methanotrophic microbiome has identified a quick succession of the initial subsurface communities, by aerobic methane-consuming populations (the Methylococcales bacteria), followed by anaerobic methanotrophic communities (the ANME archaea). Over years of development, it has been recorded that the mud volcano communities grew in diversity and complexity while maintaining both aerobic and anaerobic methanotroph populations (Ruff et al. 2019).

Diversity analyses of 16S rRNA and functional genes for methane oxidation-related functions, in combination with carbon isotopic analysis of methane and bicarbonate, in samples collected from a high-pressure environment hosting communities associated with methane hydrates on the seafloor, incubated under high pressure in the laboratory, revealed activities of both aerobic and anaerobic methanotrophs, driven by members of the Methylococcales and the ANME archaea, respectively. In this experiment, the aerobic methanotrophy was observed upon the addition of oxygen, and the anaerobic processes subsequently occurred after oxygen depletion (Case et al. 2017).

In Arctic and sub-Arctic lakes, archaeal methanogenesis and (an)aerobic methane oxidation by the *Methylococcaceae* have been observed in the very same water layers (Schütte et al. 2016; Martinez-Cruz et al. 2018). The overlapping patterns between methanogen and methanotroph (*Methylobacter*) populations were also documented in a suboxic area of a meromictic lake (Biderre-Petit et al. 2018).

7 Community Function in Methanotrophy

Recently, correlations have been noticed for the co-occurrence of specific methanotrophs, such as *Methylococcaceae* and non-methanotrophic methylotrophs, such as *Methylophilaceae* (Kalyuzhnaya et al. 2008; Beck et al. 2013; Oshkin et al. 2015; Hernandez et al. 2015). These observations supported the prior observations of the (aerobic) methanotrophs supporting communities apparently feeding off the nutrients released by the methanotrophs (Modin et al. 2007; Kalyuzhnaya et al. 2013). However, the omics data suggested that the relationships between the methanotrophs and the satellite species may not be random, as specific partnerships were observed. For freshwater environments, partnerships between the *Methylococcaceae* and the *Methylophilaceae* have been noted (Kalyuzhnaya et al. 2008; Beck et al. 2013; Schütte et al. 2016; Martinez-Cruz et al. 2017; Arora-Williams et al. 2018; Biderre-Petit et al. 2018). In marine environments, the Methylococcales appeared to be partnered with both the *Methylophaga* and the *Methylophilaceae* species (Ruff et al. 2013; Paul et al. 2017; Deng et al. 2019). These co-occurrence patterns do not support a hypothesis of the methanotrophs releasing organic compounds as free-for-all goods (Yu et al. 2017).

The specificity of partnerships in methane utilization has been further tested via manipulation of synthetic microbial communities, by following the dynamics of such communities via a phylogenetic marker (16S rRNA gene), as well as metatranscriptomic profiling (Yu et al. 2016, 2017). The trends in dynamics of the synthetic communities, in general, followed the trends in the natural communities (Oshkin et al. 2015; Hernandez et al. 2015), in terms of rapid decrease in community complexity and in terms of the dominance of the *Methylococcaceae* and the *Methylophilaceae* (Yu et al. 2017). However, the major methanotroph players in the synthetic communities were different. In these communities, the robustly growing *Methylomonas* species outcompeted the *Methylobacter* species, and the more robust *Methylophilus* species outcompeted the *Methylotenera* species, highlighting the challenges in recreating natural behaviour in synthetic communities. However, synthetic communities do present convenient and fully controllable models for testing interspecies relationships hypotheses.

Metatranscriptomic analysis of synthetic communities incubated under the conditions of varying partial pressures of methane and dioxygen and the varying sources of nitrogen revealed interesting patterns in gene expression (Yu et al. 2017). Remarkably, some of the most differentially transcribed gene clusters were the ones encoding XoxF and MxaF, the alternative MDH enzymes, suggesting that regulatory mechanisms for their transcription exist beyond REEs. Under low pressures of dioxygen and/or methane, as well as in the presence of ammonium, transcription of the *xox* gene clusters was upregulated, while transcription of the *mxa* gene clusters was upregulated under high dioxygen and methane, as well as in the presence of nitrate (Yu et al. 2017).

In a separate study, two-species synthetic communities have been employed, testing transcriptional response of the *Methylobacter* and the *Methylotenera* models to co-culture conditions (Krause et al. 2017). In this experiment, as well, *xoxF* versus *mxaF* transcription was dramatically altered in response to the community living. While REEs were present in the medium, selecting for expression of the *xoxF* operon in pure cultures of *Methylobacter*, as expected, in the co-cultures, the *mxa* operon was highly induced. It has been hypothesized that interactions between *Methylotenera* and *Methylobacter* were responsible for the reversal of the regulation by REEs, and that expression of the MxaFI type of enzyme resulted in spillage of methanol, which then served as a substrate for *Methylotenera* (Krause et al. 2017). While this result provides one explanation for the frequently observed co-occurrence of *Methylococcaceae* and *Methylophilaceae* in natural environments, it remains unclear what gives members of the *Methylophilaceae* competitive advantage against other methanol-utilizing organisms, as has been observed in the experiments with both natural and synthetic communities (Oshkin et al. 2015; Hernandez et al. 2015; Yu et al. 2017).

8 The Importance of Model Organisms

While environmental omics provide novel, exciting and sometimes unexpected insights into the diversity, physiology and ecological function of the methanotrophs and the associated microbial communities, the importance of experimentation with model organisms, including synthetic communities composed of model organisms, should not be under-estimated. Global multi-omics approaches in model methanotrophs are now providing knowledgebase for detailed understanding and manipulation of their metabolism, akin to the knowledgebase platforms available for the iconic model organisms such as *Escherichia coli*. One of the most well developed models so far is *Methylomicrobium alcaliphilum* 20ZR, an organism that demonstrates attractive growth characteristics, and for which, in addition to an expert-annotated genome sequence, a large body of enzymatic, transcriptomic and metabolomic datasets are available, as well as a curated genome-scale metabolic model (Akberdin et al. 2018a). Availability of these tools allows for addressing very specific details of the methanotrophy metabolism, such as carbon distribution among alternative metabolic pathways for central carbon metabolism. The *M. alcaliphilum* 20ZR model has been employed recently to test *in-silico* simulations through global non-targeted metabolomics and enzymatic evidence, to demonstrate the importance of substitution of the ATP-dependent reactions in central pathways for carbon assimilation with PPi-dependent reactions, the presence of a carbon shunt from acetyl-CoA to the pentose-phosphate pathway, and the highly-branched nature of the tricarboxylic acid cycle (Akberdin et al. 2018a). In a separate study, growth of *M. alcaliphilum* 20ZR in media containing calcium versus lanthanum (an REE) has been investigated, demonstrating trade-offs between growth rate versus carbon conversion efficiencies. Three complementary global omics approaches have been implemented to investigate the metabolic bases for these trade-offs (transcriptomics, proteomics, metabolomics), implicating changes in the fluxes through major carbon processing pathways as well as electron transfer systems, in response to alternative metal availability (Akberdin et al. 2018b). Other prominent models in the methanotrophy research are *Methylomicrobium buryatense* 5G (de la Torre et al. 2015; Gilman et al. 2017) and *Methylococcus capsulatus* Bath (Lieven et al. 2018; Tanaka et al. 2018). Additional attractive models include *Methylomonas* sp. (Zheng et al. 2018) and *Methylobacter* species (Yan et al. 2016; Puri et al. 2017).

9 Conclusions and Further Questions

Omics in application to natural environments provide exciting insights into methanotrophy and methanotrophs, including discovery of novel species, novel taxa, and even novel phyla not know in pure cultures. While only a handful of *Methylobacter* species are available in pure cultures, and these do not necessarily display robust growth in laboratory, omics have demonstrated that they are globally distributed, cosmopolitan species responsible for most of the methane-oxidizing activity in freshwater environments, including hypoxic and anoxic environments. Novel taxa possessing the methanotrophy potential have been identified via MAG reconstruction, such as *Hyphomicrobiaceae*. However, ideally, existence of such novel guilds of methanotrophs should be confirmed by cultivation of such organisms, considering that MAG analysis predicts that these organisms should be able to grow in pure cultures, under laboratory conditions. Likewise, the information gained so far from the analysis of the elusive upland soil high-affinity methanotroph genomes suggests that they should also be cultivable, even though, likely, only very slow growth should be expected. Analysis of the upland soil MAGs also created a controversy as these genomes appear to only encode a single type of methane monooxygenae (pMMO), typical of the low-affinity methanotrophs. Does this suggest that all methantrophs are high-affinity methanotrophs? Or, alternatively, pMMO is not the enzyme determining high versus low affinity. The new insights into REE-dependent methanotrophy and the apparent prevalence of the XoxF-type enzymes in the presumed methylotroph genomes presents intriguing questions as to how XoxF and MxaFI enzymes differ in their redox properties and how they evolved, as well as further questions on the details of REE sensing and acquisition by bacterial cells. The scale of REE dependence in major biogeochemical processes is yet to be fully uncovered, and it will likely not be limited to methylotrophy. While the evidence is mounting, based on combinations of omics approaches, in concert with biogeochemical measurements, on the activity of the 'aerobic' methanotrophs in hypoxic and anoxic environments, the mechanisms of methane activation under these conditions remain to be elucidated. Notably, non-methanotrophic methylotrophs, such as *Methylophilaceae*, are also detected in associations with the methanotrophs under hypoxic and anoxic conditions, further questioning how organisms that show obligate dioxygen dependence in pure cultures can be active in conditions of limited oxygen availability. Could the communal metabolism be a key to answering these questions, since the activities measured *in situ* are always reflective of communal action? Which members of the communities are dispensable and which are essential to support the methane oxidation step? A combination of advanced omics studies and laboratory manipulations with select model organisms and communities should bring us closer to solving these problems. Uncovering the new details of the controls of methane oxidation, under a variety of gradients of electron acceptor molecules, will be crucial for both better delineation of the activities of disparate groups of methanotrophs in environmental niches and for identifying the potential of the respective groups in industrial applications.

Acknowledgements This material is based upon work supported by the U.S. Department of Energy, Office of Science, Office of Biological and Environmental Research under Award Number DE-SC-0016224.

Conflict of Interest The authors declare no conflicts of interest.

References

Akberdin IR, Thompson M, Hamilton R, Desai N, Alexander D, Henard CA, Guarnieri MT, Kalyuzhnaya MG (2018a) Methane utilization in *Methylomicrobium alcaliphilum* 20ZR: a systems approach. Sci Rep 8:2512

Akberdin IR, Collins DA, Hamilton R, Oshchepkov DY, Shukla AK, Nicora CD, Nakayasu ES, Adkins JN, Kalyuzhnaya MG (2018b) Rare earth elements alter redox balance in *Methylomicrobium alcaliphilum* 20ZR. Front Microbiol 9:2735

Arora-Williams K, Olesen SW, Scandella B, Delwich K, Spencer SJ, Myers E, Abraham S, Sooklal A, Preheim SP (2018) Dynamics of microbial populations mediating biogeochemical cycling in a freshwater lake. Microbiome 6:165

Auman AJ, Stolyar S, Costello AM, Lidstrom ME (2000) Molecular characterization of methanotrophic isolates from freshwater lake sediment. Appl Environ Microbiol 66:5259–5266
Beck DA, Kalyuzhnaya MG, Malfatti S, Tringe SG, Glavina del Rio T, Ivanova N, Lidstrom ME, Chistoserdova L (2013) A metagenomic insight into freshwater methane-utilizing communities and evidence for cooperation between the *Methylococcaceae* and the *Methylophilaceae*. Peer J 1:e23
Beck DA, McTaggart TL, Setboonsarng U, Vorobev A, Kalyuzhnaya MG, Ivanova N, Goodwin L, Woyke T, Lidstrom ME, Chistoserdova L (2014) The expanded diversity of *Methylophilaceae* from Lake Washington through cultivation and genomic sequencing of novel ecotypes. PLoS One 9:e102458
Biderre-Petit C, Taib N, Gardon H, Hochart C, Debroas D (2018) New insights into the pelagic microorganisms involved in the methane cycle in the meromictic Lake Pavin through metagenomics. FEMS Microbiol Ecol 95(3):fiy183. https://doi.org/10.1093/femsec/fiy183
Butterfield CN, Li Z, Andeer PF, Spaulding S, Thomas BC, Singh A, Hettich RL, Suttle KB, Probst AJ, Tringe SG, Northen T, Pan C, Banfield JF (2016) Proteogenomic analyses indicate bacterial methylotrophy and archaeal heterotrophy are prevalent below the grass root zone. Peer J 4: e2687
Case DH, Ijiri A, Morono Y, Tavormina P, Orphan VJ, Inagaki F (2017) Aerobic and anaerobic methanotrophic communities associated with methane hydrates exposed on the seafloor: a high-pressure sampling and stable isotope-incubation experiment. Front Microbiol 8:2569
Chistoserdova L (2011) Modularity of methylotrophy, revisited. Environ Microbiol 13:2603–2622
Chistoserdova L (2015) Methylotrophs in natural habitats: current insights through metagenomics. Appl Microbiol Biotechol 99:5763–5779
Chistoserdova L (2016) Lanthanides: new life metals? World J Microbiol Biotechnol 32:138
Chistoserdova L (2017) Omics approaches to studying methylotrophs and methylotroph communities. Curr Issues Mol Biol 24:119–142
Chistoserdova L (2019) New pieces to the lanthanide puzzle. Mol Microbiol 111(5):1127–1131. https://doi.org/10.1111/mmi.14210
Chistoserdova L, Kalyuzhnaya MG (2018) Current trends in methylotrophy. Trends Microbiol 26:703–714
Chistoserdova L, Lidstrom ME (2013) Aerobic methylotrophic prokaryotes. In: Rosenberg E, EF DL, Thompson F, Lory S, Stackebrandt E (eds) The prokaryotes, 4th edn. Springer, Berlin, pp 267–285
Chu F, Lidstrom ME (2016) XoxF acts as the predominant methanol dehydrogenase in the Type I methanotroph *Methylomicrobium buryatense*. J Bacteriol 198:1317–1325
Conrad R (2009) The global methane cycle: recent advances in understanding the microbial processes involved. Environ Microbiol Rep 1:285–292
de la Torre A, Metivier A, Chu F, Laurens LM, Beck DA, Pienkos PT, Lidstrom ME, Kalyuzhnaya MG (2015) Genome-scale metabolic reconstructions and theoretical investigation of methane conversion in *Methylomicrobium buryatense* strain 5G(B1). Microb Cell Factories 14:188
Deng Y, Gui Q, Dumont M, Han C, Deng H, Yun J, Zhong W (2019) *Methylococcaceae* are the dominant active aerobic methanotrophs in a Chinese tidal marsh. Environ Sci Pollut Res Int 26:636–646
Edwards CR, Onstott TC, Miller JM, Wiggins JB, Wang W, Lee CK, Cary SC, Pointing SB, Lau MCY (2017) Draft genome sequence of uncultured upland soil cluster gammaproteobacteria gives molecular insights into high-affinity methanotrophy. Genome Announc 5:e00047-17
Ettwig KF, Butler MK, Le Paslier D, Pelletier E, Mangenot S, Kuypers MMM, Schreiber F, Dutilh BE, Zedelius J, de Beer D, Gloerich J, Wessels HJ, van Alen T, Luesken F, Wu ML, van de Pas-Schoonen KT, Op den Camp HJ, Janssen-Megens EM, Francoijs KJ, Stunnenberg H, Weissenbach J, Jetten MS, Strous M (2010) Nitrite-driven anaerobic methane oxidation by oxygenic bacteria. Nature 464:543–548
Evans PN, Boyd JA, Leu AO, Woodcroft BJ, Parks DH, Hugenholtz P, Tyson GW (2019) An evolving view of methane metabolism in the Archaea. Nat Rev Microbiol 17:219–232. https://doi.org/10.1038/s41579-018-0136-7

Gilman A, Fu Y, Hendershott M, Chu F, Puri AW, Smith AL, Pesesky M, Lieberman R, Beck DAC, Lidstrom ME (2017) Oxygen-limited metabolism in the methanotroph *Methylomicrobium buryatense* 5GB1C. Peer J 5:e3945

Guerrero-Cruz S, Cremers G, van Alen TA, Op den Camp HJM, MSM J, Rasigraf O, Vaksmaa A (2018) Response of the anaerobic methanotroph "*Candidatus* Methanoperedens nitroreducens" to oxygen stress. Appl Environ Microbiol 84:e01832-18

Hernandez ME, Beck DA, Lidstrom ME, Chistoserdova L (2015) Oxygen availability is a major factor in determining the composition of microbial communities involved in methane oxidation. Peer J 3:e801

Kalyuzhnaya MG, Lapidus A, Ivanova N, Copeland AC, McHardy AC, Szeto E, Salamov A, Grigoriev IV, Suciu D, Levine SR, Markowitz VM, Rigoutsos I, Tringe SG, Bruce DC, Richardson PM, Lidstrom ME, Chistoserdova L (2008) High-resolution metagenomics targets specific functional types in complex microbial communities. Nat Biotechnol 26:1029–1034

Kalyuzhnaya MG, Yang S, Rozova ON, Smalley NE, Clubb J, Lamb A, Gowda GA, Raftery D, Fu Y, Bringel F, Vuilleumier S, Beck DA, Trotsenko YA, Khmelenina VN, Lidstrom ME (2013) Highly efficient methane biocatalysis revealed in a methanotrophic bacterium. Nat Commun 4:2785

Kalyuzhnaya MG, Lamb AE, McTaggart TL, Oshkin IY, Shapiro N, Woyke T, Chistoserdova L (2015) Draft genomes of gammaproteobacterial methanotrophs isolated from Lake Washington sediment. Genome Announc 3:e00103–15

Karl DM, Beversdorf L, Björkman KM, Church MJ, Martinez A, DeLong EF (2008) Aerobic production of methane in the sea. Nat Geosci 1:473–478

Keppler F, Hamilton JTG, Braß M, Röckmann T (2006) Methane emissions from terrestrial plants under aerobic conditions. Nature 439:187–191

Knittel K, Boetius A (2009) Anaerobic oxidation of methane: progress with an unknown process. Annu Rev Microbiol 63:311–334

Krause SMB, Johnson T, Samadhi Karunaratne Y, Fu Y, Beck DAC, Chistoserdova L, Lidstrom ME (2017) Lanthanide-dependent cross-feeding of methane derived carbon is linked by microbial community interactions. Proc Natl Acad Sci U S A 114:358–363

Lieven C, Petersen LAH, Jørgensen SB, Gernaey KV, Herrgard MJ, Sonnenschein N (2018) A genome-scale metabolic odel for *Methylococcus capsulatus* (Bath) suggests reduced efficiency electron transfer to the particulate methane onooxygenase. Front Microbiol 9:2947

Martinez-Cruz K, Leewis MC, Herriott IC, Sepulveda-Jauregui A, Anthony KW, Thalasso F, Leigh MB (2017) Anaerobic oxidation of methane by aerobic methanotrophs in sub-Arctic lake sediments. Sci Total Environ 607–608:23–31

Martinez-Cruz K, Sepulveda-Jauregui A, Casper P, Anthony KW, Smemo KA, Thalasso F (2018) Ubiquitous and significant anaerobic oxidation of methane in freshwater lake sediments. Water Res 144:332–340

McTaggart TL, Benuska G, Shapiro N, Woyke T, Chistoserdova L (2015) Draft genomes of five new strains of *Methylophilaceae* isolated from Lake Washington sediment. Genome Announc 3: e01511-14

Modin O, Fukushi K, Yamamoto K (2007) Denitrification with methane as external carbon source. Water Res 41:2726–2738

Murrell JC, McGovan V, Cardy DLN (1993) Detection of methylotrophic bacteria in natural samples by molecular probing techniques. Chemosphere 26:1–11

Ochsner AM, Hemmerle L, Vonderach T, Nüssli R, Bortfeld-Miller M, Hattendorf B, Vorholt JA (2019) Use of rare-earth elements in the phyllosphere colonizer *Methylobacterium extorquens* PA1. Mol Microbiol 111(5):1152–1166. https://doi.org/10.1111/mmi.14208

Oshkin IY, Beck DA, Lamb AE, Tchesnokova V, Benuska G, McTaggart TL, Kalyuzhnaya MG, Dedysh SN, Lidstrom ME, Chistoserdova L (2015) Methane-fed microbial microcosms show differential community dynamics and pinpoint taxa involved in communal response. ISME J 9:1119–1129

Oswald K, Graf JS, Littmann S, Tienken D, Brand A, Wehrli B, Albertsen M, Daims H, Wagner M, Kuypers MM, Schubert CJ, Milucka J (2017) *Crenothrix* are major methane consumers in stratified lakes. ISME J 11:2124–2140

Padilla CC, Bertagnolli AD, Bristow LA, Sarode N, Glass JB, Thamdrup B, Stewart FJ (2017) Metagenomic binning recovers a transcriptionally active gammaproteobacterium linkinbg methanotrophy to partial denitrification in an anoxic oxygen minimum zone. Front Mar Sci 4:23

Paul BG, Ding H, Bagby SC, Kellermann MY, Redmond MC, Andersen GL, Valentine DL (2017) Methane-oxidizing bacteria shunt carbon to microbial mats at a marine hydrocarbon seep. Front Microbiol 8:186

Puri AW, Schaefer AL, Fu Y, Beck DA, Greenberg EP, Lidstrom ME (2017) Quorum sensing in a methane-oxidizing bacterium. J Bacteriol 199:e00773-16

Ruff SE, Arnds J, Knittel K, Amann R, Wegener G, Ramette A, Boetius A (2013) Microbial communities of deep-sea methane seeps at Hikurangi continental margin (New Zealand). PLoS One 8:e72627

Ruff SE, Felden J, Gruber-Vodicka HR, Marcon Y, Knittel K, Ramette A, Boetius A (2019) In situ development of a methanotrophic microbiome in deep-sea sediments. ISME J 13:197–213

Schütte UM, Cadieux SB, Hemmerich C, Pratt LM, White JR (2016) Unanticipated geochemical and microbial community structure under seasonal ice cover in a dilute, dimictic Arctic lake. Front Microbiol 7:1035

Sherwood Lollar B, Lacrampe-Couloumea G, Slatera GF, Warda J, Moserb DP, Gihringb TM, Linc L-H, Onstottc TC (2006) Unravelling abiogenic and biogenic sources of methane in the Earth's deep subsurface. Chem Geol 226:328–339

Singleton CM, McCalley CK, Woodcroft BJ, Boyd JA, Evans PN, Hodgkins SB, Chanton JP, Frolking S, Crill PM, Saleska SR, Rich VI, Tyson GW (2018) Methanotrophy across a natural permafrost thaw environment. ISME J 12:2544–2558

Skennerton CT, Ward LM, Michel A, Metcalfe K, Valiente C, Mullin S, Chan KY, Gradinaru V, Orphan VJ (2015) Genomic reconstruction of an uncultured hydrothermal vent gammaproteobacterial methanotroph (family *Methylothermaceae*) indicates multiple adaptations to oxygen limitation. Front Microbiol 6:1425

Smith GJ, Angle JC, Solden LM, Borton MA, Morin TH, Daly RA, Johnston MD, Stefanik KC, Wolfe R, Gil B, Wrighton KC (2018) Members of the genus *Methylobacter* are inferred to account for the majority of aerobic methane oxidation in oxic soils from a freshwater wetland. MBio 9:e00815-18

Tanaka K, Yokoe S, Igarashi K, Takashino M, Ishikawa M, Hori K, Nakanishi S, Kato S (2018) Extracellular electron transfer via outer membrane cytochromes in a methanotrophic bacterium *Methylococcus capsulatus* (Bath). Front Microbiol 9:2905

Tindall BJ (2019) On the nomenclatural types of *Methylothermus thermalis* Tsubota et al. 2005, *Methylothermus* Tsubota et al. 2005 and *Methylothermaceae* Hirayama et al. 2014, their status under the International Code of Nomenclature of Prokaryotes and valid publication of the names *Methylothermus* gen. nov., *Methylothermus subterraneus* Hirayama et al. and *Methylothermaceae* Hirayama et al. Int J Syst Evol Microbiol. https://doi.org/10.1099/ijsem.0.003220

Trotsenko YA, Murrell JC (2008) Metabolic aspects of aerobic obligate methanotrophy. Adv Appl Microbiol 6:183–229

Valentine DL, Reeburgh WS (2000) New perspectives on anaerobic methane oxidation. Environ Microbiol 2:477–484

Vekeman B, Speth D, Wille J, Cremers G, De Vos P, Op den Camp HJ, Heylen K (2016) Genome characteristics of two novel Type I methanotrophs enriched from North Sea sediments containing exclusively a lanthanide-dependent XoxF5-type methanol dehydrogenase. Microb Ecol 72:503–509

Versantvoort W, Guerrero-Cruz S, Speth DR, Frank J, Gambelli L, Cremers G, van Alen T, Jetten MSM, Kartal B, Op den Camp HJM, Reimann J (2018) Comparative genomics of *Candidatus methylomirabilis* species and description of *Ca. Methylomirabilis lanthanidiphila*. Front Microbiol 9:1672

Ward N, Larsen Ø, Sakwa J, Bruseth L, Khouri H, Durkin AS, Dimitrov G, Jiang L, Scanlan D, Kang KH, Lewis M, Nelson KE, Methé B, Wu M, Heidelberg JF, Paulsen IT, Fouts D, Ravel J, Tettelin H, Ren Q, Read T, DeBoy RT, Seshadri R, Salzberg SL, Jensen HB, Birkeland NK, Nelson WC, Dodson RJ, Grindhaug SH, Holt I, Eidhammer I, Jonasen I, Vanaken S, Utterback T, Feldblyum TV, Fraser CM, Lillehaug JR, Eisen JA (2004) Genomic insights into methanotrophy: the complete genome sequence of *Methylococcus capsulatus* (Bath). PLoS Biol 2:e303

Yan X, Chu F, Puri AW, Fu Y, Lidstrom ME (2016) Electroporation-based genetic manipulation in type I methanotrophs. Appl Environ Microbiol 82:2062–2069

Yu Z, Krause SM, Beck DA, Chistoserdova L (2016) A synthetic ecology perspective: how well does behavior of model organisms in the laboratory predict microbial activities in natural habitats? Front Microbiol 7:946

Yu Z, Beck DAC, Chistoserdova L (2017) Natural selection in synthetic communities highlights the roles of *Methylococcaceae* and *Methylophilaceae* and suggests differential roles for alternative methanol dehydrogenases in methane consumption. Front Microbiol 8:2392

Yu Z, Zheng Y, Huang J, Chistoserdova L (2019) Systems biology meets enzymology: recent insights into communal metabolism of methane and the role of lanthanides. Curr Issues Mol Biol 33:183–196

Zheng Y, Huang J, Zhao F, Chistoserdova L (2018) Physiological effect of XoxG(4) on lanthanide-dependent methanotrophy. mBio 9(2):e02430-17

Diversity, Physiology, and Biotechnological Potential of Halo(alkali)philic Methane-Consuming Bacteria

Snehal Nariya and Marina G. Kalyuzhnaya

Contents

Abstract Nature supplies us with a large array of microbes, known as methanotrophs, as efficient factories for methane capturing and conversions. Among them, the halo(alkali)philic methanotrophs stand out as the most favorable systems for industrial explorations as new sources of salt/pH stable enzymes and as native producers of amino acids, sugars, and osmoprotectants. In recent years, the array of halo(alkali)philic methanotroph applications has been extended to fuels and chemicals, fostering thorough investigations of their physiology, genetics, genomics and systems biology. Here we summarize four decades of research on halo(alkali)philic methanotrophic bacteria, as well as provide our vision of further developments.

S. Nariya
Department of Biology, San Diego State University, San Diego, CA, USA
e-mail: snariya@sdsu.edu

M. G. Kalyuzhnaya (✉)
Department of Biology, San Diego State University, San Diego, CA, USA

Viral Information Institute, San Diego State University, San Diego, CA, USA
e-mail: mkalyuzhnaya@sdsu.edu

E. Y. Lee (ed.), *Methanotrophs*, Microbiology Monographs 32,
https://doi.org/10.1007/978-3-030-23261-0_5

1 Introduction

Methane (CH_4), a colorless and odorless gas, is the primary hydrocarbon in natural gas and biogas. Methane plays several key roles in the natural carbon cycle as the most abundant organic compound on Earth, as a by-product of biomass decomposition, as a dangerous greenhouse gas, and as a fugitive by-product of growing anthropogenic activities (Shindell et al. 2012; IPCC 2014).

From a biological standpoint, methane is an exceptional source of carbon and energy for methane-consuming, ubiquitous organisms, known as methanotrophs (Hanson and Hanson 1996; Semrau et al. 2010). By utilizing methane, methanotrophs contribute to local carbon pools and play a vital role in reducing methane emissions. Since methane utilization is light independent, it can bring life to the most exotic and extreme places on Earth, including deep-sea thermal vents and high-pH or hypersaline ecosystems (Boetius and Wenzhöfer 2013; Khmelenina et al. 1997, 1999; Sorokin et al. 2000; Heyer et al. 2005; Kalyuzhnaya et al. 2008).

Over the years many halo(alkali)philic methanotroph strains have been isolated. Investigation of their physiology, biochemistry, and genetics has led to discoveries about how these microbes withstand environmental stresses (Trotsenko and Khmelenina 2002; Kalyuzhnaya 2016a). They also serve as robust microbial systems for exploring the metabolic flexibility and capabilities of C1-metabolism (Kalyuzhnaya et al. 2015), pyrophosphate metabolism (Khmelenina et al. 2018), sucrose and ectoine biosynthesis (Reshetnikov et al. 2006; But et al. 2015), and for improving our understanding of sterol and sterol-like lipid biosynthesis in microbes and higher organisms (Banta et al. 2015; Li et al. 2015). Studying their proteins and membranes has facilitated our understanding of the enzymatic principles of methane oxidation (Ro et al. 2018; Deng et al. 2018). Furthermore, halo(alkali)philic methanotrophs are now being converted into microbial factories for industrial purposes (Strong et al. 2015; Cantera et al. 2018). This review summarizes current knowledge of the diversity and physiology of methanotrophic bacteria and prospects for their biotechnological application.

2 Detection, Isolation, and Diversity

Moderately halophilic methanotrophs were first isolated from oceanic waters in the late 80s. That led to a formal description of two novel clades, *Methylobacter pelagicum* (Sieburth et al. 1987) and *Methylobacter marinus* (Lidstrom 1988). Only *Methylobacter marinus* (A45, BBA5, and BBA6) is currently available in pure culture (Table 1). Over the years, the diversity of marine methanotrophs has been expanded such that it now includes seven genera: *Methylomicrobium* (now *Methylotuvimicrobium, Mtm*) (Kalyuzhnaya et al. 2008; Orata et al. 2019), *Methylosphaera* (Bowman et al. 1997), *Methylothermus* (*Mth*) (Tsubota et al. 2005; Hirayama et al. 2011), *Methylomonas* (*Mm*) (Bowman et al. 1993), *Methylomarinum* (*Mmr*) (Hirayama et al. 2013), *Methylomarinovum* (Hirayama

Table 1 List of halophilic and halotolerant methanotrophic bacteria

Organism	Strains	Whole genome sequence accession numbers	Salinity range (%) [optimum]	pH range [optimum]	Application	Genetic tools applied	Systems biology tools/ datasets	References
Methylomicrobium pelagicum		NR	0.4–5.0	5.5–9.0 [5.5]	NR	NR	NR	Sieburth et al. (1987), Bowman et al. (1993), Kalyuzhnaya et al. (2001)
Methylomicrobium alcaliphilum	$20Z^T$ and $20Z^R$, 5Z	FO082060 FO082061	0.5–10 [1–3]	7.0–10.0 [9.0]	Pharmaceuticals/ cosmetics (ectoine), 2,3-butanediol; isoprenes; microbial co-cultures, heterologous protein expression, nanomaterials	*Plasmids*: pCM184, pDO1, pDO2, pDO3, pDO4, pDO5, pDO6, pDO7, pDO8, pDO9; pETectR1-his, pRK2073, pTSGex, pCM66, pCM130, pCM62, pCR2.1, pRK2013, pEBPR01, pEBP1, pMOC1, pMO1, pSMARTLCKan, Ki/KO system, pmCherry, pDrive, pEGFP-N1, pCM433:Kan	FBA model, transcriptomic, metabolomic, proteomics	Khmelenina et al. (1997), Kalyuzhnaya et al. (2008, 2013, 2018), Ojala et al. (2011), Vuilleumier et al. (2012), Akberdin et al. (2018a, b), Nguyen et al. (2018), Song et al. (2016), Hill et al. (2017), Cantera et al. (2018)
Methylomicrobium buryatense	5G(B1), 5GB1S[a], 5GB1C[a], 4G, 6G, 7G, $5B^T$	AOTL01000000KB455575 KB455576	0–8 [0.75]	6.0–11.0 [7.5–9.5]	Biofuels, lactate	*Plasmids/cassette*: pSMARTLCKan, pmCherry, pDrive, pEGFP-N1, pCM66, pCM433kanT, pCM184, pDO1, pDO2, pDO3, pDO4, pDO5, pDO6, pDO7, pDO8, pDO9; pETectR1-his, pRK2073, pTSGex, pCR2.1, pCM130, pCM62, pEBPR01, pRK2013, pEBP1,	FBA model, transcriptomic, metabolomic	Kalyuzhnaya et al. (2001), Ojala et al. (2011), Khmelenina et al. (2013a), Puri et al. (2015), Yan et al. (2016), Henard et al. (2016, 2017), Demidenko et al. (2017), de la Torre et al. (2015), Gilman et al. (2017), Fu et al. (2017), Garg et al. (2018)

(continued)

Table 1 (continued)

Organism	Strains	Whole genome sequence accession numbers	Salinity range (%) [optimum]	pH range [optimum]	Application	Genetic tools applied	Systems biology tools/ datasets	References
						pMOC1, pMO1, pAWP78, pAWP89, pCM433-moD, pET-29b(+), pET29-M, pFC25, pCM158, p7Z6, pE-FLP, pCAH01		
Methylomicrobium kenyense	AMO1^T	NR	2.8 [NR]	9.0–10.5	NR	NR	NR	Sorokin et al. (2000), Kalyuzhnaya et al. (2008), Kalyuzhnaya (2016b), Fuse et al. (1998)
Methylomicrobium japanese	N1	NR	0.2–8.1 [2.3–4.5]	NR	NR	NR	NR	
Methylobacter marinus	A45^T, BBA5, and BBA6	ARVS00000000	0.5–1.5 [0.6]	NR	NR	NR	NR	Lidstrom (1988), Bowman et al. (1993, 1997), Flynn et al. (2016)
Methylobacter modestohalophilus	10S	NR	0.2–9 [2]	5.5–8.5 [6.5]	NR	NR	NR	Kalyuzhnaya et al. (1998)
Methylosphaera hansonii	ACAM 549	NR	Seawater		NR	NR	NR	Bowman et al. (1997)
Methylotermus thermalis	MYHT^T	NR	0–3 [0.5–1]	6.5–7.5 [6.8]	NR	NR	NR	Tsubota et al. (2005)
Methylotermus subterraneus	HTM55^T	NR	0–1 [0–0.3]	5.2–7.5 [5.8–6.3]	NR	NR	NR	Hirayama et al. (2011)
Methylomonas methanica	MC09	CP002738	1–2.5 [NR]	5.5–9.0 [NR]	Biofuel	NR	NR	Bowman et al. (1993), Boden et al. (2011), Burdette (2013)
Methylococcus thermophilus		NR	1–3 [1–2.5]	5.5–9.0 [NR]	NR	NR	NR	Bowman et al. (1993)

Methylomarinum vadi	IT-4^T	JPON00000000	1–8 [2–3]	4.5–7.0 [6.2–6.6]	NR	NR	NR	Hirayama et al. (2013), Flynn et al. (2016)
	T2-1	NR	2.5–8 [2.5–3]	5.3–8.1 [6.4–7.0]	NR	NR	NR	
Methylomarinovum caldicuralii		NR	1–5 [3]	5.3–6.9 [6.0–6.4]	NR	NR	NR	Hirayama et al. (2014)
Methylohalobius crimeensis	10Ki^T and 4Kr	ATXB01000001–ATXB01000005	0.2–2.5 [1–1.5]	6.5–7.5 [7.0]	NR	NR	NR	Heyer et al. (2005), Sharp et al. (2015)
Methyloprofundus sedimenti	WF1^T	NZ_LPUF00000000.1	1–4 [2]	6–8 [6.4–7.5]	NR	NR	NR	Tavormina et al. (2015)
Methylosarcina fibrata	AML-C10^T	ARCU00000000	0–1 [NR]	5.0–9.0 [NR]	Methane mitigation (landfills)	NR	NR	Wise et al. (2001), Hamilton et al. (2015)

NR not reported, *FBA* flux balance analyses
[a]Evolved or selected strains

et al. 2014), and *Methyloprofundus* (Tavormina et al. 2015). As a general rule, these marine microbes prefer to grow at neutral or slightly alkaline pH (7.6–8.5). They require sodium ions but display a relative narrow salt-tolerance range and typically cannot grow at salinity above 3–4%. The group includes a few thermophilic species, isolated from geothermal areas or deep-sea hydrothermal vent fields. Thermophilic microbes typically demonstrate optimum growth at 55–60 °C and have relatively high growth rates (0.2 h^{-1}) but require additional nutrient supplementation (Tsubota et al. 2005; Hirayama et al. 2011). Today various metagenomic analyses have shown that methanotrophic bacteria inhabit every niche of our ocean (both aerobic or anaerobic), water columns and sediments, cold and hot vents, and even in marine animal tissues, as the most successful symbionts. Marine methanotrophs contribute to microbial chemosynthesis, are responsible for deep-sea oxygen consumption (Boetius and Wenzhöfer 2013), and undoubtedly support deep-ocean 'oases' of biodiversity (Petersen and Dubilier 2009). In spite of their many critical roles, our knowledge of the physiology and biochemistry of marine methanotrophs is still quite limited. Specifically, almost nothing is known about their strategies for survival or their methane-consumption activity in O_2-limited or even anaerobic zones.

Terrestrial saline environments, (e.g., soda lakes, salt ponds, and marshlands) also display a great diversity of methane-consuming bacteria (Khmelenina et al. 1997; Kalyuzhnaya et al. 2001; Sorokin et al. 2000, 2004; Heyer et al. 2005). Many strains have been isolated in pure culture. They were affiliated with the genera *Methylobacter* (*Mb*), *Methylomicrobium* (currently *Methylotyvimicrobium*, *Mtm*), and *Methylohalobium* (*Mh*) (Sieburth et al. 1987; Lidstrom 1988; Khmelenina et al. 1997; Kalyuzhnaya et al. 1998, 1999; Sorokin et al. 2000; Heyer et al. 2005). While the majority of these methanotrophs are described as moderate halophiles with optimal growth at 0.75–3% NaCl, the most salt-tolerant terrestrial halophiles can grow at NaCl levels as high as 9–15%. Many halophilic methanotrophs are also alkaliphiles, growing well at pH levels above 8.5 (Kalyuzhnaya et al. 2001).

Halo(alkali)philic methanotrophic bacteria restrict carbon loss and CH_4 emission from soda/saline ecosystems, contributing to both local and global carbon cycles (Zavarzin et al. 1999; Antony et al. 2013). The biotechnological potential of halophilic methanotrophs has also been long recognized (Strong et al. 2015, 2016). Both the environmental significance and biotechnological potential have stimulated extensive investigations into the biochemical capabilities of the group. Methanotrophs belonging to genera *Methylotuvimicrobium* (previously *Methylomicrobium*) are among the best-studied halo(alkali)philic methanotrophs.

3 Methanotrophic Networks: Classical Biochemistry and -omics Prospective

Several complete and draft sequences of methanotroph genomes have been produced including *Mtm. alcaliphilum* 20Z and *Mtm. buryatense* 5G, *Mm methanica* MC09, *Methylomarinum vadi*, and *Mh. crimeensis* (Table 1). Two genome-scale metabolic models of methanotrophy have been constructed, providing a computational platform for a systems-level interrogation of methane utilization (de la Torre et al. 2015; Akberdin et al. 2018a, b). These metabolic reconstructions have been grounded on numerous genetic, metabolic, transcriptomic and proteomic datasets. A number of enzymes have been purified and characterized (see Table 2). In general, halophilic methanotrophs are very similar to non-halophilic methanotrophs with respect carbon oxidation and energy conservation. However, halophilic methanotrophs require some additional metabolic functions to sustain in the high pH/high salinity environments. In the past decade, a thorough investigation of carbon fluxes in *Methylotuvimicrobium* helped to uncover new insights into the primary oxidation and assimilation as well as osmoadaptation mechanisms. The main findings are summarized below.

3.1 Central Metabolic Pathways and Electron Transfer

The first step of methane oxidation can be carried out by one of two methane monooxygenases (MMOs) and requires an input of two electrons and two protons. The reaction relies on oxygen for activation of the CH_4 molecule and results in the production of methanol and water. Possible sources of electrons include NADH/H+ for soluble methane monooxygenase (sMMO) and periplasmic pyrroloquinoline quinone (PQQ)-dependent methanol dehydrogenase and/or ubiquinol (UQH_2) for the membrane-bound monooxygenase (pMMO) (Semrau et al. 2010). The subsequent step, methanol oxidation, results in formaldehyde generation and preservation of reducing power as reduced cytochromes. In many, if not all, gammaproteobacterial methanotrophs, formaldehyde is the first branching point of metabolism, where flux is split between linear oxidation to CO_2 or incorporation into sugar-phosphates (Fig. 1).

All current -omics data suggest that the majority of formaldehyde is directed toward the ribulose monophosphate pathway (RuMP), leading to the generation of fructose-6-phosphate (Fructose-6P, F6P). The intracellular pool of fructose-6P is the second main carbon flux-flow control knob in methanotrophs harboring the RuMP pathway. Many methanotrophs have several carbohydrate pathways: pentose phosphate rearrangements (PPR), Embden–Meyerhof–Parnas (EMP), oxidative pentose phosphate pathway (oxPPP), Entner–Doudoroff pathway (ED) and phosphoketolase (PK) pathway. These multiple interconnected points in metabolic pathways enable equilibrium between C1-assimilation and oxidation in response to cellular needs. A significant portion of F6P (3/4 of the pool) is directed toward the PPR for the regeneration of ribulose-5-phosphate. It appears that many halophilic methanotrophic bacteria employ the fructose-bisphosphate aldolase (or EMP)-variant of the ribulose monophosphate pathways (Kalyuzhnaya et al. 2013), contrary to non-halophilic species, such as *Methylococcus capsulatus*, which predominantly uses the ED-variant. The majority of methanotrophic bacteria employ several pyrophosphate (PPi)-dependent enzymes, including PPi-phosphofructokinase, which

Table 2 Purified and characterized enzymes from halophilic methanotrophs

Enzyme	Cofactor	V_{max} (U mg^{-1} of protein)	Km (mM)	ID	Structure (MMDB ID)	References
Methanol dehydrogenase	PQQ, Ca-dependent	4.21	0.81	EC 1.1.99.8	–	Kim and Kim (2006)
Particulate methane monooxygenase	NADH	14.5 ± 1.2	NR	EC 1.14.18.3	162148	Ro et al. (2018)
XoxF	PQQ, La-dependent	0.18 ± 0.1	NR	EC 1.1.99.8	166379	Deng et al. (2018)
Hexulose phosphate synthase (HPS)	Ru5P	172	0.98	EC 4.1.2.43	–	Rozova et al. (2017)
HPS-PHI (phosphohexuloisomerase)	Ru5P	22	0.64	EC 4.1.2.43	–	Rozova et al. (2017)
Pyruvate kinase 2	ADP	200.1 ± 11.6	NR	EC 2.7.1.40	–	Kalyuzhnaya et al. (2013)
Acetate kinase	ADP	1288.6 ± 30.7	0.11 ± 0.01	EC 2.7.2.1	–	Rozova et al. (2015a, b)
Pyrophosphate-dependent 6-phosphofructokinase	F6P/PPi	577	0.64 ± 0.02	EC 2.7.1.90	–	Rozova et al. (2010)
Glucokinase	ATP	216.7 ± 8.4	0.32 ± 0.05	EC 2.7.1.1; EC 2.7.1.2	–	Mustakhimov et al. (2017)
Diaminobutyrate acetyltransferase	–	200	0.46	EC 2.3.1.178	–	Reshetnikov et al. (2005)
Serine-glyoxylate aminotransferase	–	13.0 ± 0.3	2.6 ± 0.3	EC 2.6.1.45	–	But et al. (2018)
Sucrose phosphate phosphatase	–	18.9	0.036	EC 3.1.3.24	–	But et al. (2015)
Amylosucrase						
Transglycosylation Hydrolysis	–	0.06 0.053	6 11	EC 2.4.1.4	–	But et al. (2015)
ATP-dependent fructokinase	ATP	141 ± 6	1.3 ± 0.2	EC 2.7.1.4	–	But et al. (2012)
Aspartokinase	ATP	105	NR	EC 2.7.2.4	–	Reshetnikov et al. (2006)
Aspartate-semialdehyde dehydrogenase	–	70	NR	EC 1.2.1.11	–	Reshetnikov et al. (2006)

DABA-transaminase	–	35	NR	EC 2.6.1.76	–	Reshetnikov et al. (2006)
Ectoine synthase	–	60	NR	EC 4.2.1.108	–	Reshetnikov et al. (2006)
Sterol reductase	NADPH/H+	NR	NR	EC 1.3.1.70	123654	Li et al. (2015)
Hydroxypyruvate reductase	NADH/H+ NADPH/H+	26 ± 7 41 ± 2	0.6 ± 0.3 0.17 ± 0.03	EC 1.1.1.81	–	But et al. (2017)
Malate dehydrogenase						
Malate Oxaloacetate	NAD+ NADH/H+	15.1 20.8	0.11 (0.45)[a] 0.34 (0.03)[a]	EC 1.1.1.37	–	Rozova et al. (2015a, b)

NR not reported
[a]Km(mM) for cofactor

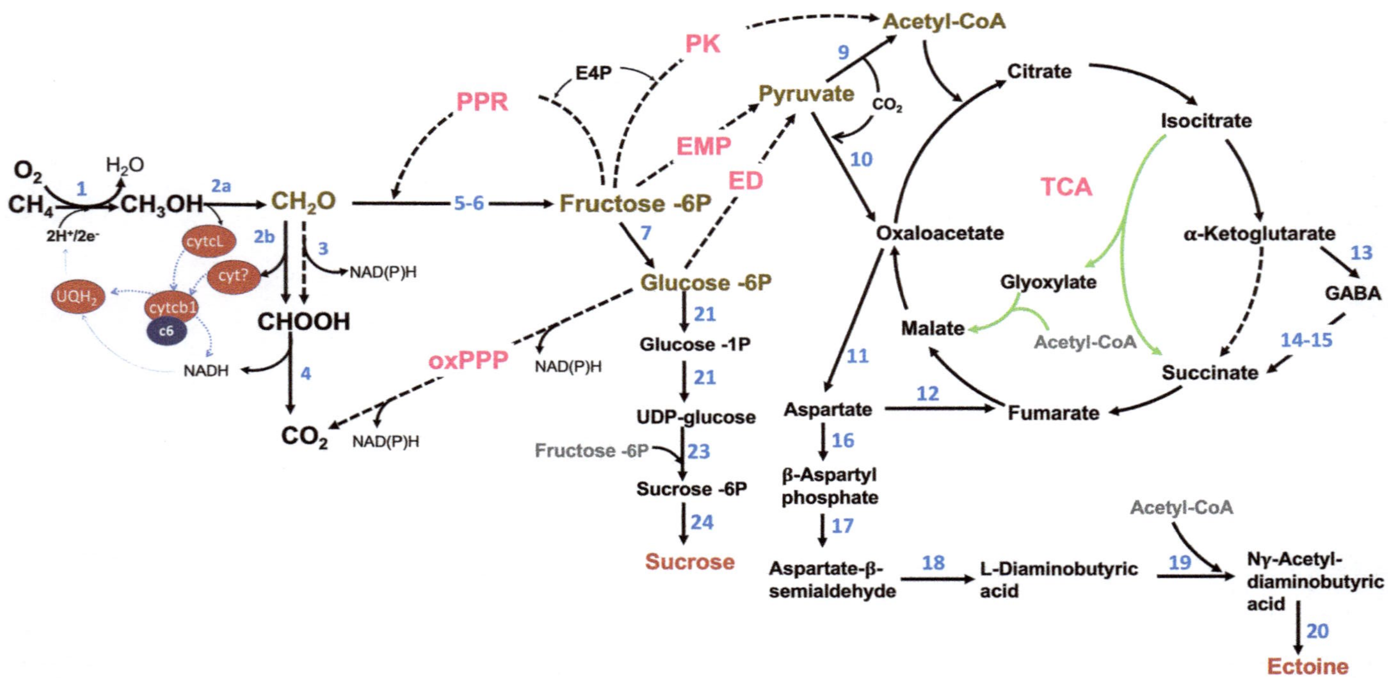

Fig. 1 Overview of central metabolic pathways in *Mtm. Alcaliphilum 20Z*^{R}. **1**—particulate methane monooxygenase; **2a**—PQQ-linked methanol dehydrogenase (MxaFI); **2b**—PQQ-linked methanol and formaldehyde dehydrogenase (XoxF); **3**—H_4FP, folate-linked C_1 transfer/H_4MTP, methanopterin-linked C_1 transfer; **4**—NAD-dependent formate dehydrogenase; **5–6**—hexulose phosphate synthase and isomerase; **7**—glucose-6-phosphate isomerase; **8**—phosphotransacetylase; **9**—pyruvate dehydrogenase; **10**—pyruvate carboxylase; **11**—oxaloacetate transaminase; **12**—aspartate ammonia-lyase; **13**—**α**-ketoglutarate decarboxylase; **14–15**—4-aminobutanoate:2-oxoglutarate aminotransferase and succinate-semialdehyde dehydrogenase; **16**—aspartokinase; **16**—aspartate **β**-semialdehyde dehydrogenase; **17**—diaminobutyric acid transaminase; **18**—diaminobutyric acid

acetyltransferase; **19**—ectoine synthase; **20**—phophoglucomutase; **21**—UDP glucose pyrophosphorylase; **22**—sucrose phosphate synthase; **23**—sucrose phosphate phosphatase; abbreviations: **PPR**—pentose phosphate pathway rearrangements; **oxPPP**—oxidative pentose phosphate pathway; **EMP**—Embden–Meyerhof–Parnas pathway; **ED**—Entner–Doudoroff pathway; PK, phoshoketolase pathway (Bifidobacterial shunt); **TCA**—citric acid cycle; **E4P**—erythrose 4-phosphate; **GABA**—diaminobutyric acid; PQQ—pyrroloquinoline quinone; UQH_2—ubiquinol; *solid lines*—enzymatic reaction; *broken lines*—multistep pathway; *green lines*—glyoxylate regeneration pathway, a metabolic alteration which improved carbon conversion efficiency. Note: Acetyl-CoA could be produced via partial serine cycle (not shown)

makes the EMP pathway the more efficient route for energy recovery. It has been proposed that the pyrophosphate-dependent enzyme plays a role in regulating carbon assimilation (Khmelenina et al. 2018). When cells are actively growing, there is an input of pyrophosphate (PPi) from biosynthesis which stimulates conversion of F6P into fructose-bisphosphate via PPi-dependent phosphofructokinase, directing carbon toward the glycolytic EMP pathway. If growth become energy limited, lack of PPi input restricts the activity of phosphofructokinase, and the excess of F6P enters oxPPP (for NAD(P)H production) and/or the ED-pathway.

Glycolytic pathways provide only a portion of the ATP needed for cell growth, and respiration complements the additional energy needs. The electron transfer reactions in methanotrophs have been recently refined (Akberdin et al. 2018a, b). Both, -omics data (gene expression and protein abundance) and genetic data suggest that the most prevalent system for respiration in the methanotroph includes complex I (NQR, NADH: ubiquinone oxidoreductase), cytochrome aa3 oxidase, and the cytochrome bc1 complex. While NQR and cytochrome aa3 contribute to proton motive force generation, the cytochrome bc1 complex is predicted to drive reverse electron transfer, further balancing energy/redox pools.

Some steps of methane oxidation metabolism have been refined, including a lanthanide-induced switch for methanol oxidation (Chu and Lidstrom 2016), the activity of serine cycle enzymes (But et al. 2018), and the complete TCA cycle (Fu et al. 2017; Akberdin et al. 2018a, b). Two TCA variants have been proposed—a canonical TCA cycle with α-ketoglutarate dehydrogenase (Fu et al. 2017) and highly branched TCA cycle with an additional GABA shunt and aspartate-fumarate loop (Akberdin et al. 2018a, b).

3.2 Fermentation and Anaerobic Respiration

Many halophilic methanotrophs employ anaerobic pathways for growth when O2-resources are limited. Fermentation pathways were first described in *Mtm. alcaliphilum* 20Z (Kalyuzhnaya et al. 2013). The strain can grow at remarkably low oxygen tensions. However, upon extreme oxygen limitation (at DO $< 1\%$), the growth slows, and pathways for mixed acid fermentation are activated (Fig. 1). The main product of fermentation is acetate. It was further discovered that a portion of F6P could be directed towards acetyl-P, via a phosphoketolase reaction, and subsequently towards acetate via an acetate kinase for regeneration of ATP (Akberdin et al. 2018b). The presence of an active phosphoketolase pathway was supported by metabolic simulation studies and confirmed by proteomics and enzymatic studies (Rozova et al. 2015a; Akberdin et al. 2018a, b). It remains to be discovered how *Mtm. alcaliphilum* 20Z cells can partition oxygen between methane oxidation and respiration.

Fermentation pathways have been also investigated in *Mtm. byryatenses* 5GB1, a derivative of 5G strain (Gilman et al. 2017). Despite being a close relative of *Mtm. alcaliphilum* 20Z, *Mtm. byryatenses* 5GB1 displays different phenotypes upon exposure to O_2-limitation. First, *Mtm. byryatenses* 5GB1 cells show O_2-limitation at higher levels of oxygen (at 5% of dissolved oxygen, DO). Both fermentation and respiration support the O_2-limited growth, and while cells also produce acetate as the primary fermentation product, it is predominantly formed from pyruvate (Gilman et al. 2017). Overall, the complexity of O_2-limiting responses suggests a variety of adaptation strategies for different species of methane-consuming bacteria.

3.3 Osmoadaptation

Environments high in salinity present an osmotic challenge for microbial growth. A few pathways linked to adaptation to high-salt environments have been characterized (Detkova and Boltyanskaya 2007; Galinski and Truper 1982; Galinski et al. 1985). Most of the knowledge in this area has resulted from research carried out with *Mtm. alcaliphilum* 20Z. It is now clear that, in response to high levels of salinity, microorganisms synthesize and accumulate osmoprotectants-inorganic and organic solutes-within their cells to prevent water loss and maintain the osmotic pressure (Khmelenina et al. 1999, 2000). The low-molecular-weight organic osmoprotectants produced by halotolerant methanotrophs include glutamate, ectoine (1,4,5,6-tetrahydro-2-methyl-4-pyrimidine carboxylic acid), sucrose, and 5-oxo-1-proline (pyrrolidone-5-carboxylic acid).

Metabolic pathways for ectoine and sucrose production have been refined via biochemical and genetic studies (Fig. 1). Sucrose biosynthesis involves the intermediates of the RuMP pathway, UDP-glucose, and fructose-6-phosphate. Sucrose phosphate synthase initiates the synthesis of sucrose from UDP-glucose and fructose-6-phosphate to produce sucrose-6-phosphate (an intermediate), which is then dephosphorylated by sucrose phosphate phosphatase to form sucrose (Trotsenko and Khmelenina 2002). Ectoine is synthesized from aspartate via the activity of several enzymes including the products of the ectA, ectB, and ectC genes. First, aspartate kinase phosphorylates aspartate to form β-aspartyl phosphate, which is then converted into L-aspartate-β-semialdehyde by aspartate semialdehyde dehydrogenase. The product of *ectB* gene, diaminobutyric acid transaminase, converts L-aspartate-β-semialdehyde to L-2,4-diaminobutyric acid (DABA). DABA is converted to Nγ-acetyl-L-2,4-diaminobutyric acid via the action of L-2,4-diaminobutryic acid acetyltransferase (EctA). Lastly, ectoine synthase (EctC) converts Nγ-acetyl-L-2,4-diaminobutyric acid to ectoine (Reshetnikov et al. 2006). In *Mtm. alcaliphilum* 20Z cells, the synthesis of nitrogen-containing osmoprotectants, 5-oxo-1-proline and ectoine, is dependent on the availability of nitrogen in the growth medium (Khmelenina et al. 2000). In nitrogen-deficient environments, *Mtm. alcaliphilum* 20Z cells opt for sucrose synthesis (Khmelenina et al. 2000).

In addition to producing osmoprotectants, some halotolerant methanotrophs employ structural changes to their cell membranes as a way to maintain cell turgor. Previous studies have shown that, as an adaption to high salt concentrations, moderately halophilic methanotrophs increase the proportion of negatively charged phospholipids

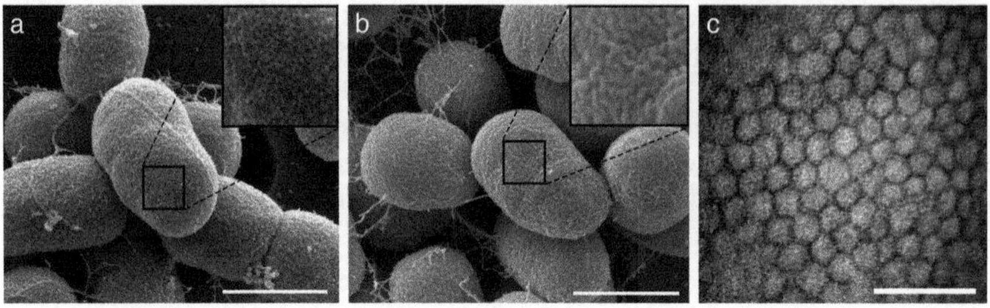

Fig. 2 EM imaging of the S-layers of *Mtm. buryatense* (**a** and **b**) and *Mtm. alcaliphilum* 20Z, 1 μm scale bars. (**a**) SEM image shows S-layer phenotype in wild strain as pits across the cell surface. (**b**) Deletion mutant of a putative S-layer protein METBUDRAFT_4471 results in a naked cell membrane phenotype where the pitting of the cell surface is replaced with ridges on the surface. The mutant strains were constructed by F. Chu (University of Washington). (**c**) SEM image of the cell surface of a *M. alcaliphilum* 20Z cell. The images are courtesy of D. Collins and M.G. Kalyuzhnaya (San Diego State University)

in their cell membranes. The negatively charged phospholipids enhance the selectivity of cation permeability and stabilize the cell membrane (Kates 1986; Trotsenko and Khmelenina 2002). The cell membranes of *M. alcaliphilum* 20Z cells grown at high salinity exhibit a decrease in phosphatidylserine (PS) and phosphatidylethanolamine (PE) levels and an increase in phosphatidylglycerol (PG) and phosphatidylcholine (PC) levels. In contrast, the cell membranes of *M. modestohalophilus* 10S cells show an increase in PG levels and a decrease in PC levels in response to high salinity environments (Khmelenina et al. 1999).

The formation of dedicated surface layers (S-layers) has been also associated with growth at high salinity (Fig. 2; Khmelenina et al. 1999). Bacterial S-layers are glycoprotein layers that cover the external surface of the cell wall. Several S-layer proteins were identified years ago (Shchukin et al. 2011; Khmelenina et al. 2013b); however, the genes encoding the core S-layer proteins were only recently identified (Kalyuzhnaya et al. 2018). Mutation of S-layer proteins has not yet been associated with any growth defects in regular growth media, suggesting that the structures might carry additional yet-to-be-discovered functions (Collins and Kalyuzhnaya unpublished data).

4 Biotechnological Applications

Compared to well-studied, biotechnology workhorses such as *Escherichia coli*, the availability of genetic tools for methanotrophic bacteria is still limited. Nevertheless, a variety of broad-host-range (BHR) vectors and metabolic engineering tools have been developed for methanotrophs (Ojala et al. 2011; Puri et al. 2015; Henard et al. 2016; Mustakhimov et al. 2010; Yan et al. 2016). The most recent advances in metabolic engineering of methanotrophs were summarized by Henard and Guarnieri

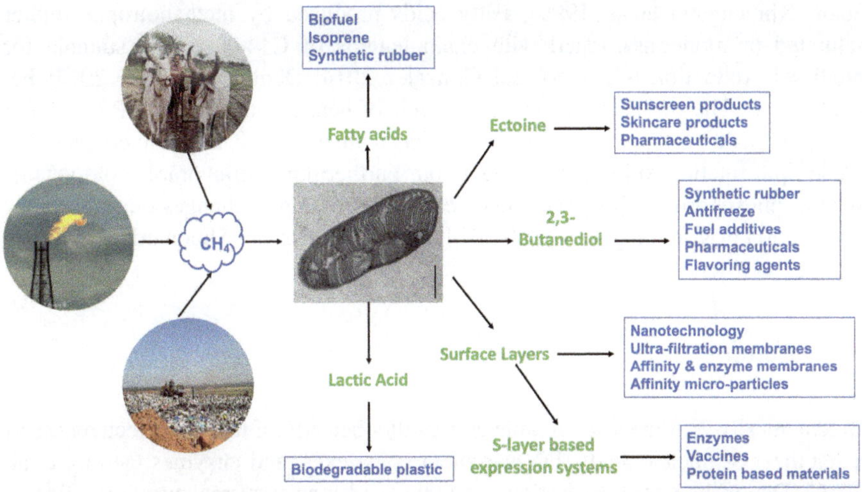

Fig. 3 Summary of biotechnological applications described for *Mtm. alcaliphilum*, a model alkaliphilic methanotrophs

(2018). In general, for genetic manipulation via conjugation, a donor strain such as *E. coli* S17-1 containing the BHR vector of interest is mixed with the recipient methanotrophic strain in a mating medium, followed by transfer of the mated cells to selective medium (Ojala et al. 2011). Optimized electroporation protocols have additionally streamlined genetic alterations in methanotrophic bacteria and opened up an array of novel genetic applications, such as genomic library construction and transposon mutagenesis (Yan et al. 2016; Henard and Guarnieri 2018). Genetic engineering of methanotrophs has further been enhanced by flux-balance modeling (Akberdin et al. 2018a, b; Nguyen et al. 2018). Today, biological engineering and systems biology provide new opportunities for metabolic system modulation and give new optimism for the concept of a methane-based bio-industry. The use of methanotrophs for the production of valuable commercial products (summarized in Fig. 3 and described below) is a very active area of current research.

4.1 *Biofuel*

In some methanotrophs the lipid fraction can make up more than 20% of cell dry weight, potentially serving as a lucrative precursor for biofuel production (Strong et al. 2015, 2016; Kalyuzhnaya et al. 2001). Lipid fractions typically used for biofuel production are rich in triglycerides; however, the lipid fractions from methanotrophs mainly consist of phospholipids from intracytoplasmic membranes. Methanotrophs synthesize two major classes of phospholipids—phosphatidylglycerol (PG) and phosphatidylethanolamine (PE) and its derivatives phosphatidylmethylethanolamine (PME) and phosphatidyldimethylethanolamine (PDME) (Fang et al. 2000; Fei et al.

2014, Khmelenina et al. 1999). Fatty acids produced by methanotrophs (either saturated or monounsaturated with chain lengths of C14–C18) are suitable for biodiesel production (Conrado and Gonzalez 2014; Demidenko et al. 2017; Fei et al. 2018). The presence of increased levels of heteroatoms (namely P and N) in the lipid fraction can pose problems for downstream extraction and conversion of the crude lipid fraction to biodiesel production. Furthermore, unfavorable components such as phosphorus, sulfur, and sugars create processing challenges due to catalyst inactivation or gumming (Fei et al. 2014, 2018; Handler and Shonnard 2018).

4.2 Ectoine

Biochemically, ectoine is a valuable commodity because of its high effectiveness as a stabilizer of nucleic acids, DNA–protein complexes, and enzymes (Strong et al. 2016). Due to its strong hydration properties and long-term moisturizing efficacy, ectoine is also being investigated as an active ingredient for the treatment of moderate atopic dermatitis (Graf et al. 2008; Marini et al. 2014). Furthermore, the protective effects of ectoine are also being evaluated for use in nanoparticle-induced neutrophilic lung inflammation (Sydlik et al. 2009; Unfried et al. 2016). In the pharmaceutical industry, the retail value of ectoine is estimated at US$1000/kg, and its consumption at a global scale is approximated at 15,000 tones per year (Strong et al. 2016). The majority of tested halophilic methanotrophs produced ectoine, a cyclic amino acid, in response to high salt concentrations in the environment (Khmelenina et al. 1999). Currently, the ectoine production is being explored mostly in *M. alcaliphilum* 20Z which can naturally accumulate up to 8% ectoine DCW (Cantera et al. 2018).

4.3 Lactic Acid

Group I methanotrophs can synthesize lactic acid from pyruvate using lactate dehydrogenase (LDH). Lactic acid is a monomeric precursor of biodegradable and renewable plastic polymers called polylactic acids (PLAs) (Rasal et al. 2010). Recently, Henard et al. achieved the bioconversion of methane to lactate by heterologous overexpression of L-LDH from *Lactobacillus helveticus* in *M. buryatense* (Henard et al. 2016). A 5 L fermentation produced 0.8 g lactate/L and yielded 0.05 g lactate/g methane with a productivity of 0.008 g/L/h. Although the maximum lactate titer achieved (at 1.3 g lactate/L) was 100-fold lower than the lactate produced by utilizing pure sugars and metabolically-engineered microbes, the lactate production achieved in this study was higher than the previous attempts using engineered methanotrophic cells. Nonetheless, further metabolic engineering is warranted for conversion of methane to lactate by *M. buryatense* (Henard et al. 2016; Garg et al. 2018).

4.4 2,3-Butanediol

2,3-Butanediol (2,3-BDO), a colorless and odorless liquid with a very high boiling point and low freezing point, is a valuable chemical owing to its wide-range of industrial applications. Derivatives of 2,3-BDO can be used for the production of a wide number of products including synthetic rubber, antifreeze (levo-isomer), fuel additives, cosmetic products, pharmaceutical products, and flavoring agents in food products (Syu 2001; Bialkowska 2016). Microbial production of 2,3-BDO could reduce our dependence on crude oil if methanotrophs could be engineered to produce this valuable chemical from methane. Recent, Nguyen et al. (2018) metabolically engineered *Mtm. alcaliphilum* 20Z for optimal production of 2,3-BDO. The optimized *Mtm. alcaliphilum* 20Z strain has been constructed by knocking out three genes-ldh, ack, and mdh-and introducing 2,3-BDO synthesis genes from *Klebsiella pneumoniae* (budA) and *Bacillus subtilis* (budB) under the control of the Tac promoter. A maximal 2,3-BDO concentration of 86.2 mg/L and a yield of 0.0318 g 2,3-BDO/g methane have been achieved under O_2-limited conditions (Nguyen et al. 2018). However, the maximum titer achieved by Nguyen et al. in a methane-based fed culture was much smaller compared to the approximately 150 g/L titers produced by using sugar-based fed cultures. Thus, further metabolic engineering of methanotrophs to enhance the production of 2,3-BDO from methane is required.

4.5 S-Layers as a New Expression Platform

As described above, it has been speculated that one of the functions of S-layers is to provide extra rigidity to the cell wall to prevent cell lysis in high salinity environments (Strong et al. 2016). Isolated S-layer glycoproteins have an intrinsic ability to self-assemble and recrystallize to form porous semi-permeable membranes (Strong et al. 2015, 2016). This property of S-layers can be exploited for a variety of applications in nanotechnology and biotechnology including the development of ultrafiltration membranes, enzyme sensors, affinity and enzyme membranes, affinity micro-particles, vaccines, and diagnostic devices (Strong et al. 2016; Sleytr and Sara 1997). The genes encoding S-layer proteins and their exporting apparatus in halophilic methanotrophs have been recently identified. Thus far, these genetic discoveries led to the description of a novel platform for heterologous protein expression in methanotrophic bacteria, which offers a reliable and inexpensive route for production of various enzymes and protein-based materials (Kalyuzhnaya et al. 2018).

5 Final Remarks

Halophilic methanotrophic bacteria serve as model systems for improving our knowledge of methane metabolism and bacterial osmoadaptation strategies. New fundamentals of cell structure and metabolic-function interplays have emerged, enhancing our understanding of as well as providing new ideas for biotechnological applications. In addition to being a model system for understanding salt tolerance, *Mtm. alcaliphilum* 20Z also serves as a reliable microbial platform for industrial applications or offsetting the impact of anthropogenic activities. Exciting breakthroughs have been made; however, there are several limiting factors that must be overcome before this technology can be fully implemented. They include identification of the parameters for supporting the activity of methanotrophic bacteria over long periods to allow efficient methane bio-capturing. Industrial applications also await additional developments in mass-transfer optimization and new approaches for oxygen delivery to bioreactors. There are several new initiatives to overcome these challenges, including new strategies for cell immobilization, new reactor designs and the possibility of co-culturing methanotrophs and photosynthetic organisms (Mühlemeier et al. 2018; Hill et al. 2017). Additional molecular engineering of methanotrophs, as well as coupling with electrocatalysis, might potentially provide the answers we are looking for as well.

Acknowledgments The work was supported by the DOE under FOA DE-FOA-0001085 and by NSF-CBET award 1605031. Authors thank David Collins for providing EM images.

References

Akberdin IR, Collins DA, Hamilton R, Oshchepkov DY, Shukla AK, Nicora CD, Nakayasu ES, Adkins JN, Kalyuzhnaya MG (2018a) Rare earth elements alter redox balance in *Methylomicrobium alcaliphilum* $20Z^R$. Front Microbiol 9:2735

Akberdin IR, Thompson M, Hamilton R, Desai N, Alexander D, Henard CA, Guarnieri MT, Kalyuzhnaya MG (2018b) Methane utilization in *Methylomicrobium alcaliphilum* $20Z^R$: a systems approach. Sci Rep 8:2512

Antony CP, Kumaresan D, Hunger S, Drake HL, Murrell JC, Shouche YS (2013) Microbiology of Lonar Lake and other soda lakes. ISME J 7:468–476

Banta AB, Wei JH, Welander PV (2015) A distinct pathway for tetrahymanol synthesis in bacteria. Proc Natl Acad Sci U S A 112:13478–13483

Bialkowska AM (2016) Strategies for efficient and economical 2,3-butanediol production: new trends in this field. World J Microbiol Biotechnol 32(12):200

Boden R, Cunliffe M, Scanlan J, Maussard H, Kits KD, Koltz MG, Jetten MS, Vuilleumier S, Han J, Peters L, Mikhailova N, Teshima H, Tapia R, Kyrpides N, Ivanova N, Pagani I, Cheng JF, Goodwin L, Han C, Hauser L, Land ML, Lapidus A, Lucas S, Pitluck S, Woyke T, Stein L, Murrell JC (2011) Complete genome sequence of the aerobic marine methanotroph *Methylomonas methanica* MC09. J Bacteriol 193(24):7001–7002

Boetius A, Wenzhöfer F (2013) Seafloor oxygen consumption fuelled by methane from cold seeps. Nat Geosci 6:725–734

Bowman JP, Sly LI, Nichols PD, Hayward AC (1993) Revised taxonomy of the methanotrophs: description of *Methylobacter* gen. nov., emendation of *Methylococcus*, validation of *Methylosinus* and *Methylocystis* species, and a proposal that the family *Methylococcaceae* includes only the group I methanotrophs. Int J Syst Bacteriol 43:735–753

Bowman JP, McCammon SA, Skerratt JH (1997) *Methylosphaera hansonii* gen. nov., sp. nov., a psychrophilic, group I methanotroph from Antartic marine-salinity, meromictic lakes. Microbiology 143:1451–1459

Burdette MD (2013) Production of biodiesel-like components by the Type I methanotroph *Methylomonas methanica*. Master's thesis. Clemson University, Clemson, SC

But SY, Rozova ON, Khmelenina VN, Reshetnikov AS, Trotsenko YA (2012) Properties of recombinant ATP-dependent fructokinase from the halotolerant methanotroph *Methylomicrobium alcaliphilum* 20Z. Biochem Mosc 77(4):372–377

But SY, Khmelenina VN, Reshetnikov AS, Mustakhimov II, Kalyuzhnaya MG, Trotsenko YA (2015) Sucrose metabolism in halotolerant methanotroph *Methylomicrobium alcaliphilum* 20Z. Arch Microbiol 197(3):471–480

But SY, Egorova SV, Khmelenina VN, Trotsenko YA (2017) Biochemical properties and phylogeny of hydroxypyruvate reductases from methanotrophic bacteria with differenent C_1-assimilation pathways. Biochem Mosc 82(11):1295–1303

But SY, Egorova SV, Khmelenina VN, Trotsenko YA (2018) Serine-glyoxylate aminotransferases from methanotrophs using different C1-assimilation pathways. Antonie Van Leeuwenhoek 112 (5):741–751. https://doi.org/10.1007/s10482-018-1208-4

Cantera S, Sánchez-Andrea I, Lebrero R, García-Encina PA, Stams AJM, Muñoz R (2018) Multi-production of high added market value metabolites from diluted methane emissions via methanotrophic extremophiles. Bioresour Technol 267:401–407

Chu F, Lidstrom ME (2016) XoxF acts as the predominant methanol dehydrogenase in the type I methanotroph *Methylomicrobium buryatense*. J Bacteriol 198(8):1317–1325

Conrado RJ, Gonzalez R (2014) Envisioning the bioconversion of methane to liquid fuels. Science 343(6171):621–623

De la Torre A, Metivier A, Chu F, Laurens LM, Beck DA, Pienkos PT, Lidsrom ME, Kalyuzhnaya MG (2015) Genome-scale metabolic reconstructions and theoretical investigation of methane conversion in *Methylomicrobium buryatense* strain 5G(B1). Microb Cell Factories 14:188

Demidenko A, Akberdin IR, Allemann M, Allen EE, Kalyuzhnaya MG (2017) Fatty acid biosynthesis pathways in *Methylomicrobium buryatense* 5G(B1). Front Microbiol 7:2167

Deng YW, Ro SY, Rosenzweig AC (2018) Structure and function of the lanthanide-dependent methanol dehydrogenase XoxF from the methanotroph *Methylomicrobium buryatense* 5GB1C. J Biol Inorg Chem 23(7):1037–1047

Detkova EN, Boltyanskaya YV (2007) Osmoadaptation of haloalkaliphilic bacteria: role of osmoregulators and their possible practical application. Microbiology 76(5):581–593

Fang J, Barcelona MJ, Semrau JD (2000) Characterization of methanotrophic bacteria on the basis of intact phospholipid profiles. FEMS Microbiol Lett 189(1):67–72

Fei Q, Guarnieri MT, Tao L, Laurens LM, Dowe N, Pienkos PT (2014) Bioconversion of natural gas to liquid fuel: opportunities and challenges. Biotechnol Adv 32(3):596–614

Fei Q, Puri AW, Smith H, Dowe N, Pienkos PT (2018) Enhanced biological fixation of methane for microbial lipid production by recombinant *Methylomicrobium buryatense*. Biotechnol Biofuels 11:129

Flynn JD, Hirayama H, Sakai Y, Dunfield PF, Klotz MG, Knief C, Op den Camp HJ, Jetten MS, Khmelenina VN, Trotsenko YA, Murrell JC, Semrau JD, Svenning MM, Stein LY, Kyrpides N, Shapiro N, Woyke T, Bringel F, Vuilleumier S, DiSpirito AA, Kalyuzhnaya MG (2016) Draft genome sequences of gammaproteobacterial methanotrophs isolated from marine ecosystems. Genome Announc 4(1):e01629–15

Fu Y, Li Y, Lidstrom M (2017) The oxidative TCA operates during methanotrophic growth of the Type I methanotroph *Methylomicrobium buryatense* 5GB1. Metab Eng 42:43–51

Fuse H, Ohta M, Takimura O, Murakami K, Inoue H, Yamaoka Y, Oclarit JM, Omori T (1998) Oxidation of trichloroethylene and dimethyl sulfide by a marine *Methylomicrobium* strain containing soluble methane monooxygenase. Biosci Biotechnol Biochem 62(10):1925–1931

Galinski EA, Truper HG (1982) Betaine, a compatible solute in the extremely halophilic phototrophic bacterium *Ectothiorhodospira halochloris*. FEMS Microbiol Lett 13:357–360

Galinski EA, Pfeiffer H, Truper HG (1985) 4,5,6-Tetrahydro-2-methyl-4-pyrimidinecarboxylic acid. Eur J Biochem 149(1):135–139

Garg S, Clomburg JM, Gonzalez R (2018) A modular approach for high-flux lactic acid production from methane in an industrial medium using engineered *Methylomicrobium buryatense* 5GB1. J Ind Microbiol Biotechnol 45(6):379–391

Gilman A, Fu Y, Hendershott M, Chu F, Puri AW, Smith AL, Pesesky M, Lieberman R, Beck DAC, Lidstrom ME (2017) Oxygen-limited metabolism in the methanotroph *Methylomicrobium buryatense* 5GB1C. Peer J 5:e3945

Graf R, Anzali S, Buenger J, Pfluecker F, Driller H (2008) The multifunctional role of ectione as a natural cell protectant. Clin Dermatol 26(4):326–333

Hamilton R, Kits KD, Ramonovskaya VA, Rozova ON, Yurimoto H, Iguchi H, Khmelenina VN, Sakai Y, Dunfield PF, Klotz MG, Knief C, Op den Camp HJ, Jetten MS, Bringel F, Vuilleumier S, Svenning MM, Shapiro N, Woyke T, Trotsenko YA, Stein LY, Kalyuzhnaya MG (2015) Draft genomes of gammaproteobacterial methanotrophs isolated from terrestrial ecosystems. Genome Announc 3(3):e00515-15

Handler RM, Shonnard DR (2018) Environmental life cycle assessment of methane biocatalysis: key considerations and potential impacts. In: Kalyuzhnaya MG, Xing XH (eds) Methane biocatalysis: paving the way to sustainability. Springer, Berlin, pp 253–270

Hanson RS, Hanson TE (1996) Methanotrophic bacteria. Microbiol Rev 60(2):439–471

Henard CA, Guarnieri MT (2018) Metabolic engineering of methanotrophic bacteria for industrial biomanufacturing. In: Kalyuzhnaya MG, Xing XH (eds) Methnane biocatalysis: paving the way to sustainability. Springer, Berlin, pp 117–132

Henard CA, Smith H, Dowe N, Kalyuzhnaya MG, Pienkos PT, Guarnieri MT (2016) Bioconversion of methane to lactate by an obligate methanotrophic bacterium. Sci Rep 6:21585

Henard CA, Smith HK, Guarnieri MT (2017) Phosphoketolase overexpression increases biomass and lipid yield from methane in an oligate methanotrophic biocatalyst. Metab Eng 41:152–158

Heyer J, Berger U, Hardt M, Dunfield PF (2005) Methylohalobius crimeensis gen. nov., sp. nov., a moderately halophilic, methanotrophic bacterium isolated from hypersaline lakes of Crimea. Int J Syst Evol Microbiol 55:1817–1826

Hill EA, Chrisler WB, Beliaev AS, Bernstein HC (2017) A flexible microbial co-culture platform for simultaneous utilization of methane and carbon dioxide from gas feedstocks. Bioresour Technol 228:250–256

Hirayama H, Suzuki Y, Abe M, Miyazaki M, Makita H, Inagaki F, Uematsu K, Takai K (2011) *Methylothermus subterraneus* sp. nov., a moderately thermophilic methanotroph isolated from a terrestrial subsurface hot aquifer. Int J Syst Evol Microbiol 61:2646–2653

Hirayama H, Hiroyuki F, Abe M, Miyazaki M, Nakamura T, Nunoura T, Furushima Y, Yamamoto H, Takai K (2013) Methylomarinum vadi gen. nov., sp. nov., a methanotroph isolated from two distinct marine environments. Int J Syst Evol Microbiol 63:1073–1082

Hirayama H, Abe M, Miyazaki M, Nunoura T, Furushima Y, Yamamoto H, Takai K (2014) *Methylomarinovum caldicuralii* gen. nov., sp. nov., a moderately thermophilic methanotroph isolated from a shallow submarine hydrothermal system, and proposal of the family *Methylothermaceae* fam. nov. Int J Syst Evol Microbiol 64:989–999

IPCC (2014) Climate change 2014: impacts, adaptation, and vulnerability. http://www.ipcc.ch/report/ar5/wg2/

Kalyuzhnaya MG (2016a) Methane biocatalysis: selecting the right microbe. In: Eckert C, Trinh CT (eds) Biotechnologies for biofuel production and optimization. Elsevier, Oxford, pp 353–383

Kalyuzhnaya MG (2016b) Methylomicrobium. Bergey's manual of systematics of archaea and bacteria. pp 1–12. https://doi.org/10.1002/9781118960608.gbm01182.pub2

Kalyuzhnaya MG, Khmelenina VN, Starostina NG, Baranova SV, Suzina NE, Trotsenko YA (1998) New moderately halophilic methanotrophic of the genus *Methylobacter*. Microbiology 67:438–444

Kalyuzhnaya MG, Khmelenina VN, Lysenko AM, Suzina NE, Trotsenko YA (1999) New methanotrophic isolates from soda lakes of the southern Transbaikal region. Microbiology 68 (5):677–685

Kalyuzhnaya MG, Khmelenina V, Eshinimaev B, Suzina N, Nikitin D, Solonin A, Lin JL, McDonald I, Murrell C, Trotsenko Y (2001) Taxonomic characterization of new alkaliphilic and alkalitolerant methanotrophs from soda lakes of the southeastern Transbaikal region and description of *Methylomicrobium buryatense* sp. nov. Syst Appl Microbiol 24(2):166–176

Kalyuzhnaya MG, Khmelenina V, Eshinimaev B, Sorokin D, Fuse H, Lidstrom M, Trotsenko Y (2008) Classification of halo (alkali) philic and halo (alkali) tolerant methanotrophs provisionally assigned to the genera *Methylomicrobium* and *Methylobacter* and emended description of the genus *Methylomicrobium*. Int J Syst Evol Microbiol 58:591–596

Kalyuzhnaya MG, Puri AW, Lidstrom ME (2015) Metabolic engineering in methanotrophic bacteria. Metab Eng 29:142–152

Kalyuzhnaya MG, Yang S, Rozova ON, Smalley NE, Clubb J, Lamb A, Gowda GA, Raftery D, Fu Y, Bringel F, Vuilleumier S, Beck DA, Trotsenko YA, Khmelenina VN, Lidstrom ME (2013) Highly efficient methane biocatalysis revealed in a methanotrophic bacterium. Nat Commun 4:2785

Kalyuzhnaya MG, Demidenko O, Collins DA (2018) Compositions and methods using methanotrophic S-layer proteins for expression of heterologous proteins. U.S. provisional patent application no. USSN 62/551:502

Kates M (1986) Influence of salt concentration on membrane lipids of halophilic bacteria. FEMS Microbiol Rev 39(1–2):95–101

Khmelenina VN, Kalyuzhnaya MG, Starostina NG, Suzina NE, Trotsenko YA (1997) Isolation and characterization of halotolerant alkaliphilic methanotrophic bacteria from Tuva Soda lakes. Curr Microbiol 35(5):257–261

Khmelenina VN, Kalyuzhnaya MG, Sakharovsky VG, Suzina NE, Trotsenko YA, Gottschalk G (1999) Osmoadaptation in halophilic and alkaliphilic methanotrophs. Arch Microbiol 172 (5):321–329

Khmelenina VN, Sakharovskii VG, Reshetnikov AS, Trotsenko YA (2000) Synthesis of osmoprotectants by halophlic and alkaliphilic methanotrophs. Microbiology 69(4):381–386

Khmelenina VN, Beck DAC, Munk C, Davenport K, Daligault H, Erkkila T, Goodwin L, Gu W, Lo CC, Scholz M, Teshima H, Xu Y, Chain P, Bringel F, Vuilleumier S, Dispirito A, Dunfield P, Jetten MS, Klotz MG, Knief C, Murrell JC, Op den Camp HJ, Sakai Y, Semrau J, Svenning M, Stein LY, Trotsenko YA, Kalyuzhnaya MG (2013a) Draft genome sequence of *Methylomicrobium buryatense* strain 5G, a haloalkaline-tolerant methanotrophic bacterium. Genome Announc 1(4):e00053-13

Khmelenina VN, Suzina NE, Trotsenko YA (2013b) Surface layers of methanotrophic bacteria. Microbiology 82(5):515–527

Khmelenina VN, Rozova ON, Akberdin IR, Kalyuzhnaya MG, Trotsenko YA (2018) Pyrophosphate-dependent enzymes in methanotrophs: new findings and views. In: Kalyuzhnaya MG, Xing XH (eds) Methane biocatalysis: paving the way to sustainability. Springer, Berlin, pp 83–98

Kim HG, Kim SW (2006) Purification and characterization of a methanol dehydrogenase derived from *Methylomicrobium* sp. HG-1 cultivated using a compulsory circulation diffusion system. Biotechnol Bioprocess Eng 11:134–139

Li X, Roberti R, Blobel G (2015) Structure of an integral membrane sterol reductase from *Methylomicrobium alcaliphilum*. Nature 517(7532):104–107. https://doi.org/10.1038/nature13797

Lidstrom ME (1988) Isolation and characterization of marine methanotrophs. Antonie Van Leeuwenhoek 54:189–199

Marini A, Reinelt K, Krutmann J, Bilstein A (2014) Ectoine-containing cream in the treatment of mild to moderate atopic dermatitis: a randomised, comparator-controlled, intra-individual double-blind, multi-center trial. Skin Pharmacol Physiol 27(2):57–65

Mühlemeier MI, Speight R, Strong PJ (2018) Biogas, bioreactors and bacterial methane oxidation. In: Kalyuzhnaya MG, Xing X (eds) Methnane biocatalysis: paving the way to sustainability. Springer, Berlin, pp 213–235

Mustakhimov II, Reshetnikov AS, Glukhov AS, Khmelenina VN, Kalyuzhnaya MG, Trotsenko YA (2010) Identification and characterization of EctR1, a new transcriptional regulator of the ectoine biosynthesis genes in the halotolerant methanotroph *Methylomicrobium alcaliphilum* 20Z. J Bacteriol 192(2):410–417

Mustakhimov II, Rozova ON, Solntseva NP, Khmelenina VN, Reshetnikov AS, Trotsenko YA (2017) The properties and potential metabolic role of glucokinase in halotolerant obligate methanotroph *Methylomicrobium alcaliphilum* 20Z. Antonie Van Leeuwenhoek 110 (3):375–386

Nguyen AD, Hwang IY, Lee OK, Kim D, Kalyuzhnaya MG, Mariyana R, Hadiyati S, Kim MS, Lee EY (2018) Systematic metabolic engineering of *Methylomicrobium alcaliphilium* 20Z for 2,3-butanediol production form methane. Metab Eng 47:323–333

Ojala DS, Beck DA, Kalyuzhnaya MG (2011) Genetic systems for moderately halo(alkali)philic bacteria of the genus *Methylomicrobium*. Methods Enzymol 495:99–118

Orata FD, Meier-Kolthoff JP, Sauvageau D, Stein LY (2019) Phylogenomic analysis of the gammaproteobacterial methanotrophs (order *Methylococcales*) calls for the reclassification of members at the genus and species levels. Front Microbiol 9:3162

Petersen JM, Dubilier N (2009) Methanotrophic symbioses in marine invertebrates. Environ Microbiol Rep 1(5):319–335

Puri AW, Owen S, Chu F, Chavkin T, Beck DA, Kalyuzhnaya MG, Lidstrom ME (2015) Genetic tools for the industrially promising methanotroph *Methylomicrobium buryatense*. Appl Environ Microbiol 81:1775–1781

Rasal RM, Janorkar AV, Hirt DE (2010) Poly(lactic acid) modifications. Prog Polym Sci 35:338–356

Reshetnikov AS, Mustakhimov II, Khmelenina VN, Trotsenko YA (2005) Cloning, purification, and characterization of diaminobutyrate acetyltransferase from the halotolerant methanotroph *Methylomicrobium alcaliphilum* 20Z. Biochemistry 70(8):878–883

Reshetnikov AS, Khmelenina VN, Trotsenko YA (2006) Characterization of the ectoine biosynthesis genes of haloalkalotolerant obligate methanotroph "*Methylomicrobium alcaliphilum* 20Z". Arch Microbiol 184(5):286–297

Ro SY, Ross MO, Deng YW, Batelu S, Lawton TJ, Hurley JD, Stemmler TL, Hoffman BM, Rosenzweig AC (2018) From micelles to bicelles: effect of the membrane on particulate methane monooxygenase activity. J Biol Chem 293(27):10457–10465

Rozova ON, Khmelenina VN, Vuilleumier S, Trotsenko YA (2010) Characterization of recombinant pyrophosphate-dependent 6-phophofructokinase from halotolerant methanotroph *Methylomicrobium alcaliphilum* 20Z. Res Microbiol 161(10):861–868

Rozova ON, Khemelenina VN, Gavletdinova JZ, Mustakhimov II, Trotsenko YA (2015a) Acetate kinase-an enzyme of the postulated phosphoketolase pathway in *Methylomicrobium alcaliphilum* 20Z. Antonie Van Leeuwenhoek 108(4):965–974

Rozova ON, Khmelenia VN, Bocharova KA, Mustakhimov II, Trotsenko YA (2015b) Role of NAD+-dependent malate dehydrogenase in the metabolism of *Methylomicrobium alcaliphilum* 20Z and *Methylosinus trichosporium* OB3b. Microorganisms 3(1):47–59

Rozova ON, But SY, Khmelenina VN, Reshetnikov AS, Mustakhimov II, Trotsenko YA (2017) Characterization of two recombinant 3-hexulose-6-phophate synthases from the halotolerant obligate methanotroph *Methylomicrobium alcaliphilum* 20Z. Biochemistry 82(2):176–185

Semrau JD, DiSpirito AA, Yoon S (2010) Methanotrophs and copper. FEMS Microbiol Rev 34:496–531

Sharp CE, Smirnova AV, Kalyuzhnaya MG, Bringel F, Hirayama H, Jetten MS, Khmelenina VN, Klotz MG, Knief C, Kyrpides N, Op den Camp HJ, Reshetnikov AS, Sakai Y, Shapiro N, Trotsenko YA, Vuilleumier S, Woyke T, Dunfield PF (2015) Draft genome sequence of the moderately halophilic methanotroph *Methylohalobius crimeensis* strain 10Ki. Genome Announc 3(3):e00644-15

Shchukin VN, Khmelenina VN, Eshinimayev BT, Suzina NE, Trotsenko YA (2011) Primary characterization of dominant cell surface proteins of halotolerant methanotroph *Methylomicrobium alcaliphilum* 20Z. Microbiology 80(5):595–605

Shindell D, Kuylenstierna JCI, Vignati E, Dingenen R, Amann M, Klimont Z, Anenberg S, Muller N, Janssens-Maenhout G, Raes F, Schwartz J, Faluvegi G, Pozzoli L, Kupiainen K, Höglund-Isaksson L, Emberson L, Streets D, Ramanathan V, Hicks K, Oanh NT, Milly G, Williams M, Demkine V, Fowler D (2012) Simultaneously mitigating near-term climate change and improving human health and food security. Science 335(6065):183–189

Sieburth JM, Johnson PW, Eberhardt MA, Sieracki ME, Lidstrom M, Laux D (1987) The first methane-oxidizing bacterium from the upper mixing layer of the deep ocean: *Methylomonas pelagica* sp. nov. Curr Microbiol 14(5):285–293

Sleytr WB, Sara M (1997) Bacterial and archaeal S-layer proteins: structure-function relationships and their biotechnological applications. Trends Biotechnol 15(1):20–26

Song J, Cho KK, Lee KS, La YH, Kalyuzhnaya M (2016) Method for producing isoprene using recombinant halophilic methanotroph. Patent no. US9994869B2

Sorokin DY, Jones BE, Kuenen JG (2000) An obligate methylotrophic, methane-oxidizing *Methylomicrobium* species form a highly alkaline environment. Extremophiles 4(3):145–155

Sorokin DY, Gorlenko VM, Namsaraev BB, Namsaraev ZB, Lysenko AM, Eshinimaev BT, Khmelenina VN, Trotsenko YA, Kuenen JG (2004) Prokaryotic communities of the north-eastern Mongolian soda lakes. Hydrobiologia 522(1–3):235–248

Strong PJ, Xie S, Clarke WP (2015) Methane as a resource: can the methanotrophs add value? Environ Sci Technol 49(7):4001–4018

Strong PJ, Kalyuzhnaya M, Silverman J, Clarke WP (2016) A methanotroph-based biorefinery: potential scenarios for generating multiple products from a single fermentation. Bioresour Technol 215:314–323

Sydlik U, Gallitz I, Albrecht C, Abel J, Krutmann J, Unfried K (2009) The compatible solute ectoine protects against nanoparticle-induced neutrophilic lung inflammation. Am J Respir Crit Care Med 180(1):29–35

Syu MJ (2001) Biological production of 2,3-butanediol. Appl Microbiol Biotechnol 55(1):10–18

Tavormina PL, Hatzenpichler R, McGlynn S, Chadwick G, Dawson KS, Cannon SA, Orphan VJ (2015) *Methyloprofundus sedimenti* gen. nov., sp. nov., an obligate methanotroph from ocean sediment belonging to the 'deep sea-1' clade of marine methanotrophs. Int J Syst Evol Microbiol 65:251–259

Trotsenko YA, Khmelenina VN (2002) The biology and osmoadaptation of haloalkaliphlic methanotrophs. Microbiology 71(2):123–132

Tsubota J, Eshinimaev BT, Khmelenina VN, Trotsenko YA (2005) *Methylothermus thermalis* gen. nov., sp. nov., a novel moderately thermophilic obligate methanotroph from a hot spring in Japan. Int J Syst Evol Microbiol 55:1877–1884

Unfried K, Kramer U, Sydlik U, Autengruber A, Bilstein A, Stolz S, Marini A, Schikowski T, Keymel S, Krutmann J (2016) Reduction of neutrophilic lung inflammation by inhalation of the compatible solute ectoine: a randomized trial with elderly individuals. Int J Chron Obstruct Pulmon Dis 11:2573–2583

Vuilleumier S, Khmelenina VN, Bringel F, Reshetnikov AS, Lajus A, Mangenot S, Rouy Z, Op den Camp HJ, Jetten MS, Dispirito AA, Dunfield P, Klotz MG, Semrau JD, Stein LY, Barbe V, Médigue C, Trotsenko YA, Kalyuzhnaya MG (2012) Genome sequence of the haloalkaliphilic methanotrophic bacterium *Methylomicrobium alcaliphilum* 20Z. J Bacteriol 194(2):551–552

Wise MG, McArthur JV, Shimkets LJ (2001) *Methylosarcina fibrata* gen. nov., sp. nov. and *Methylosarcina quisquiliarum* sp. nov., novel type I methanotrophs. Int J Syst Evol Microbiol 51:611–621

Yan X, Chu F, Puri AW, Fu Y, Lidstrom ME (2016) Electroporation-based genetic manipulation in type I methanotrophs. Appl Environ Microbiol 82:2062–2069

Zavarzin GA, Zhilina TN, Kevbrin VV (1999) The alkaliphilic microbial community and its functional diversity. Microbiology 68(5):503–552

Metabolic Engineering of Methanotrophs for the Production of Chemicals and Fuels

Ok Kyung Lee, Diep T. N. Nguyen, and Eun Yeol Lee

Contents

Abstract Methane is a promising next-generation carbon feedstock for industrial biotechnology because it is inexpensive and abundant carbon. Biological conversion of methane to valuable products can reduce greenhouse gas (GHG) emissions caused by methane. Recently, genetic manipulation techniques and systems biology has provided new opportunities for metabolic engineering of methanotrophs and engineered strains have been employed as potential industrial strains for methane gas fermentation. For commercialization of the production of chemicals and fuels from methane, methanotrophs need to be further engineered based on rational metabolic engineering strategy to enhance carbon conversion yield, titer, and productivity. In this chapter, recent advances on metabolic engineering of

O. K. Lee · D. T. N. Nguyen · E. Y. Lee (✉)
Department of Chemical Engineering, Kyung Hee University, Yongin-si, South Korea
e-mail: okblessyou@khu.ac.kr; eunylee@khu.ac.kr

E. Y. Lee (ed.), *Methanotrophs*, Microbiology Monographs 32,
https://doi.org/10.1007/978-3-030-23261-0_6

methanotrophs, including genetic tool development, strategy to enhance carbon pool for product conversion, practical example of methane bioconversion, and prospect on the engineered methanotrophic cells as a cell-factory platform are discussed.

1 Introduction

Methane is a powerful greenhouse gas and has 84 times more global warming potential than carbon dioxide (Greenhouse Gas Emissions 2017). On the other hand, methane, the primary component of natural gas and biogas has attracting lots of attention as a next-generation feedstock for industrial biotechnology due to its abundance, low price and a high degree of reduction per carbon (Hwang et al. 2015). The chemical conversion process of methane requires a lot of energy input because of a very high activation energy of the C–H bond, and hence, the installation cost per unit process is very high (Haynes and Gonzalez 2014). On the contrary, biological methane conversion using microorganisms or enzymes as biocatalysts is considered to be more sustainable than chemical methods because bioprocess is operated under ambient temperature and pressure. In addition, the carbon conversion efficiency of methane to methanol oxidation using chemical process was less than 20–50%, but the conversion efficiency using methanotrophs was over 50% (Table 1), representing that biological methane conversion has advantages over chemical methane conversion. For this reason, methanotrophs have recently been considered as a promising industrial strain for bioconversion of methane into

Table 1 Comparison of the carbon conversion efficiencies for biocatalysis and conventional chemical catalysis

	Catalyst	Pressure (atm)	Temperature (°C)	Oxidant	Methane conversion (%)
Chemical conversion	Hg(II)	–	180	O_2	50
	Au + additives ($H_2SeO_4/SeO_4/O_2$)	27 bar	180	SO_3	3–28
	$PMo_{11}V$	1 atm	700–750	O_2	3–13
	$PMo_{11}Fe$				4–23
	$SiMo_{11}Fe$				4–32
	ZSM-5	30.5 bar	50	H_2O_2	0.3
Bioconversion	*M. trichosporium* OB3b	1	30	O_2	64
	Methanotrophic consortium	1	30	O_2	43–80
	M. trichosporium OB3b	1	30	O_2	73.8–75.2

Taken from Lee et al. (2016). Copyright 2016 Wiley, England

Table 2 Pros and cons of methane bioconversion using methanotroph and non-native methanotroph

Classification	Methanotroph	Non-native methanotroph	Technical issues	References
Methane conversion	High efficiency	Not available	Heterologous expression of MMO, p450 oxidase	Lloyd et al. (1999), Zilly et al. (2011), Meinhold et al. (2005)
Methanol conversion	High efficiency	Low efficiency	High K_m and low k_{cat} value of NAD^+-dependent methanol dehydrogenase, thermodynamic constraint depending on temperature	Van Ophem et al. (1993), Anthony and Ghosh (1998), Witthoff et al. (2015), Whitaker et al. (2017)
Formaldehyde assimilation pathway	RuMP, serine pathway	RuMP, formolase based pathway	Limitation of C5 sugar regeneration in a non-native host	Siegel et al. (2015), Wang et al. (2017), Muller et al. (2015), Whitaker et al. (2015, 2017)
Genetic tools and knowledge-based system	Limited	Well-developed	Lack of vector system and gene transformation method, existence of a knowledge gap	Baani and Liesack (2008), Crombie and Murrell (2011), Puri et al. (2015)

Taken from Hwang et al. (2018). Copyright 2018 Springer, Germany

chemicals and fuels. Nonetheless, slow growth rate, lack of efficient genetic engineering tools, and poor understanding of the physiology and metabolic regulation of C1 assimilation hamper their application as biocatalysts to produce chemicals and fuels on a commercial scale (Lee et al. 2016).

As another possible option to convert methane and methanol into value-added products, metabolic reconstructions of methane and methanol oxidation steps in non-native hosts such as *Escherichia coli*, *Corynebacterium glutamicum* and *Saccharomyces cerevisiae*, have been conducted in order to take advantages of their fast growth rate and availability of highly efficient genetic manipulation techniques (Zilly et al. 2011; Meinhold et al. 2005; Witthoff et al. 2015; Whitaker et al. 2017; Siegel et al. 2015; Wang et al. 2017; Muller et al. 2015). However, the unsuccessful attempts to demonstrate the methane oxidation in non-native hosts and the low efficiency of methanol assimilation have been a major barrier for methane conversion using synthetic methanotrophy (Table 2). Thus, with respect to methane bioconversion, employment of methanotrophs as a biocatalyst is being considered a realistic option for commercialization, based on the fact that methane can only be activated by methane monooxygenase (MMO).

Over the past decade, much effort has been devoted to understand the physiology and metabolic regulation based on genome sequencing analysis and systems biology-assisted analyses of transcripts, metabolites, and fluxomes in type I and II methanotrophs (Matsen et al. 2013; Yang et al. 2013; Vorobev et al. 2014; Larsen and Karlsen 2015). Type I methanotrophs, *Methylomicrobium buryatense* 5GB, *Methylomicrobium alcaliphilum* 20Z and *Methylomonas* sp. DH-1 have attracted growing interest as fascinating model strains for methane bioconversion because of their relatively rapid doubling time (2–4 h) compared to other methanotrophs (Garg et al. 2018; Nguyen et al. 2018a). In addition, *M. buryatense* 5GB, *M. alcaliphilum* 20Z and *Methylomonas* sp. DH-1 have been used as platform strains to produce various valuable products thanks to the development of genetic manipulation tools (Yan et al. 2016; Nguyen et al. 2018b, 2019) and genome scale model (GSM) (Torre et al. 2015; Akberdin et al. 2018). Recent reports on the production of lactic acid and 2,3-butanediol (2,3-BDO) using engineered *M. buryatense* 5GB1 and *M. alcaliphilum* 20Z have demonstrated that methanotrophs have potential as promising industrial hosts for the production of valuable products from methane (Henard et al. 2016; Nguyen et al. 2018b).

In this chapter, metabolic engineering of methanotrophs for the production of chemicals and fuels from methane is comprehensively overviewed. First, we review genetic tool development for gene manipulation of methanotrophs and basics on systems biology-assisted understanding of their physiology and metabolic regulation. The methane utilization route and the target product biosynthesis one are different depending on the types of methanotroph species. In addition, there is a difference in the types of transformation method and vector systems that can be employed in different types of methanotrophs. Thus, it is necessary to determine an appropriate methanotrophic host strain according to target product, availability of the vector system, and transformation method. To achieve high carbon conversion efficiency and high productivity, methanotrophs need to be metabolically engineered using redesigned biosynthetic pathways and strategies to enhance the synthesis of metabolic precursors that can be used for target products. In order to address these points, the second section of this chapter focuses on how to enhance the accumulation of pyruvate and acetyl-CoA as the starting materials for target products, and on how to control TCA cycle and methylerythritol 4-phosphate (MEP) nodes that can supply many important precursors to be readily converted to other valuable chemicals through the rational metabolic redesign. Finally, we have summarized various examples of the production of valuable methane-derived products using metabolically engineered methanotrophs to date. The examples include various organic acids, alcohols, biodiesel, carotenoids, etc. and they are classified based on key precursors and metabolic pathways.

2 Genetic Tools for Metabolic Engineering of Methanotrophs

Development of superior methnotrophic strain requires efficient genetic tools for easy and quick metabolic engineering of strains. Over the past 20 years, many researchers have developed various vector systems and gene transfer techniques for genetic engineering of methanotrophs. A variety of case are shown below to demonstrate how various genetic tools can be employed to perform metabolic engineering of methanotrophs.

2.1 Gene Transfer Techniques

Although efficient genetic manipulation for methanotrophs is still limited as compared to that for *E. coli*, some genetic tools for methanotrophs have been developed and used already for decades. Foreign DNA is transferred to methanotrophs using broad-host-range (BHR) vectors via conjugation and electroporation. Conjugation is the most commonly used method for gene transfer in methanotrophs. The BHR vector in an *E. coli* donor strain, such as S17-1, is transferred to methanotroph recipient strains by conjugation on the mating plate followed by selective plates. Helper strains containing the plasmid with the transferred gene, such as pRK2013, can be used for triparental mating. Depending on the strains and plasmids used, as shown in Table 3, different transfer frequencies from 10 to 10^9 have been observed. Conjugation has been successfully applied to genetic manipulation regardless of the type I and type II methanotroph, but this method requires time-consuming screening to remove *E. coli* after conjugation (Puri et al. 2015). In contrast, the introduction of foreign DNA into methanotrophs via electroporation offers many advantages over conjugation; procedure is simple and rapid, and homologous recombination by transformation of linear DNA into methanotrophs is possible. Previously, electroporation tools have been mainly developed and used for type II methanotrophs, such as *Methylocystis* sp. SC2 and *Methylocella silvestris* BL2 (Baani and Liesack 2008; Crombie and Murrell 2011). Recently, electroporation techniques with high transformation efficiency have been developed for four different species of type I methanotrophs: *M. buryatense* 5G(B1), *Methylomonas* sp. LW13, *Methylobacter tundripaludum* 21/22, and *M. alcaliphilum* 20Z (Yan et al. 2016; Nguyen et al. 2018b). Transformation time of *Methylosinus trichosporium* OB3b was shortened by optimizing the medium composition and electroporation conditions (Ro and Rosenzweig 2018). Gene deletion/integration, site-specific recombination, and sacB counterselection have been successfully performed using electroporation tools. The gene transfer techniques and efficiencies for each strain are shown in Table 3.

2.2 Vector Systems for Methanotrophs

Replication vectors for the incompatibility (Inc) group IncP-1, commonly referred to IncP, have been used for heterologous gene expression in methanotrophs (Table 4), as well as pBBR and IncQ replication vectors. Plasmids containing a reporter gene have been successfully used for promoter probing in some methanotrophs. Based on

Table 3 Efficiency comparison of genetic transfer technologies

	Species	Plasmid	Selection marker	Transfer frequency	References
Conjugation	*Methylomonas 16a*	pSUSMK derives	Kan	N/A	Ye and Kelly (2012), Ye et al. (2007)
		pRK2013	Kan	N/A	Tao et al. (2007)
		pBHR1 and derives	Kan	N/A	Tao et al. (2007)
		pDCQ333	Orange (red) color	N/A	Sharpe et al. (2007)
	M. album or	pVK100, pVK100Sc	Kan, Tet	1.2×10^7	Lloyd et al. (1999)
	M. parvus	pVK104		9.2×10^9	
	M. capsulatus	pDSK509	Kan	10^3	Murrell (1992)
		pKPS	Gen, Kan	N/A	Csáki et al. (2003)
		pULB113	Kan	10^6	Murrell (1992)
	M. trichosporium	pGSS33	Kan	10^5	
		pVK100	Tet, Kan	10^3	Murrell (1992)
		pBR325	Kan	5×10^8	Martin and Murrell (1995)
		pTJS175	Spe, Str, Kan	N/A	Smith and Murrell (2011), Smith et al. (2002)
	M. silvestris	pMHA203	Kan	N/A	Crombie and Murrell (2011), Theisen et al. (2005)
Electroporation	*Methylomonas sp. LW13*	pAWP89 (linear DNA fragments)	Kan	1.1×10^5	Yan et al. (2016)
	M. tundripaludum 21/22	pAWP89 (linear DNA fragments)	Kan	2.3×10^3	
	M. buryatense 5G	pAWP89 (linear DNA fragments)	Kan	N/D	
	M. alcaliphilum 20Z	pAWP89 (linear DNA fragments)	Kan	1.2×10^2	Nguyen et al. (2018b)

(continued)

Table 3 (continued)

	Species	Plasmid	Selection marker	Transfer frequency	References
	M. silvestris BL2	pCM184 (linear DNA fragments)	Kan	9.6×10	Crombie and Murrell (2011)
	M. trichosporium	linear DNA fragments	Gen	N/A	Ro and Rosenzweig (2018)

this system, constitutive promoters such as *tac*, *lac*, *mxaF*, *hps*, and *rpoD* in methanotrophs have been characterized. An inducible vector with a tetracycline promoter/operator was constructed for heterologous gene expression in *M. buryatense*, and was used for the production of lactate (Henard et al. 2016) and fatty acids (Demidenko et al. 2017) from methane. Gene integration and deletion for methanotrophs were developed using either the homologous recombination method or the SacB-based selection system. Homologous recombination vectors such as pCM184 contained an antibiotic-resistant cassette with the loxP flanked sequence, and the antibiotic markers were removed by Cre recombinase (Marx and Lidstrom 2001). Unmarked allelic exchanges using the SacB counterselection system have been also used for methanotrophs such as *Methylococcus capsulatus* Bath, *Methylomonas* sp. 16a, *M. buryatense* and *M. alcaliphilum* (Csaki et al. 2003; Ye and Kelly 2012; Puri et al. 2015; Nguyen et al. 2018b).

2.3 Systems Biology-Based Understanding of Methanotrophs' Physiology

Systems biology tools are important to ensure insight into the basic physiology of strains before creating the metabolic engineered ones. Fragmental genetic manipulation without sufficient understanding of the overall metabolism can result in metabolic imbalance or cell growth inhibition due to depletion of important precursors, redox imbalance, or accumulation of toxic intermediates. High-throughput omics analysis, such as genome, transcriptome, metabolome, and fluxome, aid in the characterization of methanotroph on multi levels, and therefore provide a "debugging" capability for metabolic engineering of methanotrophs. Genomics and transcriptomics data provide significant information on key enzymes and core metabolic pathways for methane utilization, including transcription levels of particulate and soluble methane monooxygenase (sMMO), periplasmic pyrroloquinoline quinone (PQQ)-dependent methanol dehydrogenase (MDH), the tetrahydromethanopterin (H_4MPT) pathway, NAD-dependent formate dehydrogenase (FDH), all the essential genes in the ribulose monophosphate (RuMP), serine cycle, TCA cycle, and in the ethylmalonyl-CoA (EMC) pathway (Fig. 1). In addition, a clear difference in the core metabolic fluxes

Table 4 Selected vectors used for metabolic engineering in methanotrophs

	Vector name	Replicon	Selectable marker	Species	Note	References
Replication	pVK100	IncP	Tc, Km	*M. trichosporium* OB3b; *Methylomicrobium album* BG8		Lloyd et al. (1999)
						Nguyen and Chan (2003)
	pHM1	IncQ	Sm, Km	*M. trichosporium* OB3b		Lloyd et al. (1999)
	pBHR1	pBBR1	Km, Cm	*Methylomonas* sp. 16a		Sharpe et al. (2007)
	pSRK-Km	pBBR1	Km	*M. capsulatus* Bath		Welander and Summons (2012)
	pMHA200	IncP	Km	*M. silvestris* BL2		Theisen et al. (2005)
	pAWP78	IncP	Km	*M. buryatense* 5G	Minimized IncP-based vector	Puri et al. (2015)
Expression	pAWP89	IncP, pTac promoter	Km	*M. buryatense* 5G	Promoter probe containing dTomato reporter gene	Puri et al. (2015)
	pAWP88	IncP, lac promoter	Km	*M. buryatense* 5G	Promoter probe containing dTomato reporter gene	
	pAWP87	IncP, mxaF promoter	Km	*M. buryatense* 5G	Promoter probe containing dTomato reporter gene	
	pAWP92	IncP, rpoD promoter	Km	*M. buryatense* 5G	Promoter probe containing dTomato reporter gene	
	pMHA200_Phps	IncP, hps promoter	Km	*M. buryatense* 5G	Promoter probe containing GFP reporter gene	Mustakhimov et al. (2016)
	pCAH01::emGFP	IncP, pTetA promoter	Km	*M. buryatense* 5G	Tetracycline-inducible promoter probe containing emGFP	Henard et al. (2016)
	pMHA203	IncP, sigma54 promoter	Km	*M. silvestris* BL2	Promoter probe containing GFP reporter gene	Theisen et al. (2005)

	pJN105_Peredox-mCherry	Broad-host range, P_{BAD} promoter	Gen	*M. capsulatus* Bath	Arabinose-inducible promoter containing peredox-mCherry	Ishikawa et al. (2017)
Insertion/deletion	pK18mobsacB		Km	*M. capsulatus* Bath	SacB counterselection for unmarked allelic exchange	Csaki et al. (2003)
	pJQ200SK		Gm	*M. capsulatus* Bath	SacB counterselection for unmarked allelic exchange	Welander and Summons (2012)
	pSUKSM		Km	*Methylomonas* sp. 16a	SacB-based integration vector	Ye et al. (2007)
	pUTmTn5		Ap	*Methylomonas* sp. 16a	Promoterless transposable element based on the Tn5 transposon	Sharpe et al. (2007)
	pCM184		Km	*M. silvestris* BL2	Cre/lox system for gene knockout using marker exchange	Crombie and Murrell (2014)
				M. alcaliphilum 20Z		Ojala et al. (2011)
	pCM433kanT		Km	*M. buryatense* 5G	SacB counterselection for unmarked allelic	Puri et al. (2015)

Adapted from Hwang et al. (2018). Copyright 2018 Springer, Germany

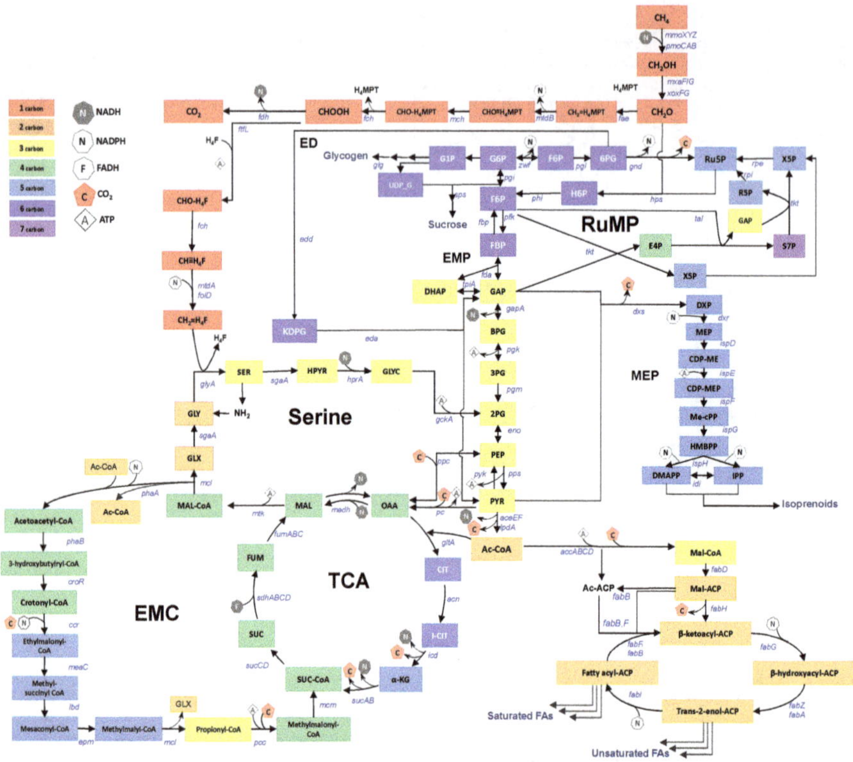

Fig. 1 Core metabolic pathway and key enzymes of methanotrophs. Endogenous metabolic pathway of methanotroph mapping products (metabolites in boxes) and the genes (italicized) of the enzymes responsible for the reactions based on type I and II methanotroph. The RuMP and PPP pathway are present in type I methanotroph. The EMC pathway is only found in II methanotroph. The enzymes name of the genes are as follows: *mmoXYZ*, soluble methane monooxygenase subunit; *pmoCAB*, particulate methane monooxygenase gene cluster; *mxaFIG*, *xoxFG*, methanol dehydrogenase; *fae*, formaldehyde activating enzyme; *mtdB*, methylene tetrahydromethanopterin dehydrogenase; *mch*, methenyl tetrahydromethanopterin cyclohydrolase; *ftr*, formyltransferase; *fmd*, formylmethanofuran dehydrogenase; *ftfL*, formate-tetrahydrofolate ligase; *fch*, methylenetetrahydrofolate dehydrogenase/methenyl-tetrahydrofolate cyclohydrolase; *mtdA/folD*, methylenetetrahydrofolate dehydrogenase; *ald*, aldehyde dehydrogenase; *fdh*, formate dehydrogenase; *hps*, 3-hexulose-6-phosphate synthase; *phi*, 6-phospho-3-hexuloisomerase; *pfk*, 6-phosphofructokinase; *fbp*, *fda*, fructose-bisphosphate aldolase; *tpiA*, triosephosphate isomerase; *gapA*, glyceraldehyde-3-phosphate dehydrogenase; *pgk*, phosphoglycerate kinase; *pgm*, phosphoglucomutase; *eno*, enolase; *pps*, phosphoenolpyruvate synthase; *pgi*, glucose-6-phosphate isomerase; *zwf*, glucose-6-phosphate 1-dehydrogenase; *pgl*, 6-phosphogluconolactonase; *gnd*, 6-phosphogluconate dehydrogenase; *edd*, phosphogluconate dehydratase; *eda*, KDPG aldolase; *tkt*, transketolase; *tal*, transaldolase; *rpi*, Ribose-5-phosphate isomerase; *rpe*, ribulose-phosphate 3-epimerase; *pyk*, pyruvate kinase; *aceE*, *aceF*, *lpdA*, pyruvate dehydrogenase complex; *pc*, pyruvate carboxylase; *gltA*, citrate synthase; *acnA*, aconitase; *icd*: isocitrate dehydrogenase; *sucAB*, 2-oxoglutarate dehydrogenase complex; *sucCD*, succinyl-CoA ligase; *sdhABCD*, succinate dehydrogenase; *fumABC*, fumarate hydratase; *madh*, malate dehydrogenase; *glyA*, serine hydroxymethyltransferase; *sgaA*, serine-glyoxylate aminotransferase; *hprA*, glycerate dehydrogenase; *gckA*, hydroxypyruvate reductase; *ppc*, phosphoenolpyruvate carboxylase; *mtk*, malate thiokinase; *accABCD*, acetyl-CoA carboxylase; *fabD*, malonyl CoA-acyl carrier protein transacylase; *fabH*, 3-oxoacyl-[acyl-carrier-protein] synthase; *fabG*, 3-oxoacyl-[acyl-carrier-protein] reductase; *fabZ*, *A*, 3-hydroxyacyl-[acyl-carrier-protein]

of type I and II methanotrophs can be identified, which can be used to select strains for the production of value-added compounds and to identify engineering targets for rational metabolic engineering. Type I methanotroph, *M. buryatense* 5G possesses relatively highly expressed RuMP cycle genes (Torre et al. 2015). Type II methanotrophs have a relatively high expression of serine cycle-related genes such as serine-glyoxylate aminotransferase (*sgaA*), serine hydroxymethyl transferase (*glyA*), and malyl-CoA/beta-methylmalyl-CoA lyase (*mclA*) genes, together with others serine, EMC, and TCA pathway-involved genes that are closely related to each other (Matsen et al. 2013). This information indicates that the target products should be different for type I and type II methanotrophs. Valuable final products need to be linked with metabolic intermediates derived from relatively highly expressed core metabolic genes.

Recently, Nguyen et al. (2018a) reported that the new isolated type I methanotroph, *Methylomonas* DH-1 has a complete set of TCA cycle enzymes, including functional 2-oxyglutarate dehydrogenase (*sucA*) based on genome analysis and enzyme activity analysis. Transcriptome analyses (GEO accession number: GSE101494) also revealed that most of the genes in TCA cycle are expressed at a relatively intermediate level. In addition, the conserved phosphoenolpyruvate carboxylase (PEPC) and pyruvate carboxylase (PC) genes in *Methylomonas* DH-1 may enhance the TCA flux by direct conversion of phosphoenolpyruvate (PEP) or pyruvate to oxaloacetate (OAA). Combined analyses of genomic and transcriptomic data from *Methylomonas* sp. DH-1 demonstrated that the DH-1 strain has high potential as an excellent succinate producer.

Vorobev et al. (2014) suggested that the core metabolic fluxes were changed depending on type of carbon source. When cultured on ethanol, transcription data of *Methylocystis* sp. strain SB2 showed a marked decrease in the expression of methane oxidation and serine cycle, and a marked increase in the expression of TCA cycle-related genes.

Metabolome and fluxome analyses can identify the core metabolites or altered metabolic flux distribution, providing the basis for understanding the physiology of methanotrophs and gene targets for metabolic engineering. Type I methanotrophs have been reported to have an incomplete TCA cycle, but recently, the complete TCA cycle of *M. buryatense* 5G on methane was identified by ^{13}C analysis (Fu et al. 2017). Kalyuzhnaya et al. (2013) have attempted a system-level approach including genome-based transcriptomic studies, metabolomics, and ^{13}C-labeled distribution

Fig. 1 (continued) dehydratase; *fabI*, enoyl-[acyl-carrier-protein] reductase; *fabF*, *B*, *F*, 3-oxoacyl-[acyl-carrier-protein] synthase; *dxs*, 1-deoxy-D-xylulose 5-phosphate synthase; *dxr*, 1-deoxy-D-xylulose 5-phosphate reductoisomerase; *ispD*, 2-C-methyl-D-erythritol 4-phosphate cytidylyltransferase; *ispE*, 4-(cytidine 5′-diphospho)-2-C-methyl-D-erythritol kinase; *ispF*, 2-C-methyl-D-erythritol 2,4-cyclodiphosphate synthase; *ispG*, 4-hydroxy-3-methylbut-2-enyl diphosphate synthase; *ispH*, 4-hydroxy-3-methylbut-2-enyl diphosphate reductase; *phaA*, β-ketothiolase; *phaB*, acetoacetyl-CoA reductase; *croR*, crotonase; *ccr*, crotonyl-CoA carboxylase/reductase; *meaC*, mesaconyl-CoA hydratase; *ibh*, methylsuccinyl-CoA dehydrogenase; *epm*, mesaconyl-CoA epimerase; *mcl*, methylmalyl-CoA lyase; *pcc*, propionyl-CoA carboxylase; *mcm*, methylmalonyl-CoA mutase

analysis for comprehensive understanding of C1 assimilation in *M. alcaliphilum* strain 20Z. The distribution of the downstream intermediates of the RuMP pathway was identified by using ^{13}C-labeled methane. Pyruvate derived from the EMP pathway was six-fold higher than the pyruvate derived from the Entner–Doudoroff (EDD) pathway in *M. alcaliphilum* strain 20Z. In addition, a novel mode of fermentation process has been discovered in producing organic acids such as formate, acetate, succinate, and lactate from methane under oxygen-limiting conditions. This result might be due to the decreased ATP production in response to redox imbalance under oxygen limitation, thus resulting in a change in the carbon flux from biomass to organic acid production. In this way, multi-omics information on methanotrophs can provide basic inputs for generating metabolic engineering strategy, including identification of key metabolic precursors, metabolic bottlenecks, and the selection of a suitable strain.

2.4 Genome-Scale Metabolic Models for Methanotrophs

The metabolic models at genome scale and *in silico* simulations are important predictive tools in metabolic engineering as the engineering point of view changes from cell components to system-level behavior. Genome sequencing and automatic annotation tools were used to construct a genome-scale metabolic model for three kinds of methanotrophs such as *M. buryatense* 5G, *M. alcaliphilum* 20Z, and *M. capsulatus* (Table 5) (Torre et al. 2015; Demidenko et al. 2017; Akberdin et al. 2018; Lieven et al. 2018). The constraint-based reconstruction and analysis (COBRA) is used to predict metabolic flux redistribution after genetic manipulation, or to predict other cell functions such as changes in expression profiles, substrate preference and adaptive evolution outcomes. It is useful for redesigning the pathway to get the desired target product. The iMb5GB1 genome-scale model (GSM) was used to improve yields of fatty acid biosynthesis (Demidenko et al. 2017). This model successfully simulated gene overexpression and multiple gene knockout, and an engineered strain with 20% higher fatty acid content was developed. Recently, the i20ZR-BDO model was reconstructed by including heterogeneous reactions of 2,3-BDO biosynthesis into the GSM model of *M. alcaliphilum* 20Z (Nguyen et al. 2018b). By solving the flux balance analysis (FBA), the maximum theoretical yield of 2,3-BDO was calculated to be 0.809 g/g. Computational framework based on FBA and MOMA, Optknock was developed to identify gene deletion targets of unnecessary competitive pathways to improve 2,3-BDO productivity. Acetate kinase (ACK), lactate dehydrogenase (LDH), pyruvate dehydrogenase (PDH), and malate dehydrogenase (MADH) knockouts were expected to increase 2,3-BDO productivity. ACK, LDH, and MADH-deficient triple mutants produced 68.8 mg/L 2,3-BDO from methane in flask culture with 20% improved productivity. The difference between the simulation result and the experimental data for 2,3-BDO production might be due to the limitation of mass transfer of methane. Currently, most of the GSMs for methanotrophs have stoichiometric information, but they do

Table 5 Summary of GSMs for methane-utilizing bacteria

Organism	Model	Genes	Reactions	Metabolites	Use	References
M. buryatense 5G	iMb5G(B1)	–	841	–	Investigation of methane oxidation process by flux balance analysis in *M. buryatense* 5G	Torre et al. (2015)
M. buryatense 5G	iMb5GB1 update	314	402	403	Update the genome-scale metabolic network of *M. buryatense* 5G to use for determining the control point of fatty acid biosynthesis pathway	Demidenko et al. (2017)
M. alcaliphilum 20Z	iIA407	407	433	423	Highlighted the importance of substitution of ATP-linked steps with PPi-dependent reactions, the presence of a carbon shunt from Acetyl-CoA to the pentose-phosphate and highly branched TCA cycle in *M. alcaliphilum* 20Z	Akberdin et al. (2018)
M. capsulatus	iMcBath	730	898	877	Theoretical investigation of methane utilization network in *M. capsulatus*	Lieven et al. (2018)

Taken from Kabimoldayev et al. (2018). Copyright 2018 Oxford University Press, England

not include kinetic and regulatory information; thus, the integration of this information can improve the accuracy and predictability of these models.

3 Metabolic Engineering Strategies for Core Methane Metabolism in Methanotrophs

To date, the production of valuable chemicals from methane using methanotrophs show low volumetric productivity. A rational metabolic engineering strategy needs to be generated for more efficient methane bioconversion using methanotrophs as biocatalysts. The elimination of by-product formation pathways and the redesign of more efficient synthetic routes that can improve the carbon conversion efficiency of methane through optimal redirection of carbon fluxes need to be accomplished.

In order to achieve high titer, productivity, and yield of the target product, the strain must accumulate metabolic precursors corresponding to the desired target product, through the deletion of unnecessary pathways and by reconstructing more efficient pathway. In this part, this approach is discussed based on the categorization of important metabolic precursors such as pyruvate and acetyl-CoA and important metabolic pathways such as the RuMP cycle and MEP ones.

3.1 Redesign Strategy of RuMP Pathway to Improve Carbon Conversion Efficiency

Increase of carbon flux through phosphoketolase (PKT) pathway has been proposed as an effective approach for metabolic engineering. The reconstruction of this pathway has been successfully used to increase the yield of acetyl-CoA-derived products in microbial biocatalysts, including *E. coli* (Bogorad et al. 2013), *C. glutamicum* (Chinen et al. 2007), and *S. cerevisiae* (Henard et al. 2015; Bergman et al. 2016; Meadows et al. 2016). PKT is an enzyme that generates acetyl phosphate (AcP) from fructose 6-phosphate (F6P) or xylulose-5-phosphate (X5P) with concurrent formation of erythrose-4-phosphate (E4P) or glyceraldehyde-3-phosphate (G3P), respectively. Many of PKTs exhibit dual substrate specificity, but these PKT variants have been typically reported to have a higher activity for X5P. The intermediate AcP can be converted to acetyl-CoA by phosphate acetyltransferase (PTA) or acetate kinase (ACK) and acetyl-CoA synthetase (ACS) (Fig. 2). Usefully, the genomes of methanotrophs often include *pkt*, *ackA*, *acs*, and *pta* genes, but no substrate specificity of PKT has been reported yet. The incorporation of PKT in methanotrophs has several advantages. The RuMP cycles of methanotrophs use ribose-5-phosphate (Ru5P) to assimilate formaldehyde, and E4P or G3P generated by PKT is then rearranged to regenerate Ru5P, resulting in an increased Ru5P pool. Another advantage is that the RuMP-PKT pathway avoids the use of ATP. This

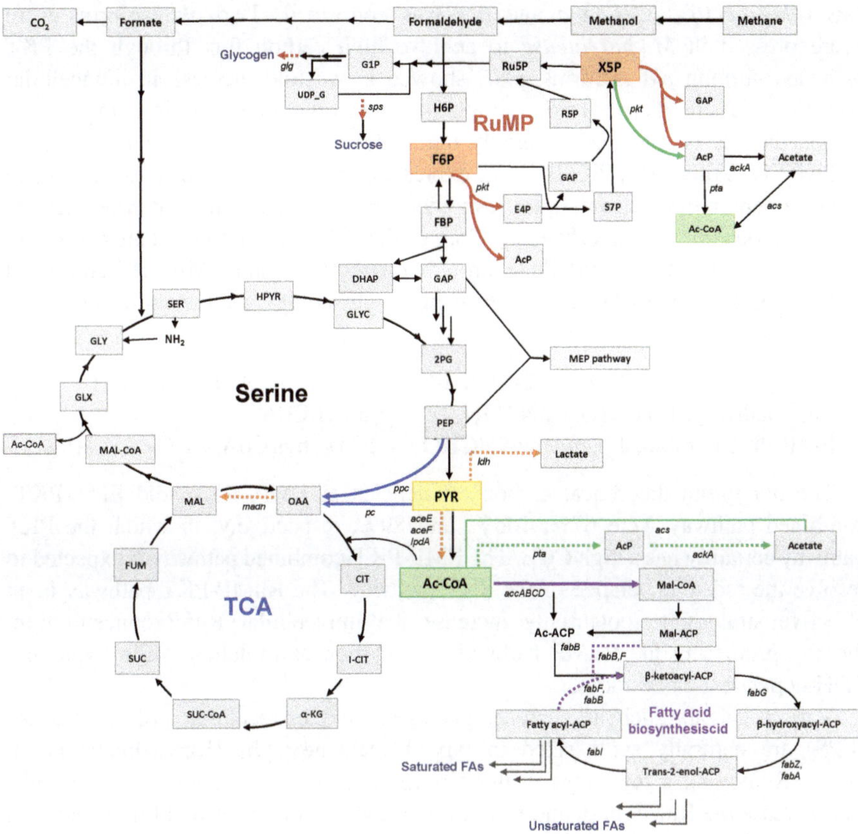

Fig. 2 Metabolic engineering strategies to achieve high carbon conversion efficiency and high productivity of target products from central carbon metabolism of methane utilization in methanotrophs. Red arrows represent strategies for improving carbon conversion efficiency. Orange and green arrows represent strategies for accumulation of pyruvate and acetyl-CoA, respectively. Blue and purple arrows indicate strategies for enhancing the accumulation of the TCA cycle and malonyl-CoA, respectively. Solid arrows indicate overexpression, and dotted arrows indicate knock-out

pathway avoids the decarboxylation of EMP-derived pyruvate and thus achieves the generation of acetyl-CoA without carbon loss. Torre et al. (2015) first conducted a simulation experiment to confirm whether the introduction of X5P/F6P PKT into a methanotroph could affect carbon conversion efficiency using an *in-silico* model (ΔPDH knockout). In this model, all of the acetyl-CoA was assumed to be produced through the PKT pathway, resulting in a deficiency of intracellular NADH. Total carbon conversion efficiency (CCE) decreases as C1 carbon is oxidized to CO_2 to satisfy NADH cellular demand. Overexpression of *pkt* was attempted to improve the biomass and lipid yield in *M. buryatense* (Henard et al. 2017). When *M. buryatense* grew on methane or methanol, the expression of a complete PKT pathway including *pkts* (*pktA*, *pktB*), *ackA*, *pta* and *acs* was confirmed. Two *pkt* isoforms were overexpressed in *M. buryatense* to achieve high carbon flux through the PKT pathway. Among *pkt* isoforms, *pktB* showed a two-fold increase in intracellular acetyl-CoA. As predicted *in silico* simulation, the oxidation process from methane to carbon dioxide is enhanced to supply intracellular NADH when a complete PKT pathway is expressed. pPKTB strain expressing *pktB* also led to a 2.6-fold yield enhancement in biomass and lipids from methane as compared to wild-type, increasing the carbon conversion efficiency. Garg et al. (2018) have proposed the maximum theoretical yield of crotonyl-CoA through EMP, PKT, and EMP-PKT-combined pathway based on knowledge of *M. buryatense* physiology and metabolic networks, as follows:

- EMP pathway: 6HCHO → 1Crotonyl-CoA + $2CO_2$ + 2ATP + 3NADH
- PKT pathway: 12HCHO + 3NADH → 3Crotonyl-CoA
- EMP-PKT combined pathway: 9HCHO → 2Crotonyl-CoA + CO_2 + ATP

The maximum theoretical carbon efficiencies of EMP, PKT, and EMP-PKT-combined pathway were 67%, 100%, and 89%, respectively, in which the PKT pathway certainly has a high CCE. The EMP-PKT-combined pathway is expected to resolve the redox imbalance of the PKT pathway. The RuMP-PKT pathway is an effective strategy to potentially increase the intracellular Ru5P concentration, thereby promoting the RuMP pathway and carbon assimilation while bypassing PDH-dependent carbon loss.

Various sugars such as sucrose, glycogen, and extracellular polysaccharides (EPS) are naturally synthesized in type I methanotrophs (Kalyuzhnaya et al. 2015). Removing a few steps in the biosynthesis pathway of these carbon sinks can increase the RuMP pool and turn it into a carbon source needed for biofuel and chemical biosynthesis. Among type I methanotrophs, *Methylomicrobium* species can accumulate up to 35% glycogen content under high cell density culture conditions when using methane or methanol as a carbon source (Gilman et al. 2015; Fei et al. 2018). Under these growth conditions, the carbon flux for cell division and membrane synthesis of *Methylomicrobium* species was redirected to glycogen accumulation. Carbon flux could be used for the biomass or target products by limiting the synthetic route of by-products such as glycogen (Henard et al. 2018). 5GB1 Δ*glgA*1 Δ*glgA*2 mutant with the removal of glycogen synthesis in *M. buryatense* 5GB1 was constructed with no effect on cell growth, as compared to the wild type under low density culture conditions (Puri et al. 2015). However, 6.4% of glycogen was still produced at high cell density culture conditions. Since methanotrophs can synthesize glycogen through sucrose degradation, a new strain AP18 was created by deleting the gene encoding the sucrose-6-phosphate synthase (*sps*) in addition to the knockout of glycogen synthase genes of *glgA*1 and *glgA*2 (Fig. 2; Fei et al. 2018). The AP18 strain accumulated no detectable glycogen and showed a 90% increase in lipid content compared to wild type was observed. Thus, the carbon for glycogen synthesis in the wild-type was redirected for biomass production of the AP18 strain using pathway redesign.

To achieve high product yield and high CCE with engineered methanotrophs, strains should accumulate metabolic precursors corresponding to the desired target compounds. For example, the accumulation of pyruvate is suitable for lactate and 2,3-BDO. The accumulation of acetyl-CoA is suitable for 1,4-BDO and succinate. The strategies for enhancing important precursor accumulation are closely related to the productivity improvement of target products. The following section focuses on the strategy of increasing the core metabolic precursors, such as pyruvate, acetyl-CoA, and deoxy-D-xylulose 5-phosphate (DXP), which can be used as starting materials for valuable chemicals.

3.2 Redesign Strategy of Enhancing Pyruvate Pool

Pyruvate is an intermediate metabolite in the RuMP-EMP pathway in methanotrophs and an important chemical compound for chemical synthesis and biosynthesis of valuable products. Pyruvate can be converted to lactate, acetate, or acetyl-CoA by the native pathway, and it can also be transformed to useful chemicals such as acetoin, 2,3-BDO via a non-native synthetic pathway. Enhancement of pyruvate has been one of the main subjects in industrial biotechnology so far. The simplest strategy for earning metabolic engineering strains with improved pyruvate production is to eliminate genes of another competitive pathway, such as the acetyl-CoA, lactate, formate, or the acetate pathway (Fig. 2).

Pyruvate-derived products can be efficiently produced by enhancing pyruvate pool. Type I methanotrophs, *M. alcaliphilum* 20Z, naturally produce lactate from pyruvate by LDH. Henard et al. (2018) reported that an increase in lactate flux was achieved by reducing pyruvate conversion to acetyl-CoA by PDH. The PDH mutant increased the lactate titer and specific productivity by two- to threefold and fourfold when compared to wild-type *M. alcaliphilum* 20Z, respectively. Another example is 2,3-BDO production using engineered *M. alcaliphilum* 20Z. Nguyen et al. (2018b) constructed a triple mutant based on *in silico* knockout simulations to direct more flux from the pyruvate node into 2,3-BDO formation. The mutant strain with no LDH, ACK, and MADH improved the 2,3-BDO productivity by 1.68-fold compared to the wild-type. The elimination of LDH and ACK could improve the pyruvate node, and the deletion of LDH and MADH could effectively increase 2,3-BDO productivity by the supply from the NADH pool.

3.3 Redesign Strategy of Enhancing Acetyl-CoA Pool

Acetyl-CoA is a key metabolic intermediate generated by pyruvate dehydrogenase complex (PDHc). Acetyl-CoA can be converted into a variety of industrial biotechnology products, including fatty acids, organic acid, polyhydroxyalkanoate, 1-butanol, and vitamins. To improve the yield of these products, strategies for increasing the acetyl CoA node are necessary. Acetyl-CoA node can be increased through overexpression of genes contributing to the acetyl-CoA synthesis pathway, which has been successfully applied to *E. coli* and yeast. The increase of flux from pyruvate to acetyl-CoA by overexpression of PDH complex (*aceE*, *aceF*, *lpdA*) in *E. coli* improved the yield of 1-butanol. The expression of *pkt* in yeast improved the acetyl-CoA, which resulted in a 1.6-fold improvement in FAME yield as compared to the reference strain. Studies on the improvement of acetyl-CoA by PDH overexpression in methanotrophs have not been reported yet, but the improvement in acetyl-CoA by *pkt* overexpression has been successfully applied (Fig. 2). The *M. buryatense* expressing *pkt*B isoform increased the intracellular acetyl-CoA concentration by two-fold, and therefore, enhanced fatty acid production. Particularly, the fatty acids produced by wild-type and pPKT strains were 4.2% and 9% of the biomass, respectively. Obviously, the incorporation of PKT could improve both acetyl-CoA and fatty acid productivity.

Another strategy to improve the acetyl-CoA node is the deletion of a competitive pathway, such as the PTA-ACK pathway (Fig. 2). Although deletion of the PTA-ACK pathway provides an increase in cellular levels of acetyl-CoA, this mutation inhibited cell growth. The overexpression of *acs* has been used to increase the synthesis of acetyl-CoA from acetate. These strategies were applied to *M. buryatense* 5GB1C for crotonate production. Garg et al. (2018) attempted to synthesize crotonate from acetyl-CoA node using an engineered reverse β-oxidation pathway in *M. buryatense* 5GB1C. The carbon flux lost by acetate was redirected to improve crotonate productivity. The PTA elimination inhibited acetate formation, but did not improve crotonate productivity. Overexpression of *acs* led to a twofold increase in productivity by recycling acetate to acetyl-CoA (Fig. 2).

Acetyl-CoA is a molecule for the TCA cycle through condensation with OAA. Under sufficient oxygen-supply conditions, acetyl-CoA is mainly oxidized through the TCA cycle to generate energy and to produce organic acids and various amino acids. The increase in carbon flux through acetyl-CoA alone in the TCA cycle does not expand the pool of constituent metabolites such as OAA, malate, and succinate. Thus, TCA cycle requires the supply of carbon flux from other supply routes, PEP and pyruvate. Enhancement of TCA flux can be achieved by overexpression of *ppc* and *pc* by conversion of PEP or pyruvate to OAA (Fig. 2). Methanotrophs have partial serine or complete serine cycles, and the serine cycle includes PPC, which is an advantage for improving the TCA cycle (Torre et al. 2015). *Methylomonas* sp. DH-1 naturally has both *ppc* and *pc*, which can support an increase of the TCA cycle flux and thus, is considered as a potent strain for the production of TCA cycle-derived products (Nguyen et al. 2018a).

3.4 Understanding of MEP Pathway of Methanotrophs

Secondary metabolite production has attracted much attention to produce pharmaceuticals, next-generation fuels, and etc. Microorganisms have either of MEP or mevalonate (MVA) pathway. Many types of methanotrophs have MEP pathway, which contains seven enzymes such as 1-deoxy-D-xylulose 5-phosphate (DXP) synthase (DXS, dxs), DXP reductoisomerase (DXR, *IspC*), 4-diphosphocytidyl-2-C-methyl-D-erythritol synthase (CDP-MES, *IspD*), 4-disphophocytidyl-2-C-methyl-D-erythritol kinase (CDP-MEK, *IspE*), 2-C-methyl-D-erythritol 2,4-cyclodiphosphate synthase (or ME-cPPS, *IspF*), 1-hydroxy-2-methyl-2-(E)-butenyl 4-diphosphate synthase or HMBPP synthase (HMBPPS, *IspG*), 1-hydroxy-2-methyl-butenyl 4-diphosphate reductase or HMBPP reductase (HMBPPR, *IspH*), and isopentenyl diphosphate isomerase (IPPI, *idi*) (Fig. 1; Lee et al. 2016). Pyruvate and G3P are condensed to DXP, subsequently converted to isopentenyl diphosphate (IPP) via a multi-step reaction, which serves as a basic isoprene building block for the synthesis of isoprenoids. A sufficient pool of key precursors can lead to enhancement in the MEP node. Intracellular concentrations of pyruvate, G3P, and acetyl-CoA of *M. alcaliphilum* 20Z have been reported (Akberdin et al. 2018). The pools of pyruvate and G3P were approximately 150- and 30-fold higher than that of acetyl-CoA pool. The theoretical maximum IPP yield was reported to be higher in the MEP pathway than in the MVA pathway based on glucose (Niu et al. 2017). Methanotrophs are also expected to generate higher IPP yield through the MEP pathway. This suggests that the MEP pathway in methanotrophs is a suitable route to produce isoprenoid-related products. In contrast, the acetyl-CoA pool of *E. coli* is much higher than those of pyruvate and G3P, and thus reconstruction of heterologous MVA pathway in *E. coli* resulted in higher diterpene yield as compared to the native MEP pathway. The production of isoprenoids using MEP pathway of methanotrophs is described in the following section.

4 Metabolic Engineering of Methanotrophs and Application for the Production of Chemicals and Fuels

With the use of rational metabolic engineering strategies, engineered methanotrophs are being developed for the production of chemicals and biofuels. In this section, developments of metabolic engineered methanotrophs are reviewed based on pyruvate, acetyl-CoA, TCA, and MEP pathway-derived products.

4.1 Pyruvate-Derived Products

Pyruvate is an important precursor for biosynthesis of valuable compounds. Recently, several approaches have been successfully developed for the production of valuable products using pyruvate as a starting precursor in methanotrophs.

4.1.1 Production of Lactic Acid

Lactic acid is one of the most important organic acids being used in a variety of industrial and biotechnological applications, such as matrices for drug delivery systems, environmental-friendly biodegradable plastics, food and health applications (Hyon 2000; Vroman and Tighzert 2009; Rydz et al. 2014; Pawar et al. 2014; Patel and Prajapat 2013). Therefore, metabolic engineering of methanotrophic bacteria for lactate production from methane is pivotal for industrial application.

Methanotrophs can produce lactic acid using LDH for converting pyruvate into lactate. Calysta Co. has issued a patent which focused on engineering methanotrophic bacteria for producing lactic acid from methane or methanol. Several bacterial LDHs were cloned into vector pMS3 and expressed in the methanotrophic host to verify the functional expression of LDH (Saville et al. 2016). Vector pMS3 contained *mxaF* promoter from *M. capsulatus* Bath with native and mutated ribosome binding site (RBS) were used for LDH expression, these vectors were introduced into *M. capsulatus* Bath via conjugation with *E. coli* S17-1. Lactate production from methane was evaluated and confirmed by analyzing carbon (^{13}C) content and carbon stable isotope ratio. LDH was constitutively expressed in *M. capsulatus* Bath recombinants and a maximum detectable level of lactic acid was more than 2 mM. LDH gene was also introduced into *M. trichsporium* OB3b and *M. buryatense* 5G for production of lactate from methane in these strains. The LDH gene was cloned into a pMS10 expression vector and transferred into methanotrophs by conjugation. The OB3b-originated promoter *sga*, *Methylomonas* sp. 16a *mxaF* promoter and *Methylomonas* sp. 16a *hps* promoter were used for LDH expression in *M. trichosporium* OB3b and *M. buryatense* 5G, respectively. The recombinant strains of *M. trichosporium* OB3b and *M. buryatense* 5G were cultured ranging from 0.31 to 0.71 at OD_{600}, but lactate accumulation was very low at detectable levels (1–20 μM greater and three-fold higher concentrations of lactate compared to strains not carrying a functional LDH) (Saville et al. 2016).

As another example, lactate production from methane was reported by Subbian (2017a) using *M. capsulatus* as the host strain with heterologous expressions of LDHs from *E. coli*, *Pseudomonas aeruginosa* MTCC 424 and *Pectobacterium carotovorum* MTCC 1428 in the broad host range vector pMHA201 under the control of MDH promoter (PmxaF) or sigma 70 promoter (σ70). The recombinant methanotrophs with *E. coli* LDH under σ70 promoter resulted in highest lactate production with a titer of approximately 0.37 g/L (more than 20 times higher than wild-type control).

Recently, bioconversion of methane to lactate using obligate methanotrophic bacteria *M. buryatense* by heterologous expression of codon-optimized L-LDH from *E. coli*, *Bifidobacterium longum*, and *Lactobacillus helveticus* was achieved (Henard et al. 2016). Different L-LDH genes were expressed under the control of an inducible BHR expression vector containing the tetracycline promoter/operator, which can be used in an array of methanotrophs with oriV origin of replication. Among them, *L. helveticus* LDH expression showed highest lactate production in

M. buryatense with approximately 70-fold higher productivity than the wild-type strain. In 5 L fermentation, the engineered *M. buryatense* produced lactate with 0.05 g lactate/g methane yield, and 0.8 g/L lactate titer and 0.008 g/L/h productivity in continuous gas fermentation. Leveraging the results of Henard et al. (2016), Garg et al. (2018) performed a production of lactic acid from methane by examining the effects of various genetic factors such as inducible *vs* constitutive promoters and ribosome-binding sites of various strengths. Employing a synthetic biology-based approach, three sets of plasmids were constructed containing varying promoters and RBS. 5GB1 (pCMR3) strain represented the highest flux at all times, ranging 4- to 14-fold higher compared to previous studies, with the maximum productivity of 0.14 mmol L-lactate/g CDW/h achieved at 24 h. This is the result obtained in the Hungate tubes without medium optimization, and it is expected that productivity will be further improved if the production of lactic acid in the bioreactor is carried out using the optimum medium as in a previous study (Henard et al. 2016). Most recently, Henard et al. (2018) illustrated a significant increase in lactate production in type I methanotrophs, *M. alcaliphilum* 20Z, by inactivating PDH. The PDH mutant enhanced the lactate titer and specific productivity by two- to threefold and fourfold compared to the wild-type *M. alcaliphilum* 20Z, respectively, with a productivity of 0.027 g lactate/g DCW/h from biogas. This is the highest lactate specific productivity obtained up to now.

4.1.2 Production of C4 Alcohols: Isobutanol

Isobutanol is considered a good gasoline alternative because it is less water soluble and has energy density/octane value close to the gasoline (Rodriguez-Anton et al. 2016). Isobutanol has various applications in the paint industry and as fuel additives. The isobutanol ester derivatives, such as diisobutyl phthalate, can be used as plasticizer agents (Atsumi et al. 2008).

In bacterial metabolism, key enzymes are involved in isobutanol production from pyruvate including the acetolactate synthase (ALS, *ilvB*), ketol-acid reductoisomerase (*KARI*, *ilvC*), dihydroxy-acid dehydratase (DHAD, *ilvD*), ketoacid decarboxylase (KDC, *kdc*) and the alcohol dehydrogenase (ADH, *adh*) (Avalos et al. 2013; Atsumi et al. 2008, 2010) (Fig. 3). Metabolic engineering for isobutanol production was only recently demonstrated, the carbon source was converted to 2-ketoisovalerate, through overexpressed *ilvB* (*Bacillus subtilis*), *ilvC* (*E. coli*), and *ilvD* (*E. coli*). The resulting 2-ketoisovalerate was then converted to isobutanol using *kdc* from *L. lactis* with the *S. cerevisiae adh4*. The engineered *E. coli* strain produced more than 20 g/L of isobutanol from glucose (Atsumi et al. 2008). This strategy has also been implemented in several other organisms, among them, a *C. glutamicum* strain was successfully engineered to produce 13 g/L of isobutanol with a yield of 0.2 g/g on glucose (Blombach et al. 2011). Isobutanol was also produced by microbial fermentation of engineered *S. cerevisiae* using glycine as a substrate through the glycine degradation pathway *via* glyoxylate and α-ketoacid intermediates (Branduardi et al. 2013).

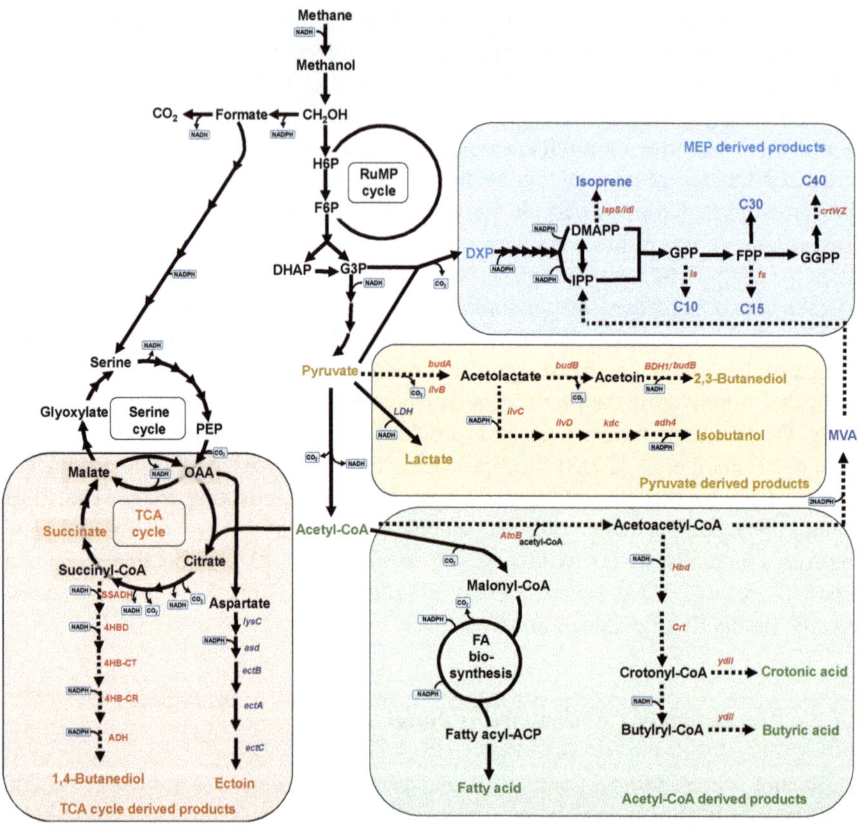

Fig. 3 Simplified metabolic pathway for production of chemicals and fuels from methane. The dotted arrows represent the non-native pathway, and the interactions mediated by the heterologous enzyme are marked in red. Enzymes (or encoding genes): LDH, lactate dehydrogenase; *ilvB*/*budA*, acetolactate synthase; *budB*, α-acetolactate decarboxylase; *BDH1*/*budA*, 2,3-butanediol dehydrogenase; *ilvC*, ketol-acid reductoisomeras; *ilvD*, dihydroxy-acid dehydratase; *kdc*, 2-keto acid decarboxylase; *adh4*, alcohol dehydrogenase; *atoB*, thiolase; *fadB*/*hbd*, 3-hydroxyacyl-CoA dehydrogenase; *fadB*/*crt*, enoyl-CoA hydratase; *ydiI*, enoyl-CoA specific thioesterase; *ectB*, aspartylsemialdehyde-glutamate aminotransferase; *ectA*, diaminobutyrate acetyltransferase; *ectC*, ectoine synthase, SSADH, CoA-dependent succinate semialdehyde dehydrogenase; 4HBD, 4-hydroxybutyrate dehydrogenase; 4HB-CT, 4-hydroxybutyryl-CoA transferase; 4HB-CR, 4-hydroxybutyryl-CoA reductase; ADH, alcohol dehydrogenase; *IspS*, isoprene synthase; *idi*, IPP:DMAPP isomerase; LS, limonene synthase; FS, farnesene synthase; *crtW*, β-carotene ketolase; *crtZ*, β-carotene hydroxylase

Methane can be converted to isobutanol by introducing the above-mentioned pathway into methanotrophs (Fig. 3). Two- and five-gene pathways for isobutanol production were introduced in *M. capsulatus* Bath (Coleman et al. 2014). For the two-gene pathway, isobutanol was produced from ketoacid by expression of *kdc* and *adh*. Several *kdc* and *adh* genes from *M. trichosporium* OB3b, *M. capsulatus* Bath and codon-optimized genes from *Lactococcus lactis*, *S. cerevisiae*, and *C. acetobutylicum* were screened and expressed under constitutive and inducible promoters. A 2 g/L of 2-ketoisovaleric acid (2-KIV) was the best for the recombinant strain growth, and the highest isobutanol production was observed in 48–72 h after 2-KIV feeding, which was about 0.22 g/L by a combination of *M. capsulatus* Bath *kdc* and *S. cerevisiae adh6* expressed under a constitutive promoter J23115. The five-gene pathway includes five complete genes for isobutanol synthesis from pyruvate such as *kdc*, *ilvB*, *ilvC*, *ilvD* gene from *M. capsulatus* Bath and *adh4* from *S. cerevisiae*. By combination expression of those genes under a modified J23100 promoter, the *M. capsulatus* host produced approximately 0.001 g/L isobutanol (Coleman et al. 2014). Recently, Intrexon Co. has engineered a methanotroph to produce isobutanol from methane in natural gas, with potential yield of approximately 70% (w/w) (Lane 2015).

4.1.3 Production of C4 Diols: 2,3-Butanediol

The compound 2,3-BDO, a four-carbon diol, is an important intermediate for the chemical industry (Coleman et al. 2014). 2,3-BDO is a promising bulk fuel biochemical with a potentially wide range of application that can be produced via biotechnological routes (Ji et al. 2011). It has a high heating value of 27,200 J/g and ability to increase the octane number of fuels so it can be used either as liquid fuel or fuel additive (Białkowska 2016; Xiao et al. 2012). Especially, 2,3-BDO exhibits lower toxicity for the microbial system, so it is very potential for producing a high titer of 2,3-BDO in microbes (Xu et al. 2014). At the commercial scale, 2,3-BDO is mainly produced or generated from petroleum and then serves as a precursor for the production of various commodity and specialty chemicals, such as the solvent methyl ethyl ketone, gamma-butyrolactone and 1,3-butadiene (Coleman et al. 2014).

In bacterial metabolism, three key enzymes are involved in the 2,3-BDO biosynthesis from pyruvate including acetolactate synthase (ALS), α-acetolactate decarboxylase (ALDC), and 2,3-BDO dehydrogenase (BDH) that provide simple engineering strategy for the production of 2,3-BDO in heterologous hosts (Ji et al. 2011) (Fig. 3). Recently, biological production of 2,3-BDO from glucose has been conducted using microbial fermentation of native hosts such as *Enterobacter aerogenes*, *Klebsiella pneumonia*, *S. cerevisiae* or non-native host such as *E. coli and Cyanobacteria* (Ji et al. 2011; Xu et al. 2014; Oliver et al. 2013). Thus, heterologous expression of 2,3-BDO biosynthesis pathway in methanotrophic bacteria could be a promising strategy using methane as major carbon and energy source for 2,3-BDO production.

In 2014, Coleman et al. have mentioned engineering pathways for production of 2,3-BDO by heterologously expressing (2R, 3R)-2,3-butanediol dehydrogenase (*BDH1*) enzyme from *S. cerevisiae* in *M. capsulatus* Bath (Fig. 3). The *BDH1* gene was codon-optimized and cloned into the BHR pCM132 vector. This invention is the first demonstration of 2,3-BDO production from a methane substrate. This biological production of 2,3-BDO can take advantage of the endogenous production of (R)-acetoin by the native metabolism of methanotrophs. The (R)-acetoin compound is produced in methanotrophs from two molecules of pyruvate, which are ultimately derived from methane. By expressing a single gene *BDH1* from *S. cerevisiae* encoding in 2,3-butanediol dehydrogenase *M. capsulatus* Bath, the (R)-acetoin is converted into 2,3-BDO (Coleman et al. 2014). In this patent, however, there was no additional information on yield and production rate.

A breakthrough of bio-production of 2,3-BDO from methane by an engineered methanotroph was recently developed based on the advances of systematic metabolic engineering and synthetic biology (Nguyen et al. 2018b). In this significant work, a 2,3-BDO pathway was designed in *M. alcaliphilum* 20Z by manipulation of three reactions catalyzed by three heterologous enzymes, or two combined heterologous enzymes from pyruvate as a central metabolic intermediate to 2,3-BDO. The design of the 5-UTR was obtained and the elimination of the byproduct pathway was manipulated. A triple mutant strain *Δldh ΔackA Δmadh* produced 68.8 mg/L 2,3-BDO from methane (Fig. 2). The productivity of triple-mutant strain was obtained with a final product concentration of 86.2 mg/L in a fed-batch stirred-tank bioreactor under O_2-limited conditions.

4.2 *Acetyl-CoA-Derived Molecules*

Among the metabolites in methanotrophs, acetyl-CoA is by far most important because it is connected with EMP, serine, TCA and EMC pathways. Many studies have been focused on production of acetyl CoA-derived target products from methane based on strategies to enhance acetyl-CoA pools.

4.2.1 Production of Fatty Acids-Derived Compounds as Diesel Precursors

Fatty acids are potential precursors for the production of liquid transportation fuels. Methanotrophs can accumulate a high level of fatty acid since their central metabolism is linked with the synthesis of intracytoplasmic membranes (ICMs) (Demidenko et al. 2017). The fatty acids in methanotrophs are saturated or monounsaturated, with different positions of double bonds and C14–C18 chain lengths, which have been highlighted in the potential to produce liquid fuel (Fei et al. 2014). Recently, the fatty acid biosynthetic pathways in *M. buryatense* 5GB1 have been characterized, and several approaches for enhancing fatty acid accumulation have also been investigated. The strain accumulated 111 ± 2 mg/g DCW of extractable fatty acids with a 20% increase in content compared to the wild type strain. Interestingly, the fatty acids biosynthesis regulation gene *farE* was identified, and its deletion resulted in an increase in the pool of C18 fatty acid methyl ester (Demidenko et al. 2017). Furthermore, a biorefinery process integrating biological conversion of methane to lipids using *M. buryatense* 5GB1 with subsequent hydrotreating has been demonstrated (Dong et al. 2017).

4.2.2 Production of C-4 Carboxylic Acids: Crotonic Acid and Butyric Acid

The C-4 carboxylic acid is employed in many industrial applications and it is mainly used as a monomer for the synthesis of copolymers which are widely applied in paints, adhesives, coatings, ceramics, and agrochemicals (Blumenstein et al. 2015).

Production of crotonic acid was demonstrated by functional expression of a modified version of reverse β-oxidation pathway in *E. coli* (Kim et al. 2016; Clomburg et al. 2012), wherein a acetoacetyl-CoA thiolase (*atoB*) initiates the pathway by condensing two molecules of acetyl-CoA to generate acetoacetyl-CoA, subsequently acetoacetyl-CoA was reduced to 3-hydroxybutyryl-CoA by a 3-hydroxyacyl-CoA dehydrogenase (*fadB*), followed by dehydration of 3-hydroxybutyryl-CoA by an enoyl-CoA hydratase (*fadB*) to generate crotonyl-CoA. Then, an enoyl-CoA specific thioesterase (*ydiI*) cleaves the thiol group of crotonyl-CoA to produce the free unsaturated fatty acid crotonic acid (Fig. 3). Applying the strategy from *E. coli*, Garg et al. (2018) reported the diversion of carbon from the acetyl-CoA to crotonic acid synthesis through a heterologous expression of *atoB*, *fadB* and *ydiI* genes from *E. coli* MG1655 in *M. buryatense* 5GB1C. To optimize the expression level of each enzyme in the heterologous pathway, RBS variants of different strengths designed using RBS calculator 2.0 program were examined, and the order of genes as gene clusters was also varied on the pAWP87 plasmid under the control of a constitutive *mxaF* promoter. Finally, up to 70 mg/L of crotonic acid in addition to the unexpected formation of 40 mg/L butyric acid from methane was obtained in engineered *M. buryatense* 5GB1C.

4.3 TCA Cycle-Derived Molecules

The intermediates of TCA cycle themselves are important products and can be used as precursors for valuable products. Metabolically engineered methanotrophs have been developed for the production of TCA cycle-derived target products via natural or synthetic pathways.

4.3.1 Production of Succinic Acid

Succinate has been considered as the top building block chemical, which can be used for agricultural, food, pharmaceutical, and cosmetic industries (Meng et al. 2016; McKinlay et al. 2007). Many molecules derived from a succinic acid by known chemical processes are 1,4-butanediol, maleic anhydride, succinimide,

2-pyrrolidinone, and tetrahydrofuran. All of them can be converted to a various range of more valuable molecules to use in polymers, industrial solvents, and specialty chemicals, such as biodegradable succinate esters (Zeikus et al. 1999; Budarin et al. 2007; Xu and Guo 2010). Around 38 kilotons of bio-based succinic acid are currently produced, and the production scale is expected to grow at an annual growth rate of approximately 18.7% from 2011 to 2016 (Nattrass et al. 2013).

Subbian et al. (2017b) reported metabolic engineering of methanotrophs for the production of succinic acid from methane in *M. capsulatus* for the first time. *M. capsulatus*-catalyzed succinic acid production by genetic manipulation, such as overexpression and/or down-regulation of many key enzymes of central carbon metabolism including MADH, PC, PEPC, phosphofructokinase (PFK), cytryl-CoA lyase, isocitrate lyase (ICL), fumarate reductase (FRD), malate synthase (MS), aspartate transaminase (AST), succinyl-CoA synthetase (SCS), pyruvate kinase (PK), and so on were performed. Production of succinate in the recombinant *M. capsulatus* was affected by overexpression and/or down-regulation of the selected genes. Succinate titer produced by the recombinant strains expressing MADH and PC was 15% and 50%, respectively higher than that of the wild-type strain. The maximum titer of succinic acid is approximately 105 mg/L by overexpression of MADH and ICL (Subbian 2017b). Recently, a type I methanotroph, *Methylomonas* sp. DH-1 was developed to produce succinate by metabolic engineering (Nguyen et al. 2019). Succinate accumulation increased by ~10-fold via deleting succinate dehydrogenase (*sdh*) in the TCA cycle. The succinate titer of ~134 mg/L in DS-GL strain was obtained by reconstructing glyoxylate shunt in the Δ *sdh* strain. The additional disruptions of competitive pathways through deletion of pyruvate formate lyase (*pfl*) and *pta-ack* genes did not improve succinate titer. The best succinate producing strain, DS-GL strain, was cultured in a fed-batch bioreactor, and high cell growth and succinate production (~195 mg/L succinate with 0.0789 g-succinate/g-methane yield) were obtained. This is the highest succinate titer of methane bioconversion by metabolic engineered methanotrophs up to date.

4.3.2 Production of Ectoine

Ectoine (1,4,5,6-tetrahydro-2-methyl-4-pyrimidine carboxylate) is a secondary metabolite which is produced by cells as a protectant against osmotic pressure (Strong et al. 2015). Ectoine attracts scientific and commercial interest as protectant of biomolecules and cells from the effects of various physico-chemical factors such as temperature, dehydration, UV radiation, and chemotherapeutic agents (Khmelenina et al. 2015). It can be used as a moisturizing and sun protecting ingredient in cosmetics (Strong et al. 2015). Extremophilic methanotrophs are promising producers of ectoine because of their specific mechanisms of haloadaptation (Khmelenina et al. 2015). Halotolerant or haloalcaliphilic methanotrophs isolated are new species of *Methylomicrobium* (*M. alcaliphilum*, *M. buryatense*, *M. kenyense*, and *M. japanense*) and also *Methylobacter marinus* and *Methylohalobius cremeensis*. In order to adapt to the high salinity of their living environment, they synthesize osmoprotectants including ectoine, sucrose, and glutamate. The total concentration of these osmolytes in the cytoplasm balances the external osmotic pressure, and ectoine is the major osmoprotectant (Khmelenina et al. 2015).

Methanotrophic and heterotrophic halophilic bacteria synthesize ectoine from aspartate (Fig. 3). This biochemical pathway is a side branch of the synthesis of aspartate family amino acids. The genes encoding three specific enzymes including diaminobutyrate-2-oxoglutarate aminotransferase (*ectB*), diaminobutyrate acetyltransferase (*ectA*), and ectoine synthase (*ectC*) typically organized as an *ectABC* cluster. An ectABC cluster also containing an additional gene of aspartokinase (ASK, *lysC*) has been found in *M. alcaliphilum* 20Z and other strains of this species. It has been speculated that the ASK isoenzyme provides independent ectoine synthesis and relatively high salt tolerance to the methanotroph *via* the catalysis of aspartate phosphorylation to aspartylphosphate in a common reaction for the biosynthesis of ectoine and other structural amino acids (Khmelenina et al. 2015; Reshetnikov et al. 2011).

The haloalkaliphile *M. alcaliphilum* 20Z accumulates ectoine up to 20% of dry cell weight, and quickly and stably grows on methane or high methanol concentrations under wide ranges of salinities (0–10% NaCl) and pH (7–10) (Khmelenina et al. 1999). Recently, by an increase of NaCl concentration in the cultivation broth, ectoine has been produced from 16.5 to 37.4 mg/g biomass during the continuous abatement of diluted emissions of methane in stirred tank reactors by *M. alcaliphilum* 20Z (Cantera et al. 2017a). Additionally, *M. alcaliphilum* 20Z can efficiently synthesize and excrete ectoine of 70.4 ± 14.3 mg/g biomass from methane when cultivating at a NaCl concentration of 6% through a tailored bio-milking process (Cantera et al. 2017b).

4.3.3 Production of C4 Diol: 1,4-Butanediol

1,4-Butanediol (BDO) is a precursor in the production of important polymers such as polyethers, polyurethanes, and polyester such as polybutylene terephthalate. It can be used for the production of tetrahydrofuran and as an intermediate of spandex and gamma-butyrolactone (Hunter et al. 2006; Hwang et al. 2011). It was presented that 1,4-BDO is a major commodity chemical used to make over 2.5 million tons of plastics, polyesters and spandex fibers annually (Yim et al. 2011). Furthermore, 1,4-BDO-based polymers possess a greater stability and property in comparison with polymers made from 1,2-propanediol or ethylene glycol (Cheng et al. 2013). Nowadays, the global 1,4-BDO market size was valued at USD 6.19 billions in 2015 and is expected to grow at an estimated CAGR of 7.7% from 2016 to 2025 (Grand View Research 2017).

The 1,4-BDO production has attracted significant investments for the use of sustainable sources such as succinate (Yim et al. 2011). In 1,4-BDO biosynthetic pathway, succinate is converted into succinyl-CoA catalyzed by succinate-CoA ligase (SCS, *sucCD*), and to 4-hydroxybutyrate through subsequent two steps by CoA-dependent succinate semialdehyde dehydrogenase (SSADH, *sucD*) and

4-hydroxybutyrate dehydrogenase (4HBD, *4hbD*). 4-Hydroxybutyrate is further converted to 1,4-BDO by 4-hydroxybutyryl-CoA transferase (4HB-CT), 4-hydroxybutyryl-CoA reductase (4HB-CR), and alcohol dehydrogenase (ADH) with a three-step reaction (Fig. 3).

Yim et al. (2011) presented a metabolic engineering approach enabling new bioprocesses for producing chemicals from engineered *E. coli* which are not produced naturally. The research indicated the first direct biocatalytic routes for the production of 1,4-BDO from renewable carbohydrate feedstocks, resulting in an *E. coli* strain capable of producing 18 g/L of 1,4-BDO. The *E. coli* host was engineered to improve the operation of the oxidative TCA cycle under anaerobic condition, releasing reducing power to control the 1,4-BDO pathway. Noticeably, in addition to carbon from sugar, one carbon of 1,4-BDO directly derives from CO_2.

Based on *E. coli* system, bioconversion of methane using methanotrophs also has the potential of producing 1,4-BDO. 1,4-BDO can be produced from succinate *via* 4-hydroxybutyrate by introducing the 1,4-BDO biosynthetic pathway into methanotrophs (Fig. 3). 1,4-BDO production from methane can be an environment-friendly method and a fascinating way for commercialization; however, there is only a few report on metabolic engineering of methanotrophs for producing 1,4-BDO. Recently, Intrexon Co. has announced that an engineered microbial cell was developed to produce 1,4-BDO (Lane 2015). Compared to traditional multi-step catalytic processes with high energy and hydrogen inputs, the engineered methanotroph can produce 1,4-BDO through a single-step fermentation process at ambient temperature and pressure, but the production titer was not mentioned.

4.4 MEP-Derived Molecules

Various terpenoids via the pathway can be broadly applied for the production of next-generation fuels, pharmaceuticals and aromatic chemicals. Some examples of terpenoids production from methane has been demonstrated using metabolically engineered methanotrophs.

4.4.1 Production of Isoprene

Isoprene, referred as 2-methyl-1,3-butadiene, is a volatile 5-carbon component produced by a variety of organisms including microbes, plants, and animals (Fall and Copley 2000; Kuzuyama 2002). It is an important platform chemical in the production of polyisoprene used in tires and rubber industry; elastomers, for use in footwear, medical supplies, latex, sporting goods; adhesives; and isoprenoids for medicines (Morais et al. 2015; Gordillo et al. 2009). Isoprene polymerization products are also useful as fuels (Clement et al. 2008; Jackstell et al. 2007). However, current industrial use of isoprene is limited due to the supply shortage.

Isoprene can be produced from petroleum by cracking hydrocarbons in the naphtha portion, but the yield is low (Lidov et al. 1963).

Therefore, efforts to enable or enhance the microbial production of isoprene from abundant and inexpensive renewable resources have been increasing. Especially, recombinant microorganisms such as *E. coli*, algae, and cyanobacteria have been used to convert the raw material derived from biomass to produce isoprene, but more than half of the mass for cellulosic carbohydrate feedstocks consist of oxygen, which is a major reason of limitation in theoretical conversion efficiency and yield, since isoprene and its derivatives have much lower oxygen composition than the feedstocks (Leonard et al. 2014). For these reasons, methane, a highly-reduced carbon source, was investigated as a feedstock for isoprene production. To utilize methane as a carbon feedstock, the role of methane-consuming bacteria known as methanotrophs was investigated.

There are two pathways for isoprene biosynthesis (Fig. 3), the MVA pathway and the MEP pathway or also named the DXP pathway (Kuzuyama 2002). The enzymes for the upper step of the MEP pathway are present in many methanotrophic strains, however, methanotrophs lack of Isoprene synthase (*IspS*) for the conversion of dimethylallyl-DP (DMAPP) into isoprene.

For producing isoprene from methane using methanotrophs, heterogeneous expression of IspS in the MEP pathway for conversion of DMAPP to isoprene could lead to isoprene production. The recombinant methanotrophs produced isoprene by expression of codon-optimized *Salix* sp. *IspS* gene with MDH promoter or *Pueraria montana IspS* gene with an IPTG-inducible promoter in *M. capsulatus* Bath (Leonard et al. 2014). A 10 mg/L isoprene was produced by expression of *P. montana IspS*, more isoprene production as compared to the expression of the *Salix* sp. *IspS* in *M. capsulatus* Bath (Leonard et al. 2014). IspS and isopentenyl diphosphate isomerase (*idi*) expression vector was constructed and expressed for the production of isoprene from methane and methanol in *Methylomonas* (Donaldson et al. 2015). To create an *IspS* and *idi* expression vector for *Methylomonas*, the chloramphenicol resistance gene promoter of the pBHR1 vector was used to direct expression of a modified codon-optimized *P. alba IspS* and the *S. cerevisiae idi*. After introducing the expression vector into *Methylomonas* by triparental mating, the recombinant *Methylomonas* produced 56.4 ng/mL and 53.3 ng/mL of isoprene from methane and methanol, respectively (Donaldson et al. 2015).

4.4.2 Production of C10 Isoprenoid: Limonene

Limonene, one of the simplest monocyclic p-menthane (1-methyl-4-isopropylcyclohexane) type monoterpenes, has been used as a flavor or fragrance with aroma value (Colby et al. 1993; Alonso-Gutierrez et al. 2013). Limonene is also an important plant monoterpene precursor of several fine chemicals, flavorings, fragrances, and pharmaceuticals such as carveol, carvone, perillyl alcohol, and menthol (Colby et al. 1993; Kirby and Keasling 2009). In addition, the hydrogenated form of limonene can be used as fuel (Tracy et al. 2009). However, the major supply of limonene is limited to plant sources currently. Although chemical processes for monoterpene synthesis have been well documented, very few pathway-engineered processes (Alonso-Gutierrez et al. 2013) have been reported so far for the production of limonene. Therefore, it is promising and attractive to develop a route of microbial synthesis for limonene production.

IPP and its isomer DMAPP are the two essential building blocks to synthesize all isoprenoids (terpenoids) (Du et al. 2014) (Fig. 3). Based on the IPP and DMAPP produced from the MEP pathway, limonene can be synthesized by the catalysis of two key enzymes, geranyl diphosphate synthase (*GPPS*) and limonene synthase (LS), which were reported to occur in *Abies grandis* (Burke and Croteau 2002) and *Mentha spicata* (Colby et al. 1993), respectively. GPPS catalyzes the condensation between IPP and DMAPP, forming a linear diphosphate intermediate, geranyl diphosphate (GPP), the precursor of all the monoterpenes. On the other hand, LS catalyzes the intramolecular cyclization of GPP to give limonene (Du et al. 2014).

Methanotrophs have been engineered for limonene production by the expression of *M. spicata* LS in a pR58 vector in *Methylomonas* sp. 16a (DiCosimo et al. 2004). The LS carries a deletion of the first 57 amino acids of the enzyme from N-terminal as truncation of LS preprotein provides a fully active 'pseudomature' form of this monoterpene cyclase and reveals the function of the amino-terminal arginine pair (Williams et al. 1998). However, the titer of limonene production in *Methylomonas* sp. 16a is very low, just 0.5 ppm (DiCosimo et al. 2004).

4.4.3 Production of C15 Isoprenoid: Farnesene

Farnesene is one of the simplest acyclic sesquiterpenes (Wang et al. 2011) and sesquiterpenes belong to a large and diverse isoprenoid family with important medical and industrial properties (Berger 2009; Muntendam et al. 2009; Dhingra et al. 1999). Farnesene was first discovered in apple peels and was found to play a role in plant defense (Huelin and Murray 1966; Pare and Tumlinson 1999; Yang et al. 2009). Farnesene has also been developed as a biofuel precursor owing to its inherent properties of low hygroscopicity and high energy density (Wang et al. 2011). Fanesane, which is reduced from farnesene, is a promising bio-jetfuel candidate (Renninger and McPhee 2008). The market was 9.35 kilotons in 2016, while the market size is growing at an annual growth rate of over 27.7%, and is expected to reach the production scale of 81.14 kilotons by 2025 (Factor and Equilibrium 2018). Thus, the metabolic engineering of organisms is an attractive route and alternative method for the production of those valuable and rare compounds (Munoz-Bertomeu et al. 2008).

Farnesene biosynthesis begins with the formation of precursors, IPP and DMAPP, generated from the MEP pathway in methanotroph (Fig. 3). IPP and DMAPP further condense to form GPP by GPPS; then, a second condensation step of GPP and IPP by farnesyl pyrophosphate synthase (*FPPS*) generates farnesyl pyrophosphate (FPP). Finally, heterologously expressed α-farnesene synthase (FS) catalyzes α-farnesene formation from the FPP precursor (Wang et al. 2011).

Intrexon Co. has developed a lab-scale bioconversion of methane to farnesene by constructing a biosynthetic pathway of farnesene in a genetically engineered methanotrophic strain. There is no report on productivity yet, and the yield of farnesene is expected to be over 70% (Lane 2015).

4.4.4 Production of C40 Carotenoids

Canthaxanthin and astaxanthin are carotenoids that have been widely commercialized (Higuera-Ciapara et al. 2006). In the aquaculture industry, they are applied as colorants in farmed fish and are important for fish growth and survival (Torrissen and Christiansen 1995). Astaxanthin (3,3′-dihydroxy-β,β′-carotene-4,4′-dione) is the most common carotenoid naturally present in marine fish and shellfish (Miki et al. 1982), and exhibits various biological activities. It has a strong antioxidant activity (Nishikawa et al. 2005), anti-photoaging effect (Arakane 2001), and enhances immune responses (Chew and Park 2004). Benefits of astaxanthin to human health have been suggested (Hussein et al. 2006). Astaxanthin protects skin against UV-induced photo-oxidation. It is used for anti-tumor therapies and treatment of neural damage due to age-related macular degeneration, Alzheimer, and Parkinson diseases (Zhang et al. 2007; Cardozo et al. 2007).

Currently, the majority of commercial astaxanthin is chemically synthesized (Rick et al. 2007), although some is still obtained from biological sources such as yeast Xanthophyllomyces dendrorhous (Visser et al. 2003; Johnson 2003), and the green algae *Haematococcus pluvialis* (Guerin et al. 2003). Recombinant bacteria can be another option for astaxanthin production.

Biosynthesis of carotenoids in methanotrophs is derived from the MEP pathway (Fig. 3). Genes, geranylgeranyl pyrophosphate synthetase (*crtE*), phytoene synthase (*crtB*), phytoene desaturase (*crtI*), and lycopene cyclase (*crtY*) can be used to produce the β-carotene from FPP. The two β-ionone rings in β-carotene are subsequently modified by hydroxylases and ketolases to produce canthaxanthin and astaxanthin. The β-carotene hydroxylase (*crtZ*) introduces the hydroxyl groups on the β-ionone rings, while the β-carotene ketolase (*crtW*) catalyzes the addition of two keto-groups. Only the β-carotene ketolase is necessary for the production of canthaxanthin, while a combination of both β-carotene hydroxylase and β-carotene ketolase is required for the biosynthesis of astaxanthin (Rick et al. 2007). Both β-carotene ketolase and hydroxylase enzymes need oxygen as a cofactor (Fraser et al. 1997). Limited activities of β-carotene ketolase and hydroxylase lead to accumulation of less oxygenated intermediates, which decrease astaxanthin production (Tao et al. 2007). As a result, adequate activity of both enzymes is required to achieve high conversion of the starting β-carotene substrate to astaxanthin (Rick et al. 2007).

Efforts were undertaken to engineer an obligate methanotrophic bacterium for astaxanthin production. Although astaxanthin is widely used, it is a challenge to produce it in bacteria because astaxanthin generally forms a small amount of the total carotenoids (Strong et al. 2015). Astaxanthin was produced by genetic engineering and the development of plasmid-free production strain of *Methylomonas* sp. 16a. The integration and expression of the canthaxanthin gene cluster in combination with *crtW* and *crtZ* encoding astaxanthin synthesis enzymes under the control of the *hps* promoter into the chromosome of *Methylomonas* sp. 16a resulted in the accumulation of astaxanthin from 1 to 2.4 mg/g DCW (Rick et al. 2007). Astaxanthin production was affected by oxygen availability, and thus, other researchers increased its production by expressing hemoglobin genes to improve oxygen accessibility (Tao et al. 2007). The *thbN*1-*crtWZ* cassette encoding hemoglobin genes and astaxanthin synthesis enzymes, respectively, was integrated into the chromosome of *Methylomonas* sp. 16a strain. When cells were grown in a 5 mL-well, the recombinant methanotroph containing the *thbN*1-*crtWZ* cassette produced astaxanthin at approximately 80% of total carotenoids (Tao et al. 2007). Total carotenoid synthesis has been increased by integrating the genes involved in the synthesis of the C40 carotenoids, canthaxanthin and astaxanthin, within the chromosome of *Methylomonas* sp. strain 16a leading to optimal expression using random transposon mutagenesis (Ye and Kelly 2012; Sharpe et al. 2007). Tn5 transposon promoter-probe vectors were used to insert the promoterless genes of the carotenoid synthetic pathway into five chromosomal locations including flagelline export (*fliCS*), modification of type II restriction system (*hsdM*), cytochrome c peroxidase (*ccp*-3), phosphoadenosine phosphosulfate reductase (*cysH*), and nitrite reductase (*nirS*). The number of total carotenoids produced by the recombinant strain increased by 10- to 20-fold when the carotenoid gene clusters were inserted at those chromosomal locations as compared to when the same carotenoid gene clusters were integrated at neutral locations under the control of a chloramphenicol resistance promoter. The titer of total carotenoid was approximately 2 mg/g DCW (Sharpe et al. 2007).

5 Perspectives on Biological Methane Conversion

Since 2008, shale gas revolution in the United States has significantly reduced natural gas price and, as a result, methane is considered a potential alternative carbon feedstock for industrial biotechnology. The theoretical carbon conversion efficiency and yield of target products of *E. coli* and methanotroph from glucose and methane are shown in Table 6, respectively. While the oxygen in glucose is lost during metabolic processes, methane substrate contains no waste oxygen, thus resulting in high theoretical conversion efficiency and yield. Despite these advantages, practical titer, productivity, and yield have been low because the research and development status of metabolic engineering of methanotrophs is still at the beginning stage.

In order to address technical challenges in methane bioconversion both in strain and process development, well-organized research programs have been recently conducted. Reducing Emissions using Methanotrophic Organisms for Transportation Energy (REMOTE), as one of the ARPA-E programs of the US Department of Energy, was launched in 2013. This program was proposed to accelerate the development of anaerobic methanotrophy for bioconversion of methane to liquid

Table 6 Comparison the carbon conversion efficiencies for *E. coli*- vs methanotroph-based cell factory

Target products	Sugar/*E. coli* Carbon conversion eq.	Theoretical CCE (%)	Theoretical Yield (%, w/w)	Experimental titer (Yield)	Methane/methanotroph Carbon conversion eq.	Theoretical CCE (%)	Theoretical Yield (%, w/w)
Lactate	$C_6H_{12}O_6 \rightarrow 2C_3H_6O_3$	100	100	138 g/L (86%)	$3CH_4 \rightarrow C_3H_6O_3$	100	188
Succinate	$C_6H_{12}O_6 \rightarrow C_4H_6O_4 + 2CO_2$	66.7	66	73–87 g/L (80–100%)	$6CH_4 \rightarrow C_4H_6O_4 + 2CO_2$	66.7	123
1,4-BDO	$C_6H_{12}O_6 \rightarrow C_4H_{10}O_2 + 2CO_2$	66.7	50	18 g/L (-)	$6CH_4 \rightarrow C_4H_{10}O_2 + 2CO_2$	66.7	94
Isobutanol	$C_6H_{12}O_6 \rightarrow C_4H_{10}O + 2CO_2$	66.7	41	50 g/L (-)	$6CH_4 \rightarrow C_4H_{10}O + 2CO_2$	66.7	77
Isoprene	$C_6H_{12}O_6 \rightarrow C_5H_8 + CO_2$	83.3	38	60 g/L (11%)	$6CH_4 \rightarrow C_5H_8 + CO_2$	83.3	71
Farnesene	$3C_6H_{12}O_6 \rightarrow C_{15}H_{24} + 3CO_2$	83.3	38	–	$18CH_4 \rightarrow C_{15}H_{24} + 3CO_2$	83.3	71

Taken from Lee et al. (2016). Copyright 2016 Wiley, England
Calculated based on type I methanotroph

fuels. The C1 gas refinery R&D program has been launched in 2015 with the financial support of the National Research Foundation of Korea funded by the Ministry of Science, Technology, and ICT. This program aims at the development of aerobic methanotrophy for the production of fuel and chemicals from methane. These kinds of research programs can inspire many microbiologists and biochemical engineers to revisit the principle and application of methane bioconversion with respect to both of biocatalysis and bioprocessing.

In order to use methanotroph as an efficient industrial strain, there are many technical issues that need to be solved. First of all, system-levels understanding of cellular behavior including specific and global gene expression regulation, and control of C1 metabolic enzyme activity and metabolic flux is needed for better design of metabolic engineering strategies. Currently, the GSMs of methanotroph are steady-state models, thus next generation GSMs integrated with various regulatory mechanisms and enzyme dynamic kinetics need to be developed to derive more quantitative strategies for metabolic engineering. In addition, synthetic biology, protein engineering and adaptive evolution engineering will certainly contribute to the development of highly efficient engineered methanotroph strains for the commercialization of methane bioconversion.

As another comment to be added, low titer and productivity are also due to mass transfer limitation of methane and oxygen, indicating that only strain development using metabolic engineering cannot be a shortcut to commercialization of methane bioconversion. Therefore, it is necessary to develop suitable methane gas fermentation systems for engineered methanotroph strains through collaboration with bioreactor engineer in order to overcome mass transfer limitation and low productivity. Lastly, metabolic engineering of methanotrophs will play a key role in methane-based C1 gas biorefinery.

Acknowledgements This work was supported by the C1 Gas Refinery Program through the National Research Foundation of Korea (NRF), funded by the Ministry of Science and ICT (2015M3D3A1A01064882).

References

Akberdin IR, Thompson M, Hamilton R, Desai N, Alexander D, Henard CA, Guarnieri MT, Kalyuzhnaya MG (2018) Methane utilization in *Methylomicrobium alcaliphilum* 20Z R: a systems approach. Sci Rep 8:2512

Alonso-Gutierrez J, Chan R, Batth TS, Adams PD, Keasling JD, Petzold CJ, Lee TS (2013) Metabolic engineering of *Escherichia coli* for limonene and perillyl alcohol production. Metab Eng 19:33–41

Anthony C, Ghosh M (1998) The structure and function of the PQQ-containing quinoprotein dehydrogenases. Prog Biophys Mol Biol 69:1–21

Arakane K (2001) Superior skin protection by astaxanthin. In: Presentation at the 15th annual meeting on carotenoid research

Atsumi S, Hanai T, Liao JC (2008) Non-fermentative pathways for synthesis of branched-chain higher alcohols as biofuels. Nature 451:86–89

Atsumi S, Wu T, Eckl E, Hawkins SD, Buelter T, Liao JC (2010) Engineering the isobutanol biosynthetic pathway in *Escherichia coli* by comparison of three aldehyde reductase/alcohol dehydrogenase genes. Appl Microbiol Biotechnol 85:651–657

Avalos JL, Fink GR, Stephanopoulos G (2013) Compartmentalization of metabolic pathways in yeast mitochondria improves the production of branched-chain alcohols. Nat Biotechnol 31:335–341

Baani M, Liesack W (2008) Two isozymes of particulate methane monooxygenase with different methane oxidation kinetics are found in *Methylocystis* sp. strain SC2. Proc Natl Acad Sci 105:10203–10208

Berger RG (2009) Biotechnology of flavours—the next generation. Biotechnol Lett 31:1651

Bergman A, Siewers V, Nielsen J, Chen Y (2016) Functional expression and evaluation of heterologous phosphoketolases in *Saccharomyces cerevisiae*. AMB Express 6:115

Białkowska AM (2016) Strategies for efficient and economical 2, 3-butanediol production: new trends in this field. World J Microbiol Biotechnol 32:200

Blombach B, Riester T, Wieschalka S, Ziert C, Youn JW, Wendisch VF, Eikmanns BJ (2011) *Corynebacterium glutamicum* tailored for efficient isobutanol production. Appl Environ Microbiol 77:3300–3310

Blumenstein J, Albert J, Shulz RP, Kohlpaintner C (2015) Crotonaldehyde and crotonic acid. In: Ullmann's encyclopedia of industrial chemistry. Wiley, Weinheim

Bogorad IW, Lin TS, Liao JC (2013) Synthetic non-oxidative glycolysis enables complete carbon conservation. Nature 502:693–697

Branduardi P, Longo V, Berterame NM, Rossi G, Porro D (2013) A novel pathway to produce butanol and isobutanol in *Saccharomyces cerevisiae*. Biotechnol Biofuels 6:68

Budarin V, Luque R, Macquarrie DJ, Clark JH (2007) Towards a bio-based industry: benign catalytic esterifications of succinic acid in the presence of water. Chem Eur J 13:6914–6919

Burke C, Croteau R (2002) Geranyl diphosphate synthase from Abies grandis: cDNA isolation, functional expression, and characterization. Arch Biochem Biophys 405:130–136
Cantera S, Lebrero R, Rodriguez E, Garcia-Encina PA, Munoz R (2017a) Continuous abatement of methane coupled with ectoine production by *Methylomicrobium alcaliphilum 20Z* in stirred tank reactors: a step further towards greenhouse gas biorefineries. J Clean Prod 152:134–141
Cantera S, Lebrero R, Rodriguez S, Garcia-Encina PA, Munoz R (2017b) Ectoine bio-milking in methanotrophs: a step further towards methane-based bio-refineries into high added-value products. Chem Eng J 328:44–48
Cardozo KH, Guaratini T, Barros MP, Falcão VR, Tonon AP, Lopes NP, Campos S, Torres MA, Souza AO, Colepicolo P (2007) Metabolites from algae with economical impact. Comp Biochem Physiol C Toxicol Pharmacol 146:60–78
Cheng K, Wang G, Zeng J, Zhang J (2013) Improved succinate production by metabolic engineering. BioMed Res Int 2013:538790
Chew BP, Park JS (2004) Carotenoid action on the immune response. J Nutr 134:257S–261S
Chinen A, Kozlov YI, Hara Y, Izui H, Yasueda H (2007) Innovative metabolic pathway design for efficient l-glutamate production by suppressing CO_2 emission. J Biosci Bioeng 103:262–269
Clement ND, Routaboul L, Grotevendt A, Jackstell R, Beller M (2008) Development of palladium–carbene catalysts for telomerization and dimerization of 1, 3-dienes: from basic research to industrial applications. Chem Eur J 14:7408–7420
Clomburg JM, Vick JE, Blankschien MD, Rodríguez-Moyá M, Gonzalez R (2012) A synthetic biology approach to engineer a functional reversal of the β-oxidation cycle. ACS Synth Biol 1:541–554
Colby SM, Alonso WR, Katahira EJ, McGarvey DJ, Croteau R (1993) 4S-Limonene synthase from the oil glands of spearmint (*Mentha spicata*). cDNA isolation, characterization, and bacterial expression of the catalytically active monoterpene cyclase. J Biol Chem 268:23016–23024
Coleman WJ, Vidanes GM, Cottarel G, Muley S, Kamimura R, Javan AF, Sun J, Groban ES (2014) Biological conversion of multi-carbon compounds from methane. US 14/206:835
Crombie A, Murrell JC (2011) Development of a system for genetic manipulation of the facultative methanotroph *Methylocella silvestris* BL2. Methods Enzymol 495:119–133
Crombie AT, Murrell JC (2014) Trace-gas metabolic versatility of the facultative methanotroph *Methylocella silvestris*. Nature 510:148–151
Csaki R, Bodrossy L, Klem J, Murrell JC, Kovacs K (2003) Genes involved in the copper-dependent regulation of soluble methane monooxygenase of *Methylococcus capsulatus* (Bath): cloning, sequencing and mutational analysis. Microbiology 149:1785–1795
Demidenko A, Akberdin IR, Allemann M, Allen EE, Kalyuzhnaya MG (2017) Fatty acid biosynthesis pathways in *Methylomicrobium buryatense* 5G(B1). Front Microbiol 7:2167
Dhingra V, Rao KV, Narasu ML (1999) Current status of artemisinin and its derivatives as antimalarial drugs. Life Sci 66:279–300
DiCosimo DJ, Koffas M, Odom JM, Wang S (2004) Production of cyclic terpenoids. US 09/938:956
Donaldson GK, Hollands K, Picataggio SK (2015) Biocatalyst for conversion of methane and methanol to isoprene. US 14/618:066
Dong T, Fei Q, Genelot M, Smith H, Laurens LM, Watson MJ, Pienkos PT (2017) A novel integrated biorefinery process for diesel fuel blendstock production using lipids from the methanotroph, *Methylomicrobium buryatense*. Energy Convers Manag 140:62–70
Du F, Yu H, Xu J, Li C (2014) Enhanced limonene production by optimizing the expression of limonene biosynthesis and MEP pathway genes in *E. coli*. Bioresour Bioprocess 1:10
Factor and Equilibrium (2018) Farnesene market report. https://www.factorandequilibrium.com/press-release/farnesene-market-size-research-report
Fall R, Copley SD (2000) Bacterial sources and sinks of isoprene, a reactive atmospheric hydrocarbon. Environ Microbiol 2:123–130
Fei Q, Guarnieri MT, Tao L, Laurens LM, Dowe N, Pienkos PT (2014) Bioconversion of natural gas to liquid fuel: opportunities and challenges. Biotechnol Adv 32:596–614

Fei Q, Puri AW, Smith H, Dowe N, Pienkos PT (2018) Enhanced biological fixation of methane for microbial lipid production by recombinant *Methylomicrobium buryatense*. Biotechnol Biofuels 11:129

Fraser PD, Miura Y, Misawa N (1997) In vitro characterization of astaxanthin biosynthetic enzymes. J Biol Chem 272:6128–6135

Fu Y, Li Y, Lidstrom M (2017) The oxidative TCA cycle operates during methanotrophic growth of the Type I methanotroph *Methylomicrobium buryatense* 5GB1. Metab Eng 42:43

Garg S, Wu H, Clomburg JM, Bennett GN (2018) Bioconversion of methane to C-4 carboxylic acids using carbon flux through acetyl-CoA in engineered *Methylomicrobium buryatense* 5GB1C. Metab Eng 48:175–183

Gilman A, Laurens LM, Puri AW, Chu F, Pienkos PT, Lidstrom ME (2015) Bioreactor performance parameters for an industrially-promising methanotroph *Methylomicrobium buryatense* 5GB1. Microb Cell Fact 14:182

Gordillo A, Pachón LD, de Jesus E, Rothenberg G (2009) Palladium-catalysed telomerisation of isoprene with glycerol and polyethylene glycol: a facile route to new terpene derivatives. Adv Synth Catal 351:325–330

Grand View Research (2017) 1,4-Butanediol market size, share 1,4 BDO industry report 2014–2025

Greenhouse Gas Emissions (2017) Understanding global warming potentials. Environmental Protection Agency

Guerin M, Huntley ME, Olaizola M (2003) Haematococcus astaxanthin: applications for human health and nutrition. Trends Biotechnol 21:210–216

Haynes CA, Gonzalez R (2014) Rethinking biological activation of methane and conversion to liquid fuels. Nat Chem Biol 10:331–339

Henard CA, Freed EF, Guarnieri MT (2015) Phosphoketolase pathway engineering for carbon-efficient biocatalysis. Curr Opin Biotechnol 36:183–188

Henard CA, Smith H, Dowe N, Kalyuzhnaya MG, Pienkos PT, Guarnieri MT (2016) Bioconversion of methane to lactate by an obligate methanotrophic bacterium. Sci Rep 6:21585

Henard CA, Smith HK, Guarnieri MT (2017) Phosphoketolase overexpression increases biomass and lipid yield from methane in an obligate methanotrophic biocatalyst. Metab Eng 41:152–158

Henard CA, Franklin TG, Youhenna B, But S, Alexander D, Kalyuzhnaya MG, Guarnieri MT (2018) Biocatalysis: methanotrophic bacterial cultivation, metabolite profiling, and bioconversion to lactic acid. Front Microbiol 9:2610

Higuera-Ciapara I, Felix-Valenzuela L, Goycoolea FM (2006) Astaxanthin: a review of its chemistry and applications. Crit Rev Food Sci Nutr 46:185–196

Huelin FE, Murray KE (1966) α-Farnesene in the natural coating of apples. Nature 210:1260–1261

Hunter SE, Ehrenberger CE, Savage PE (2006) Kinetics and mechanism of tetrahydrofuran synthesis via 1, 4-butanediol dehydration in high-temperature water. J Org Chem 71:6229–6239

Hussein G, Sankawa U, Goto H, Matsumoto K, Watanabe H (2006) Astaxanthin, a carotenoid with potential in human health and nutrition. J Nat Prod 69:443–449

Hwang DW, Kashinathan P, Lee JM, Lee JH, Lee U, Hwang J, Hwang YK, Chang J (2011) Production of γ-butyrolactone from biomass-derived 1, 4-butanediol over novel copper-silica nanocomposite. Green Chem 13:1672–1675

Hwang IY, Hur DH, Lee JH, Park C, Chang IS, Lee JW, Lee EY (2015) Batch conversion of methane to methanol using *Methylosinus trichosporium* OB3b as biocatalyst. J Microbiol Biotechnol 25:375–380

Hwang IY, Nguyen AD, Nguyen TT, Nguyen LT, Lee OK, Lee EY (2018) Biological conversion of methane to chemicals and fuels: technical challenges and issues. Appl Microbiol Biotechnol 102:3071–3080

Hyon SH (2000) Biodegradable poly (lactic acid) microspheres for drug delivery systems. Yonsei Med J 41:720–734

Ishikawa M, Tanaka Y, Suzuki R, Kimura K, Tanaka K, Kamiya K, Ito H, Kato S, Kamachi T, Hori K, Nakanishi S (2017) Real-time monitoring of intracellular redox changes in *Methylococcus capsulatus* (Bath) for efficient bioconversion of methane to methanol. Bioresour Technol 241:1157–1161

Jackstell R, Grotevendt A, Michalik D, El Firdoussi L, Beller M (2007) Telomerization and dimerization of isoprene by in situ generated palladium–carbene catalysts. J Organomet Chem 692:4737–4744

Ji X, Huang H, Ouyang P (2011) Microbial 2, 3-butanediol production: a state-of-the-art review. Biotechnol Adv 29:351–364

Johnson EA (2003) Phaffia rhodozyma: colorful odyssey. Int Microbiol 6:169–174

Kabimoldayev I, Nguyen AD, Yang L, Park S, Lee EY, Kim D (2018) Basics of genome-scale metabolic modeling and applications on C1-utilization. FEMS Microbiol Lett 365:fny241

Kalyuzhnaya MG, Yang S, Rozova ON, Smalley NE, Clubb J, Lamb A, Nagana Gowda GA, Raftery D, Fu Y, Bringel F, Vuilleumier S, Beck DAC, Trotsenko YA, Khmelenina VN, Lidstrom ME (2013) Highly efficient methane biocatalysis revealed in a methanotrophic bacterium. Nat Commun 4:2785

Kalyuzhnaya M, Puri AW, Lidstrom ME (2015) Metabolic engineering in methanotrophic bacteria. Metab Eng 29:142–152

Khmelenina VN, Kalyuzhnaya MG, Sakharovsky VG, Suzina NE, Trotsenko YA, Gottschalk G (1999) Osmoadaptation in halophilic and alkaliphilic methanotrophs. Arch Microbiol 172:321–329

Khmelenina VN, Rozova N, But CY, Mustakhimov II, Reshetnikov AS, Beschastnyi AP, Trotsenko YA (2015) Biosynthesis of secondary metabolites in methanotrophs: biochemical and genetic aspects (review). Prikl Biokhim Mikrobiol 51:140–150

Kim S, Cheong S, Gonzalez R (2016) Engineering *Escherichia coli* for the synthesis of short- and medium-chain α,β-unsaturated carboxylic acids. Metab Eng 36:90–98

Kirby J, Keasling JD (2009) Biosynthesis of plant isoprenoids: perspectives for microbial engineering. Annu Rev Plant Biol 60:335–355

Kuzuyama T (2002) Mevalonate and nonmevalonate pathways for the biosynthesis of isoprene units. Biosci Biotechnol Biochem 66:1619–1627

Lane J (2015) Methane-munching platform microbe: the digest's 2015 8-slide guide to intrexon energy. Biofuels Digest. 2017

Larsen Ø, Karlsen OA (2015) Transcriptomic profi ling of *Methylococcus capsulatus* (Bath) during growth with two different methane monooxygenases. Microbiologyopen 5:254–267

Lee OK, Hur DH, Nguyen DTN, Lee EY (2016) Metabolic engineering of methanotrophs and its application to production of chemicals and biofuels from methane. Biofuels Bioprod Biorefin 10 (6):848–863

Leonard E, Minshull J, Ness JE, Purcell TJ (2014) Compositions and methods for biological production of isoprene. U.S. Patent Application No. 14/773,118 PCT/US2014/021258

Lidov RE, Schaffel GS, White CJ (1963) Cracking of neohexene to isoprene. 24 December 1963

Lieven C, Petersen LA, Jorgensen SB, Gernaey K, Herrgard M, Sonnenschein N (2018) A genome-scale metabolic model for *Methylococcus capsulatus* predicts reduced efficiency uphill electron transfer to pMMO. bioRxiv 329714

Lloyd JS, De Marco P, Dalton H, Murrell JC (1999) Heterologous expression of soluble methane monooxygenase genes in methanotrophs containing only particulate methane monooxygenase. Arch Microbiol 171:364–370

Martin H, Murrell JC (1995) Methane monooxygenase mutants of constructed by marker-exchange mutagenesis. FEMS Microbiol Lett 127:243–248

Marx CJ, Lidstrom ME (2001) Development of improved versatile broad-host-range vectors for use in methylotrophs and other gram-negative bacteria. Microbiology 147:2065–2075

Matsen JB, Yang S, Stein LY, Beck D, Kalyuzhnaya MG (2013) Global molecular analyses of methane metabolism in methanotrophic alphaproteobacterium, *Methylosinus trichosporium* OB3b. Part I: transcriptomic study. Front Microbiol 4:40

McKinlay JB, Vieille C, Zeikus JG (2007) Prospects for a bio-based succinate industry. Appl Microbiol Biotechnol 76:727–740

Meadows AL, Hawkins KM, Tsegaye Y, Antipov E, Kim Y, Raetz L, Dahl RH, Tai A, Mahatdejkul-Meadows T, Xu L, Zhao L (2016) Rewriting yeast central carbon metabolism for industrial isoprenoid production. Nature 537:694–697

Meinhold P, Peters MW, Chen MM, Takahashi K, Arnold FH (2005) Direct conversion of ethane to ethanol by engineered cytochrome P450 BM3. Chembiochem 6:1765–1768

Meng J, Wang B, Liu D, Chen T, Wang Z, Zhao X (2016) High-yield anaerobic succinate production by strategically regulating multiple metabolic pathways based on stoichiometric maximum in *Escherichia coli*. Microb Cell Fact 15:141

Miki W, Yamaguchi K, Konosu S (1982) Comparison of carotenoids in the ovaries of marine fish and shellfish. Comp Biochem Physiol B 71:7–11

Morais AR, Dworakowska S, Reis A, Gouveia L, Matos CT, Bogdał D, Bogel-Łukasik R (2015) Chemical and biological-based isoprene production: green metrics. Catal Today 239:38–43

Murrell JC (1992) Genetics and molecular biology of methanotrophs. FEMS Microbiol Lett 88:233–248

Muller JE, Meyer F, Litsanov B, Kiefer P, Potthoff E, Heux S, Quax WJ, Wendisch VF, Brautaset T, Portais J (2015) Engineering *Escherichia coli* for methanol conversion. Metab Eng 28:190–201

Munoz-Bertomeu J, Ros R, Arrillaga I, Segura J (2008) Expression of spearmint limonene synthase in transgenic spike lavender results in an altered monoterpene composition in developing leaves. Metab Eng 10:166–177

Muntendam R, Melillo E, Ryden A, Kayser O (2009) Perspectives and limits of engineering the isoprenoid metabolism in heterologous hosts. Appl Microbiol Biotechnol 84:1003

Mustakhimov II, But SY, Reshetnikov AS, Khmelenina VN, Trotsenko YA (2016) Homo and heterologous reporter proteins for evaluation of promoter activity in *Methylomicrobium alcaliphilum* 20Z. Appl Biochem Microbiol 52:263–268

Nattrass L, Aylott M, Higson A (2013) NNFCC renewable chemicals factsheet: succinic acid. NNFCC, York

Nguyen HH, Chan S (2003) Protein and nucleic acid expression systems. US20030032141 A1

Nguyen AD, Hwang IY, Lee OK, Hur DH, Jeon YC, Hadiyati S, Kim MS, Yoon SH, Jeong H, Lee EY (2018a) Functional analysis of *Methylomonas* sp. DH-1 genome as a promising biocatalyst for bioconversion of methane to valuable chemicals. Catalysts 8:117

Nguyen AD, Hwang IY, Lee OK, Kim D, Kalyuzhnaya MG, Mariyana R, Hadiyati S, Kim MS, Lee EY (2018b) Systematic metabolic engineering of *Methylomicrobium alcaliphilum* 20Z for 2, 3-butanediol production from methane. Metab Eng 47:323–333

Nguyen DTN, Lee OK, Hadiyati S, Affifah AN, Kim MS, Lee EY (2019) Metabolic engineering of the type I methanotroph *Methylomonas* sp. DH-1 for production of succinate from methane. Metab Eng 54:170–117

Nishikawa Y, Minenaka Y, Ichimura M, Tatsumi K, Nadamoto T, Urabe K (2005) Effects of astaxanthin and vitamin C on the prevention of gastric ulcerations in stressed rats. J Nutr Sci Vitaminol 51:135–141

Niu FX, Lu Q, Bu YF, Liu JZ (2017) Metabolic engineering for the microbial production of isoprenoids: carotenoids and isoprenoid-based biofuels. Synth Syst Biotechnol 2:167

Ojala DS, Beck DA, Kalyuzhnaya MG (2011) Genetic systems for moderately halo (alkali) philic bacteria of the genus *Methylomicrobium*. Meth Enzymol 495:99–118

Oliver JW, Machado IM, Yoneda H, Atsumi S (2013) Cyanobacterial conversion of carbon dioxide to 2, 3-butanediol. Proc Natl Acad Sci 110:1249–1254

Pare PW, Tumlinson JH (1999) Plant volatiles as a defense against insect herbivores. Plant Physiol 121:325–332

Patel A, Prajapat JB (2013) Food and health applications of exopolysaccharides produced by lactic acid bacteria. Adv Dairy Res 1:107

Pawar RU, Tekale SU, Shisodia SU, Totre JT, Domb AJ (2014) Biomedical applications of poly (lactic acid). Recent Pat Regen Med 4:40–51

Puri AW, Owen S, Chu F, Chavkin T, Beck DA, Kalyuzhnaya MG, Lidstrom ME (2015) Genetic tools for the industrially promising methanotroph *Methylomicrobium buryatense*. Appl Environ Microbiol 81:1775–1781

Renninger NS, Mcphee DJ (2008) Inventors; Amyris Biotechnologies Inc, assignee. Fuel compositions comprising farnesane and farnesane derivatives and method of making and using same. United States patent US 7,399,323

Reshetnikov AS, Khmelenina VN, Mustakhimov II, Trotsenko YA (2011) 2 Genes and enzymes of ectoine biosynthesis in halotolerant methanotrophs. Methods Enzymol 495:15

Rick WY, Yao H, Stead K, Wang T, Tao L, Cheng Q, Sharpe PL, Suh W, Nagel E, Arcilla D (2007) Construction of the astaxanthin biosynthetic pathway in a methanotrophic bacterium *Methylomonas* sp. strain 16a. J Ind Microbiol Biotechnol 34:289

Ro SY, Rosenzweig AC (2018) Recent advances in the genetic manipulation of *Methylosinus trichosporium* OB3b. Methods Enzymol 605:335–349

Rodriguez-Anton LM, Gutierrez-Martin F, Doce Y (2016) Physical properties of gasoline, isobutanol and ETBE binary blends in comparison with gasoline ethanol blends. Fuel 166:73–78

Rydz J, Sikorska W, Kyulavska M, Christova D (2014) Polyester-based (bio) degradable polymers as environmentally friendly materials for sustainable development. Int J Mol Sci 16:564–596

Saville RM, Lee S, Regitsky DD, Resnick SM, Silverman JA (2016) Inventors; CALYSTA Inc, assignee. Compositions and methods for biological production of lactate from C1 compounds using lactate dehydrogenase transformants. United States patent application US 14/898,948

Sharpe PL, DiCosimo D, Bosak MD, Knoke K, Tao L, Cheng Q, Rick WY (2007) Use of transposon promoter-probe vectors in the metabolic engineering of the obligate methanotroph *Methylomonas* sp. strain 16a for enhanced C40 carotenoid synthesis. Appl Environ Microbiol 73:1721–1728

Siegel JB, Smith AL, Poust S, Wargacki AJ, Bar-Even A, Louw C, Shen BW, Eiben CB, Tran HM, Noor E (2015) Computational protein design enables a novel one-carbon assimilation pathway. Proc Natl Acad Sci U S A 112:3704–3709

Smith TJ, Murrell JC (2011) Chapter nine – Mutagenesis of soluble methane monooxygenase. In: Rosenzweig AC, Ragsdale SW (eds) Methods in enzymology methods in methane metabolism, part B: methanotrophy, vol 495. Academic Press, Cambridge, pp 135–147

Smith TJ, Slade SE, Burton NP, Murrell JC, Dalton H (2002) Improved system for protein engineering of the hydroxylase component of soluble methane monooxygenase. Appl Environ Microbiol 68:5265–5273

Strong PJ, Xie S, Clarke WP (2015) Methane as a resource: can the methanotrophs add value? Environ Sci Technol 49:4001–4018

Subbian E (2017a) Inventor; String Bio Private Limited, assignee. Production of lactic acid from organic waste or biogas or methane using recombinant methanotrophic bacteria. United States patent application US 15/303,188

Subbian E (2017b) Inventor; String Bio Private Limited, assignee. Production of succinic acid from organic waste or biogas or methane using recombinant methanotrophic bacterium. United States patent application US 15/303,184

Tao L, Sedkova N, Yao H, Rick WY, Sharpe PL, Cheng Q (2007) Expression of bacterial hemoglobin genes to improve astaxanthin production in a methanotrophic bacterium *Methylomonas* sp. Appl Microbiol Biotechnol 74:625

Theisen AR, Ali MH, Radajewski S, Dumont MG, Dunfield PF, McDonald IR, Dedysh SN, Miguez CB, Murrell JC (2005) Regulation of methane oxidation in the facultative methanotroph *Methylocella silvestris* BL2. Mol Microbiol 58:682–692

Torre A, Metivier A, Chu F, Laurens LM, Beck DA, Pienkos PT, Lidstrom ME, Kalyuzhnaya MG (2015) Genome-scale metabolic reconstructions and theoretical investigation of methane conversion in *Methylomicrobium buryatense* strain 5G(B1). Microb Cell Fact 14:188

Torrissen OJ, Christiansen R (1995) Requirements for carotenoids in fish diets. J Appl Ichthyol 11:225–230

Tracy NI, Chen D, Crunkleton DW, Price GL (2009) Hydrogenated monoterpenes as diesel fuel additives. Fuel 88:2238–2240

Van Ophem PW, Van Beeumen J, Duine JA (1993) Nicotinoprotein [NAD (P)-containing] alcohol/aldehyde oxidoreductases. Purification and characterization of a novel type from Amycolatopsis methanolica. Eur J Biochem 212:819–826

Visser H, van Ooyen AJ, Verdoes JC (2003) Metabolic engineering of the astaxanthin-biosynthetic pathway of Xanthophyllomyces dendrorhous. FEMS Yeast Res 4:221–231

Vorobev A, Jagadevan S, Jain S, Anantharaman K, Dick GJ, Vuilleumier S, Semrau JD (2014) Genomic and transcriptomic analyses of the facultative methanotroph *Methylocystis* sp. strain SB2 grown on methane or ethanol. Appl Environ Microbiol 80:3044–3052

Vroman I, Tighzert L (2009) Biodegradable polymers. Materials 2:307–344

Wang C, Yoon S, Jang H, Chung Y, Kim J, Choi E, Kim S (2011) Metabolic engineering of *Escherichia coli* for α-farnesene production. Metab Eng 13:648–655

Wang X, Wang Y, Liu J, Li Q, Zhang Z, Zheng P, Lu F, Sun J (2017) Biological conversion of methanol by evolved *Escherichia coli* carrying a linear methanol assimilation pathway. Bioresour Bioprocess 4:41

Welander PV, Summons RE (2012) Discovery, taxonomic distribution, and phenotypic characterization of a gene required for 3-methylhopanoid production. Proc Natl Acad Sci U S A 109:12905–12910

Whitaker WB, Sandoval NR, Bennett RK, Fast AG, Papoutsakis ET (2015) Synthetic methylotrophy: engineering the production of biofuels and chemicals based on the biology of aerobic methanol utilization. Curr Opin Biotechnol 33:165–175

Whitaker WB, Jones JA, Bennett RK, Gonzalez JE, Vernacchio VR, Collins SM, Palmer MA, Schmidt S, Antoniewicz MR, Koffas MA (2017) Engineering the biological conversion of methanol to specialty chemicals in *Escherichia coli*. Metab Eng 39:49–59

Williams DC, McGarvey DJ, Katahira EJ, Croteau R (1998) Truncation of limonene synthase preprotein provides a fully active 'pseudomature' form of this monoterpene cyclase and reveals the function of the amino-terminal arginine pair. Biochemistry 37:12213–12220

Witthoff S, Schmitz K, Niedenfhr S, Nh K, Noack S, Bott M, Marienhagen J (2015) Metabolic engineering of Corynebacterium glutamicum for methanol metabolism. Appl Environ Microbiol 81:2215–2225

Xiao Z, Wang X, Huang Y, Huo F, Zhu X, Xi L, Lu JR (2012) Thermophilic fermentation of acetoin and 2, 3-butanediol by a novel Geobacillus strain. Biotechnol Biofuels 5:88

Xu J, Guo B (2010) Poly (butylene succinate) and its copolymers: research, development and industrialization. Biotechnol J 5:1149–1163

Xu Y, Chu H, Gao C, Tao F, Zhou Z, Li K, Li L, Ma C, Xu P (2014) Systematic metabolic engineering of *Escherichia coli* for high-yield production of fuel bio-chemical 2, 3-butanediol. Metab Eng 23:22–33

Yan X, Chu F, Puri AW, Fu Y, Lidstrom ME (2016) Electroporation-based genetic manipulation in type I methanotrophs. Appl Environ Microbiol 82:2062–2069

Yang DS, Son KC, Kays SJ (2009) Volatile organic compounds emanating from indoor ornamental plants. Hortic Sci 44:396–400

Yang S, Matsen JB, Konopka M, Green-Saxena A, Clubb J, Sadilek M, Orphan VJ, Beck D, Kalyuzhnaya MG (2013) Global molecular analyses of methane metabolism in methanotrophic Alphaproteobacterium, *Methylosinus trichosporium* OB3b. Part II. Metabolomics and 13C-labeling study. Front Microbiol 4:70

Ye R, Yao H, Stead K, Wang T, Tao L, Cheng Q, Sharpe PL, Suh W, Nagel E, Arcilla D, Dragotta D, Miller ES (2007) Construction of the astaxanthin biosynthetic pathway in a methanotrophic bacterium *Methylomonas* strain 16a. J Ind Microbiol Biotechnol 34:289–299

Ye RW, Kelly K (2012) Construction of carotenoid biosynthetic pathways through chromosomal integration in methane-utilizing bacterium *Methylomonas* sp. Strain 16a. Methods Mol Biol 892:185–195

Yim H, Haselbeck R, Niu W, Pujol-Baxley C, Burgard A, Boldt J, Khandurina J, Trawick JD, Osterhout RE, Stephen R (2011) Metabolic engineering of *Escherichia coli* for direct production of 1, 4-butanediol. Nat Chem Biol 7:445–452

Zeikus JG, Jain MK, Elankovan P (1999) Biotechnology of succinic acid production and markets for derived industrial products. Appl Microbiol Biotechnol 51:545–552

Zhang X, Pan L, Wei X, Gao H, Liu J (2007) Impact of astaxanthin-enriched algal powder of *Haematococcus pluvialis* on memory improvement in BALB/c mice. Environ Geochem Health 29:483–489

Zilly FE, Acevedo JP, Augustyniak W, Deege A, Husig UW, Reetz MT (2011) Tuning a P450 enzyme for methane oxidation. Angew Chem 123:2772–2776

Methanobactin: A Novel Copper-Binding Compound Produced by Methanotrophs

Jeremy D. Semrau and Alan A. DiSpirito

Contents

Abstract Aerobic methanotrophs are a novel group of microorganisms that play a critical role in the global carbon cycle. Expression and activity of a key enzyme in the metabolism in these microbes, the methane monooxygenase, is controlled by the availability of copper. To sequester copper, some methanotrophs produce a ribosomally synthesized post-translationally modified polypeptide called methanobactin. This peptide, or chalkophore, has unique features, with all known forms being small (<1300 Da) with two heterocyclic rings with associated thioamide groups that bind

J. D. Semrau (✉)
Department of Civil and Environmental Engineering, University of Michigan, Ann Arbor, MI, USA
e-mail: jsemrau@umich.edu

A. A. DiSpirito
Roy J. Carver Department of Biochemistry, Biophysics and Molecular Biology, Iowa State University, Ames, IA, USA
e-mail: aland@iastate.edu

E. Y. Lee (ed.), *Methanotrophs*, Microbiology Monographs 32,
https://doi.org/10.1007/978-3-030-23261-0_7

copper quite strongly. Here we provide an overview of the known diversity of methanobactins, describe the genetics underlying the methanobactin biosynthesis, the variety of functions methanobactin has been observed to have, as well as various environmental, industrial and medical applications of methanobactin.

1 Introduction

As described throughout this volume, methanotrophs are an interesting group of microbes that can utilize methane as a carbon and/or energy source (Semrau et al. 2010). Both aerobic and anaerobic methanotrophs have been characterized, as well as methanotrophy ascribed to both Bacteria and Archaea. It is difficult, if not impossible, however, to thoroughly describe the known metabolic or phylogenetic diversity of methanotrophy in any one book chapter (and indeed, it is becoming increasingly challenging to do so in only one book). As a result, here in this chapter we will focus solely on mechanisms utilized by aerobic methanotrophs of the α-Proteobacteria for the collection of copper, a key nutrient required by these fastidious microbes (Semrau et al. 2010, 2018). We do so as these mechanisms not only are the first examples of chalkophores (or copper-binding compounds) synthesized by various microbes, but also as these compounds have a wide range of environmental, industrial and medical applications (DiSpirito et al. 2016; Semrau et al. 2018; Strong et al. 2015, 2016, 2017).

1.1 General Overview of Proteobacterial Methanotrophic Phylogeny, Physiology and Metabolism

Briefly, Proteobacterial methanotrophs are primarily defined by their ability to utilize methane as their sole carbon and energy source. A key characteristic of these methanotrophs is the production of the enzyme responsible for the initial conversion of methane to methanol, the methane monooxygenase (MMO). Many Proteobacterial methanotrophs can utilize other C1 compounds for growth (e.g., methanol) but some can also be considered to be facultative, i.e., some α-Proteobacteria can use simple multi-carbon compounds as growth substrates [i.e., compounds with C–C bonds such as C2–C4 organic acids and ethanol, (Dedysh et al. 2005; Dunfield et al. 2010; Dunfield and Dedysh 2014; Semrau et al. 2011)]. To date, most γ–Proteobacteria methanotrophs are not known to be facultative, with the possible exception of *Crenothrix polyspora* that appears able to take up acetate and glucose (Stoecker et al. 2006).

What is more germane for the discussion on the role of copper in Proteobacterial methanotrophs is that two different forms of MMO have been found, with the expression and activity of these MMOs strongly dependent on the availability of copper. That is, it is well known that most (but not all) Proteobacterial methanotrophs express a membrane-bound or particulate methane monooxygenase (pMMO). The mechanism whereby pMMO converts methane to methanol is still a subject of much discussion, but it is generally accepted that copper plays an important role in the expression and activity of pMMO (Semrau et al. 2010, 2018). Some Proteobacteria methanotrophs can express a cytoplasmic or soluble methane monooxygenase (sMMO), either in addition or in lieu of the pMMO. The fact that methane can be converted to methanol via a biotic mechanism is in itself quite remarkable given the strong C–H bonds in methane (i.e., a bond dissociation energy of 104 kcal mol^{-1}), but the finding that there are two different enzymes capable of converting methane to methanol is even more extraordinary.

1.1.1 Soluble Methane Monooxygenase

H. Dalton's group at the University of Warwick was responsible for the initial isolation and characterization of the sMMO (Colby and Dalton 1976, 1978, 1979; Green and Dalton 1985; Woodland and Dalton 1979). The isolation required the separate purification of a reductase (component C), a regulatory protein (component B) and a hydroxylase (component A). Component C, or MMOR was shown to be a FAD-dependent reductase. The hydroxylase, or MMOH was a heterodimer with a subunit molecular structure of $(\alpha\beta\gamma)_2$. Component B, or MMOB was a regulatory protein and shown to mediate electron flow from component C to component A. In the initial characterization, the hydroxylase was shown to contain 1.25 iron per hydroxylase components, suggesting the active site was composed of a single iron atom. The laboratory of J. Lipscomb at the University of Minnesota dramatically improved the activity of sMMO preparations by stabilizing the iron center of the hydroxylase component and demonstrated that the hydroxylase component actually contained two iron per hydroxylase component (Chang et al. 1999; DeWitt et al. 1991; Fontecave et al. 1998; Fox et al. 1988, 1989, 1991, 1993; Fox and Lipscomb 1988; Lipscomb 1994). They also determined the catalytic center of the sMMO to be a μ-oxo bridged di-iron center. The Lispcomb group also demonstrated the loss of activity in earlier preparations was associated with disruption of the di-iron center. In *Methylosinus trichosporium* OB3b the di-iron center is coordinated by Glu 114, Glu 144, Glu 209, Glu 243, His 146 and His 246 of the hydroxylase α-subunit (Elango et al. 1997).

Once stabilized, structural characterizations verified the methane oxidation site as a di-iron center and the catalytic cycle was determined via a number of single turnover experiments (Banerjee et al. 2013, 2015; Elango et al. 1997; Lee et al. 1993a, b; Rosenzweig et al. 1992, 1993; Wallar and Lipscomb 1996; Walters et al. 1999; Zhang et al. 2006; Fig. 1).

The catalytic cycle begins with the reduction of the diiron center (diferrous form) followed by the reductive activation of O_2 and the formation of an oxygen bridged dinuclear iron cluster (compound O). This step is followed by a peroxo-bridged di-ferric intermediate, compound P. Homolytic cleavage of the O–O bond rapidly converts the di-ferric center to the di-(μ-oxo)diferryl form, compound Q. Compound

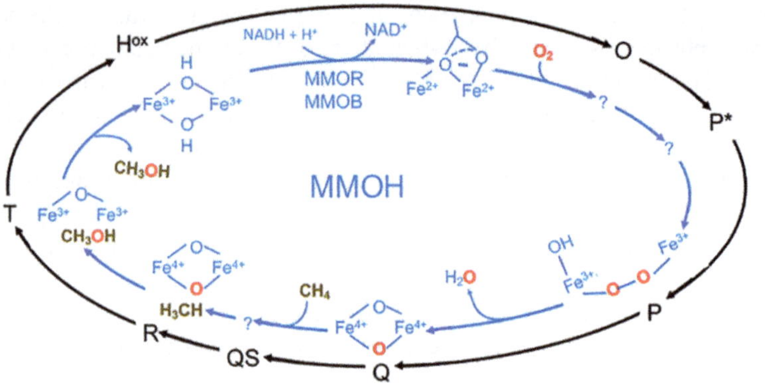

Fig. 1 Catalytic cycle of MMOH. Modified from Banerjee et al. (2013, 2015)

Q then reacts with methane and abstracts a proton from CH_4 to yield a substrate and a hydroxyl radical (intermediate R). Recombination of radicals eventually breaks down to release methanol.

1.1.2 Membrane-Associated Methane Monooxygenase (pMMO)

Current studies on the membrane associated methane monooxygenase (pMMO) are similar to the initial studies of the sMMO with a significant loss of activity during purification (Basu et al. 2003; Chan et al. 2007; Choi et al. 2003; Lieberman and Rosenzweig 2005; Lieberman et al. 2003; Nguyen et al. 1998; Smith and Dalton 1989; Takeguchi et al. 1998; Zahn and DiSpirito 1996). However, the loss of activity of the pMMO is more extreme and even the best enzyme preparations represent only 10–12% of the measured physiological rates (Martinho et al. 2007). In spite of this limitation, the enzyme has been shown to be a homotrimer $(\alpha\beta\gamma)_3$ made of PmoB (β), PmoA (β) and PmoC (γ) (Basu et al. 2003; Choi et al. 2003; Lieberman et al. 2003, 2006; Lieberman and Rosenzweig 2005; Semrau et al. 1995; Smith and Dalton 1989; Stanley et al. 1983; Zahn and DiSpirito 1996). In addition, biochemical, structural and spectral studies have identified three different metal binding/catalytic sites (Hakemian et al. 2008). One site, found in the periplasmic region of PmoB of some pMMOs, is a mononuclear copper site. The copper at this site is coordinated by two His (His 48 and His 72 in *Methylococcus capsulatus* Bath) and a Glu (Glu 404 in *M. capsulatus* Bath) (Hakemian et al. 2008; Lawton and Rosenzweig 2017; Lieberman and Rosenzweig 2005). This site is not well conserved, and thus an unlikely site of methane oxidation or electron flow to the catalytic center.

A second copper site is also present in the periplasmic region of PmoB. The copper at this site is coordinated by three His (His 33, 137 and 139 in *M. capsulatus* Bath). Based on spectral analysis this site was originally described as a mononuclear copper center (Choi et al. 2003, 2005; Semrau et al. 2010; Zahn and DiSpirito 1996). Structural studies, however, suggested this site was a bi-nuclear copper center and has generally been called the di-copper center of the pMMO (Hakemian et al. 2008; Hakemian and Rosenzweig 2007; Kenney and Rosenzweig 2012; Lawton and Rosenzweig 2017; Lieberman et al. 2006; Lieberman and Rosenzweig 2005). However, proponents of the bi-nuclear copper center have recently changed this site to a mononuclear copper center after re-examination of the original crystal structure data (Cao et al. 2018). Residues for this site are conserved in most but not all pMMO sequences (Semrau et al. 2018). As a bi-nuclear copper center, some groups have proposed this copper center is the site of methane oxidation (Balasubramanian et al. 2010; Culpepper et al. 2014; Hakemian et al. 2008; Hakemian and Rosenzweig 2007; Kenney and Rosenzweig 2012; Lawton and Rosenzweig 2017; Lieberman et al. 2006; Lieberman and Rosenzweig 2005). Theoretically, a high valence di-copper center could catalyze the initial oxidation of methane to methanol (Semrau et al. 2018). However, no model system for a mononuclear copper center is known to be capable of oxidizing the C–H bond in methane. The most compelling evidence for this site being the catalytic center of the pMMO is the reported methane oxidation by an N-terminal fragment of PmoB containing only this copper center (Balasubramanian et al. 2010). The reported methane oxidation activity, however, was approximately 2% of the estimated physiological rate. Although the activity is low, any activity in such a small peptide fragment containing one copper atom is surprising. For catalysis, this peptide fragment must oxidize duroquinol, a soluble quinone analog (Smith and Dalton 1989), or NADH, catalyze O–O bond cleavage and breakage the stable C–H bond of methane.

A third metal coordination site was found in all pMMO sequences as well as in all closely related ammonia and hydrocarbon monooxygenase sequences (Coleman et al. 2012; Semrau et al. 2008, 2010; Tumanova et al. 2008). In *M. capsulatus* Bath, this site is located within the lipid bilayer near the periplasmic membrane interface (Hakemian et al. 2008; Lieberman and Rosenzweig 2005). The metal in this site is coordinated by Asp 156, Asp 166, Asp 168, His 160 and His 173 from PmoC plus Glu 195 from PmoA and is very similar to the coordination of the diiron center of the sMMO (Lieberman and Rosenzweig 2005). In crystal structures of the pMMO, the metal associated with this site is identical to the metal content of the crystallization buffer suggesting the metal loss and the replacement, as the samples used showed little to no enzyme activity.

The most active preparations of the pMMO contain 1–2 Fe atoms and a Mössbauer spectrum (called doublet 1) similar to that observed from the di-iron center of the sMMO (Martinho et al. 2007). The direct correlation of the amount of doublet 1 iron in pMMO samples to activity suggests this is the site of methane oxidation. Other evidences for the involvement of PmoA in methane oxidation came from peptide labeling studies using the suicide substrate acetylene (Hynes and Knowles 1978; Prior and Dalton 1985). Acetylene is oxidized by the both MMOs as well as the closely related ammonia monooxygenase (AMO) to a ketene that binds presumably to amino acids in the active site. Exposure of methane or ammonia oxidizing cells, membrane fractions or purified pMMO to ^{14}C-acetylene results in the specific labeling of PmoA or the comparable peptide in AMO, AmoA (Choi et al. 2003; Gilch et al. 2009; Hyman and Arp 1990, 1992; Zahn and DiSpirito 1996). In

AmoA, the ketene is bound specifically to His 191 of AmoA which is adjacent to the Glu 200 of AmoA modeled to be involved in coordination of metals at this site. Lastly, mutagenesis studies in a related hydrocarbon monooxygenase also supports the third metal binding site is the active site of the pMMO (Coleman et al. 2012).

The possibility that both sMMO and pMMO utilize similar mechanisms for methane oxidation has some attraction given that it is difficult for the C–H bond to be cleaved. If this is true, however, what is the role of copper in pMMO activity? It may be that the one or both of the identified single copper sites serves to shuttle electrons to the di-iron site from the in vivo reductant, which is still unknown, but could be from a cytochrome, quinone, or NADH where electrons are collected from methanol, formate or formaldehyde oxidation (Leak and Dalton 1986; Myronova et al. 2006; Tonge et al. 1975; Torre et al. 2015).

2 Methanobactin: A Unique Copper Uptake System

Despite the uncertainty regarding the exact role(s) of copper in pMMO, it has been known for over 30 years that copper is a critical factor controlling the relative expression of pMMO/sMMO for those strains that can express both MMOs (Dalton et al. 1984; Stanley et al. 1983), and many Proteobacterial methanotrophs that can only express pMMO do not thrive in the absence of copper (Brantner et al. 1997; Choi et al. 2010). That is, there is a well-characterized "copper-switch" in some methanotrophs where sMMO expression and activity are predominant at low copper: biomass ratios, but decrease as the copper:biomass ratio increases, while pMMO expression and activity increase. Copper will also affect the formation of intracytoplasmic membranes (ICM), i.e., as copper availability increases, ICM content also increases (Brantner et al. 1997; Choi et al. 2003). Such a finding is not surprising as it is in the ICM that pMMO is located.

Although the response of Proteobacterial methanotrophs to copper was a well-known phenomenon, how these microbes sensed or collected copper remained a puzzle until the 1990s. Phenotypic characterization of constitutive sMMO mutants in *Methylosinus trichosporium* OB3b provided the initial clue when the results showed a dramatic increase in soluble copper (Fitch et al. 1993; Phelps et al. 1992). This increased soluble copper suggested the presence of a soluble copper-binding compound or copper-binding ligand similar to iron-binding siderophores. Isolation of the copper-binding compound from the spent media verified that methanotrophs were producing a copper-binding ligand in response to low copper conditions (DiSpirito et al. 1998; Zahn and DiSpirito 1996). Based on its antimicrobial activity the copper-binding compound was renamed methanobactin (MB) (DiSpirito et al. 2004; Johnson 2006). The unique structure of methanobactin made initial structural characterization problematic, however, a crystal structure of the methanobactin from *M. trichosporium* OB3b was eventually obtained (Kim et al. 2004). Corrections to the initial crystal structure was provided by NMR (Krentz et al. 2010) and verified by a high resolution crystal structure (El Ghazouani et al. 2011). Based on the unique structural elements, regulation and metal binding properties described below MBs were proposed as the first representative of a chalkophore (Greek for copper bearing or copper carrying) similar to siderophores (Greek for iron bearing or iron carrying) (Kim et al. 2004). Several other copper-binding/ copper acquisition compounds have recently been placed within the chalkophore group expanding our understanding of copper acquisition systems (Anttila et al. 2011; Chaturvedi et al. 2012; Kenney and Rosenzweig 2018; Koh et al. 2017; Pelludat et al. 1998).

2.1 Diversity of Methanobactins

Methanobactins (MBs) are small (<1300 Da) ribosomally synthesized post-translationally modified peptides (RiPPs) characterized by the presence of one C-terminal oxazolone group with an associated thioamide and either an N-terminal oxazolone or an N-terminal imidazolone or pyrazinedione group with an associated thioamide (Fig. 2). Based on the crystal and NMR structures, and sequence comparisons, MBs appear to fall into two structural groups. Group I methanobactins are characterized by the MBs from *Methylosinus* species. Multiple NMR structures and high resolutions crystal structures (Basle et al. 2018; Behling et al. 2008; El Ghazouani et al. 2011; Kenney et al. 2016; Kim et al. 2004) have been determined for Group I methanobactins. Group I MBs were initially characterized by an N-terminal oxazolone group, the presence of Cys in the mature peptide, two of which form a disulfide bond. Quite recently, the presence of an N-terminal pyrazinedione in the MB from *Methylosinus sporium* NR3K (Basle et al. 2018) has altered the initial description of group I MBs (DiSpirito et al. 2016) to include the presence of different N-terminal rings as observed with group II MBs. MBs from *Methylocystis* species represent the structurally characterized group II MBs (El Ghazouani et al. 2012; Krentz et al. 2010). Group II MBs lack Cys residues in the mature peptide, are smaller and less rigid due to the absence of the disulfide bond found in group I MBs, and have either an N-terminal imidazolone or pyrazinedione group in addition to the C-terminal oxazolone ring. Structurally characterized group II MBs also contain a sulfate group that aids in the formation of a tight bend in the molecule. The sulfate group also increases the affinity for Cu (El Ghazouani et al. 2012).

2.2 Genetics of Methanobactin

Discovery that MB from *M. trichosporium* OB3b (MB-OB3b) could indeed be chromosomally encoded, i.e., was a ribosomally synthesized and post-translationally modified polypeptide (a RiPP) led to the finding that the putative gene encoding for the precursor polypeptide, *mbnA*, appeared to be part of an operon, with genes potentially encoding for extrusion, uptake, and conversion of the precursor polypeptide into mature MB (Krentz et al. 2010). Initial efforts to use the *mbnA* sequence

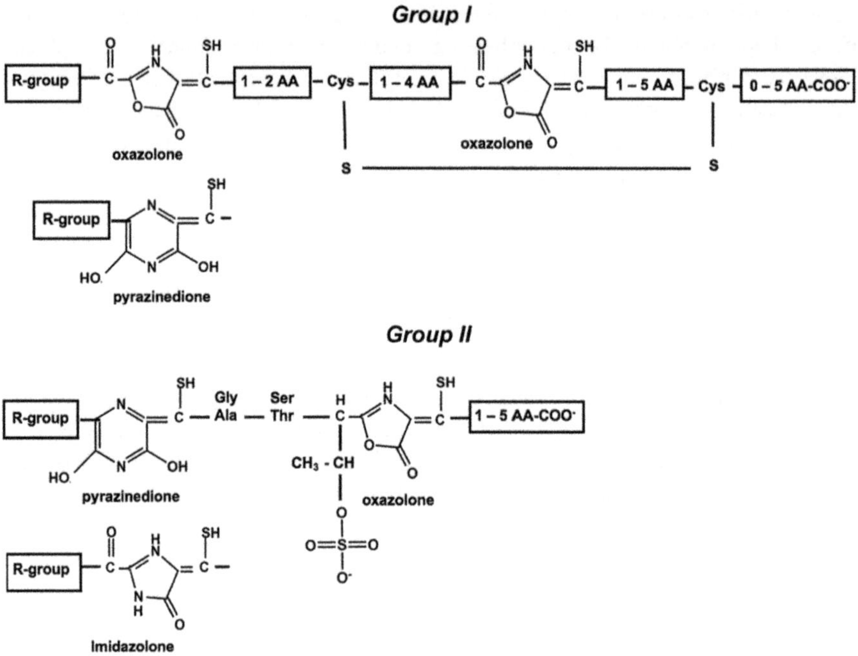

Fig. 2 Core features of Group I and II MBs. *AA* amino acid(s)

from *M. trichosporium* OB3b to screen other microbes for the ability to produce MB met with little success due to the small size of this gene and sequence variability of the core region of *mbnA*. Instead, when other genes in the putative *mbn* gene cluster were used as screens, i.e., *mbnB* or *mbnC*, additional microbes were found that appear to be able to make MB-like compounds (Arnison et al. 2013). Using such an approach, a number of microbes, including methanotrophs and non-methanotrophs, were found to have similar gene clusters, suggesting that these microbes make MB (Kenney and Rosenzweig 2013; Semrau et al. 2013). To date, however, MB has only been characterized from methanotrophs (Basle et al. 2018; El Ghazouani et al. 2012; Kenney et al. 2016; Kim et al. 2004; Krentz et al. 2010), so it cannot be stated unequivocally that non-methanotrophs produce MB-like compounds.

Despite circumstantial data indicating that *mbnA* encoded for the precursor polypeptide, definitive proof that this was indeed the case was not generated until it was knocked-out in *M. trichosporium* OB3b using marker exchange mutagenesis (Semrau et al. 2013). The resulting mutant was unable to make MB under any condition, while it was found that in *M. trichosporium* OB3b wildtype, *mbnA* expression decreased with increasing copper concentrations, suggesting that as copper becomes more available, this microbe represses expression of this copper uptake system [similar to that found for expression of siderophores when iron concentrations are varied (Neilands 1995)].

Subsequent investigation has identified the function of several (but not all genes) involved in MB biosynthesis, secretion and uptake. *mbnT*, encoding for a TonB-dependent transporter, was found via marker exchange mutagenesis to be responsible for MB uptake (Gu et al. 2016), while *mbnN*, encoding for an aminotransferase, is required for the formation of the N-terminal oxazolone ring in MB-OB3b (Gu et al. 2017). It has been long suspected that *mbnB* and *mbnC* also played critical roles in ring formation (DiSpirito et al. 2016; Gu et al. 2017; Krentz et al. 2010; Semrau et al. 2013, 2018), and that has indeed been shown to be the case (Kenney et al. 2018). The roles of other genes in MB biosynthesis or secretion have yet to be explicitly determined, but it is widely believed based on bioinformatic comparisons that *mbnM* is the means by which MB is secreted, while other genes, i.e., *mbnP* and *mbnH*, may play roles in the conversion of the precursor polypeptide to mature MB complete with heterocyclic rings or may be involved in extraction of copper after MB is internalized into the cell (Kenney and Rosenzweig 2018). Not every *mbn* gene cluster, however, has *mbnPH* (Kenney et al. 2016), so there is still some uncertainty as to its role. Finally there are two genes adjacent to *mbnT—mbnI* and *mbnR*—that putatively encode for an extracytoplasmic sigma factor and membrane sensor protein, respectively. It has been speculated that these genes play a role in controlling expression of other *mbn* genes (Kenney and Rosenzweig 2013), but experimental evidence suggests that they do not (Gu et al. 2016). It should be stressed, however, that knock-outs of these genes have not been performed to date, and should be pursued to determine what the role(s) of these genes might be.

There is significant diversity amongst characterized forms of MBs, and such differences are reflected in the identified MB gene clusters in methanotrophs (DiSpirito et al. 2016). That is, there appears to be a core group of *mbn* genes required for MB formation, secretion, uptake, and possibly regulation (i.e., *mbnABCMPHIRT*). Group I MBs have in addition an aminotransferase (encoded by *mbnN*) that is critical for the formation of the N-terminal oxazolone ring (Gu et al. 2017). As noted above, methanotrophs that produce Group II MBs (e.g., *Methylocystis rosea* SV97, *M. hirsuta* CSC1, and *M.* strains M and SB2) do not have an N-terminal oxazolone ring, but rather have a pyrazinedione/imidazolone ring. Only two of these strains have genome data available (*M. rosea* SV97 and *M.* strain SB2), and close homologs to *mbnN* are not found in the *mbn* gene clusters of these strains. Interestingly, the *mbn* gene clusters of *M. rosea* and *M.* strain SB2 include *mbnF* (encoding for a putative FAD-dependent oxidoreductase) and *mbnS* (encoding for a putative sulfotransferase). It is tempting to speculate that *mbnF* plays a role in the formation of heterocyclic rings in MB made by these methanotrophs (particularly the N-terminal pyrazinedione/imidazolone ring) and that *mbnS* is responsible for sulfonating a threonine in these MBs. More methanobactins from different methanotrophs must be characterized as well as the genomes of these methanotrophs to verify if those methanotrophs that produce MB with a N-terminal oxazolone ring have *mbnN* while those that have an N-terminal imidazolone/pyrazinedione ring have *mbnF*. In conjunction, mutational analyses should be performed to verify the role of these genes, e.g., deletion of *mbnS* in methanotrophs shown to make Group II MBs and *mbnF* in those methanotrophs that have a N-terminal imidiazolone/pyrazinedione ring. Such work would provide more definitive data, but to date only a small number of methanotrophs have validated systems for gene mutation. Alternatively, one could try to heterologous express *mbnF/S* in *Methylosinus trichosporium* OB3b (which has a validated set of genetic tools for manipulation). It is recommended that a multi-tiered approach combining bioinformatics and genetic manipulation be used to resolve this intriguing issue.

3 Metal Binding by Methanobactin

3.1 Range of Metals Bound by Methanobactin

Although MB binds copper quite strongly, it can also bind many other metals that be grouped into two general categories. One group, "Group A", include Cu(II), Ag(I), Au(III), Hg(II) and Pb(II). All these metals are bound by both heterocyclic rings of MB and are reduced after binding (Choi et al. 2006). Interestingly, although MB is believed to be a major mechanism used by methanotrophs for copper uptake, some of these Group A can compete effectively with Cu(II) for binding to MB, i.e., Hg(II) and Au(III). As a result, copper uptake can be inhibited in the presence of at least some other Group A metals, with the result that sMMO expression is possible, even in the presence of copper if other Group A metals are also present (Kalidass et al. 2015). It may be that those methanotrophs that can express both forms of MMO (i.e., "switchover" methanotrophs) have a competitive advantage over non-switchover methanotrophs in that they can express an active form of MMO (sMMO) when copper uptake is limited by the presence of other Group A metals. The second group of metals, "Group B", include Cd(II), Co(II), Fe(III), Ni(II), Zn(II), and Mn (II) (Choi et al. 2006). These metals are only bound to one of the heterocyclic rings of MB, are not reduced by MB, and copper easily displaces them. The presence of these metals had little effect on methanotrophic growth or MMO expression, indicating that they did not limit copper uptake (Kalidass et al. 2015). The fact that MB targets soft metals such as Hg(II), Cu(II), etc. (i.e., Group A metals) is not surprising when one considers that these metals are preferentially bound by N-/S-containing ligands (as found in MB) while Group B metals bind preferentially to oxygen-containing ligands that are not part of MB (Nieboer and Richardson 1980).

3.2 Detoxification of Mercury Species by Methanobactin

One interesting implication of the fact that MB binds Hg(II) even in the presence of copper is that this may be a novel biogenic mechanism of mercury detoxification. Mercury, in a variety of forms is quite toxic, i.e., inorganic mercury (Hg(II)) forms very stable complexes with sulfhydryl groups, leading to protein inactivation. Other forms of Hg, particularly organic forms are of even greater concern, e.g., methylmercury is a potent neurotoxin (Barkay and Wagner-Döbler 2005).

It has been found that MB, in addition to collecting copper important for methanotrophic activity, also detoxifies Hg(II) as this metal rapidly and strongly binds to MB (Baral et al. 2014; Vorobev et al. 2013). That is, if *M. trichosporium*

OB3b is exposed to 5 μM Hg(II) (provided as $HgCl_2$), no growth is observed. If an equimolar amount of MB from *M. trichosporium* OB3b was exogenously added, however, growth was restored. MB bound Hg(II) strongly and was apparently reduced to elemental mercury after binding, and thus became unavailable for subsequent protein inactivation. Such data also indicate that the initial amount of MB in the growth flasks was quite low (not surprisingly as cell densities were initially quite low), but that MB can and does reduce Hg(II) toxicity if present at least at an equimolar amount. Further, it was found that MB served as a general prophylactic and allowed other methanotrophs incapable of producing MB to grow in the presence of Hg(II). Such a finding is intriguing as it is known that MB is secreted after synthesis (Gu et al. 2016) and as such could serve to protect the broader microbial community from Hg(II) toxicity.

Interestingly, not only does MB play a critical role in controlling the availability (and by extension, toxicity) of Hg(II), it was later found that methanotrophs capable of expressing MB also could demethylate methylmercury and by so doing, reduce the toxicity of this potent neurotoxin (Lu et al. 2017). *merB*, encoding for the organomercurial lyase, is not found in the genome of methanotrophs such as *M. trichosporium* OB3b, indicating that another mechanism, hitherto unknown mechanism is used that involves MB. At this moment, it is still unclear how MB-producing methanotrophs degrade methylmercury, but MB is critical for such demethylation (Lu et al. 2017). It may be that MB serves to bind and deliver methylmercury in a controlled fashion into the cell where it is then attacked by some (as yet unknown) enzymes. Initial data suggests that the methanol dehydrogenase may be involved as the addition of methanol inhibited methylmercury degradation. That is, it may be that methylmercury is attacked by the methanol dehydrogenase where inorganic mercury is released and the methyl group converted either to methanol or formaldehyde depending on how many electrons are transferred. Such a mechanism is analogous to that found for methylamine oxidation by the methylamine dehydrogenase (Bishop et al. 1996; Bishop and Davidson 1998; Davidson 2001, 2005), but more work remains to be done to prove that this is indeed the case.

3.3 *Formation of Gold Nanoparticles by Methanobactin*

As mentioned earlier, Group A metals are those that are bound to both rings of MB and are reduced after binding. In particular, not only was Au(III) was reduced to elemental gold (Au(0)), substantial amounts of gold nanoparticles were formed (Choi et al. 2006). Both Group I and II MBs were found to form gold nanoparticles from Au(III), and such nanoparticles were stable, and had uniform size (typically ~2 nm) and spherical shape (Choi et al. 2006; DiSpirito et al., unpublished data). As gold nanoparticles have used for decades as biosensors and also in some medical therapies (Connor et al. 2005; Ghosh et al. 2008; Mandal et al. 2005; Saha et al. 2012), there is great interest in developing more efficient methods for their production. Current methods typically involve chemical reduction, followed by temperature-, sonic- or photo-induction and chemical stabilization (Guo and Wang 2007; Polte et al. 2010; Zhou et al. 2009). MB produces gold nanoparticles in the absence of an external reductant and does so at room temperature without induction, making its use in gold nanoparticle production a potentially viable alternative method (DiSpirito et al. 2011, 2014).

3.4 *Methanobactin Allows Methanotrophs to Exert a "Copper Monopoly"*

As stated earlier, aerobic methanotrophy is highly dependent on copper, with expression and activity of methane monooxygenases controlled by copper availability (Murrell et al. 2000; Semrau et al. 2010). Interestingly, biological removal of nitrous oxide produced by denitrifiers during the conversion of nitrate to dinitrogen is also strongly dependent on copper as the nitrous oxide reductase or NosZ (responsible for reducing N_2O to N_2) has a high copper requirement (Brown et al. 2000). Under copper stress induced with synthetic chelators, transient N_2O accumulation was found during denitrification (Moffett et al. 2012; Sullivan et al. 2013; Twining et al. 2007). Given the importance of copper availability for both methane oxidation and N_2O reduction, we hypothesized that methanotrophs may be able to exert a "monopoly" over copper through production of copper chelating agents such as MB, leading to increased rates and amounts of N_2O production by denitrifiers due to inactivation of NosZ. Indeed, when the denitrifier *Pseudomonas stutzeri* DCP-Ps1 was incubated in co-culture with *M. trichosporium* OB3b or with purified MB from *M. trichosporium*, stoichiometric N_2O production was observed from NO_3^- reduction (Chang et al. 2018). No significant N_2O accumulation was observed in co-cultures with a mutant of *M. trichosporium* OB3b defective in MB production. Additional assays showed that increased N_2O production by *P. stutzeri* DCP-Ps1 in the presence of MB was due to the inability of this microbe to collect copper. That is, in the absence of MB, cellular copper was 0.69 ± 0.14 μg Cu mg protein^{-1}, but only 0.10 ± 0.03 μg Cu mg protein^{-1} in the presence of 1 μM MB ($p < 0.01$). The reduced copper uptake led to reduced *nosZ* transcription, i.e., in the absence of MB, *nosZ* expression was 4.42 ± 1.13 *nosZ* transcript (*recA* transcript)$^{-1}$, while in the presence of MB, *nosZ* transcription decreased significantly ($p < 0.05$) to 1.59 ± 0.55 *nosZ* transcript (*recA* transcript)$^{-1}$. Similar findings of the effects of MB on N_2O production, copper uptake and gene expression were observed with three other denitrifier strains, i.e., *Dechloromonas aromatica* RCB, *Paracoccus denitrificans* ATCC17741, and *Shewanella loihica* PV-4 (Chang et al. 2018).

These data conclusively showed that methanotrophs can exert a "copper monopoly", thereby limiting the ability of other microbes, i.e., denitrifiers, to collect copper. This is particularly relevant as co-occurrence of methanotrophy and denitrification in situ is not rare. As the main substrates for aerobic methanotrophs are CH_4 and O_2, which diffuse towards the oxic-anoxic interface from opposite ends of the oxygen gradient, the largest abundance of methanotrophs is often found at this interface in subsurface environments (Reim et al. 2012). The oxic-anoxic interface is also a "hotspot" for soil denitrification in agricultural soils. That is, NO_3^-/NO_2^- formed from oxidation of ammonia by nitrifying microbes in the oxic upper soil layer leaches to this interface where it is used by denitrifying bacteria as terminal electron acceptors (Maeda et al. 2003). Hitherto unrecognized microbial competition for copper thus is a significant mechanism that affects both methane and nitrous oxide emissions. As a result, attempts to stimulate methanotrophs to enhance methane removal can inadvertently boost net N_2O production, and thus increase rather than decrease net greenhouse gas emissions. Microbial competition for copper must be more fully explored to learn how best to predict, and hopefully minimize future greenhouse gas emissions, as well as to better understand the diversity of microbial interactions that occur in situ. It may be that methanotrophs, through the production of MB not only affect NosZ activity, but also the activity of other enzymes containing copper, e.g., ammonia monooxygenase and nitrite reductase.

4 Other Functions of Methanobactin

Methanobactin clearly plays a critical role in controlling the availability and speciation of multiple metals. Interestingly, MBs also have many other intriguing properties that suggest these molecules may be considered to be "moonlighting" proteins, i.e., proteins that perform multiple functions (either biochemical and/or biophysical) and such functions are not due to gene fusions, RNA splice variants or breakdown into multiple fragments that have different properties (Jeffery 1999, 2003, 2009). Such proteins have been found in both eukaryotes and prokaryotes. Although a natural conclusion may be that moonlighting proteins are rare, there are almost 400 such proteins identified on the MoonProt database as of August 2018 (moonlightingproteins.org). Most such proteins were adventitiously found and it is difficult to predict a priori if any protein can perform multiple functions. Methanobactin is no exception to this general rule. That is, although initial studies clearly showed that MB played a critical role in copper binding and uptake, subsequent studies found that at least one form of MB, that from *M. trichosporium* OB3b (MB-OB3b) also had significant oxidase, superoxide dismutase and hydrogen peroxide reductase activity (Choi et al. 2008). Specifically, MB-OB3b (a Group I MB), after being reduced by either NADH or duroquinol, exhibited substantial oxidase activity in both the presence and absence of copper. Superoxide dismutase and hydrogen peroxide reductase activity, however, was found only when copper was present, and hydrogen peroxide reductase activity also required MB-OB3b to be first reduced. Such findings indicate the MB-OB3b is redox active, and may play an important role in scavenging potential toxic radicals formed via methane oxidation and dioxygen respiration. It should be noted, however, that similar properties for Group II MBs have not been established, thus it is not known how if all MBs are the moonlighting proteins. The finding that both groups of MB bind and reduce Group A metals suggests, however, that both can serve as effective electron carriers and as such Group II MBs may have similar properties as that observed for MB-OB3b. We hasten to add, however, that we do not know if these additional functions are physiologically relevant, i.e., are these abilities of MB merely a fortuitous accident or actively promoted by the cell?

5 Potential Medical Applications of Methanobatin

5.1 Use of Methanobactin to Treat Wilson Disease/Fatty Liver Disease

Methanobactin has been shown to be uniquely effective in treating the copper-related disease, Wilson disease (WD) (Einer et al. 2019; Lichmannegger et al. 2016; Mullert et al. 2018; Summer et al. 2011; Zischka et al. 2011). WD is an autosomal recessive disease where the Cu-transporting P-type ATPase, ATP7B, is inactivated in the liver (de Bie et al. 2007). As a result, copper export to the bile is limited, causing copper accumulation in the liver. If not corrected, such accumulation leads to sudden and complete hepatic failure due to copper induced production of oxygen radicals via Fenton and Harber-Weiss reactions (Fentona and Chemizmu 2009; Fridovick 1978; Kehrer 2000). Other symptoms of WD include serious neurological and psychiatric problems (e.g., migraine headaches, seizures, and cognitive impairment) due to spillover of copper from the liver to the brain (Ala et al. 2007; Dalyl and Padmanaban 2014; Dusek et al. 2014).

There is no cure for WD, although a small number of therapies have been developed that have some efficacy (Ala et al. 2007; Dalyl and Padmanaban 2014; Roberts 2011; Schilsky 2001, 2014). Current therapies consist of one or more of the following options: (1) use of chelating agents to remove copper via enhanced urine excretion; (2) implementation of a low copper diet, and (3) zinc dietary supplements to stimulate production of the copper-binding protein metallothionein, thereby reducing the buildup of copper in the liver (Ala et al. 2007; Dalyl and Padmanaban 2014; Schilsky 2001). Of these options, the use of chelating agents is the most commonly prescribed treatment.

Two drugs, D-penicillamine and triethylenetetramine, are currently approved by the US Food and Drug Administration and the European Medicines Agency. Another chelating agent, tetrathiolmolybdate is under consideration, but is not yet approved by either agency. Unfortunately, D-penicillamine and triethylenetetramine have limited effectiveness, must be taken multiple times a day, and are only useful when taken before the body burden of copper becomes high. D-penicillamine, triethylenetetramine and tetrathiolmolybdate have many other flaws, including that they remove excess copper via urination and not the bile (a more efficacious route), they have no value if severe liver damage has already occurred, and unfortunately for a significant fraction of patients are of no value. For situations where excessive liver damage occurs, death is almost certain unless a liver transplant is performed, which carries significant health risks (Gitlin 2003; Schilsky 2014; Walshe 2007; Weiss and Stremmel 2012). Thus, better therapies for treating WD are critical.

Using a robust rodent model for WD, MB from *M. trichosporium* OB3b (MB-OB3b) has been shown to be a very powerful de-coppering agent (Lichmannegger et al. 2016; Mullert et al. 2018; Summer et al. 2011; Zischka et al. 2011). That is, MB-OB3b reversed the buildup of copper in the liver as well as reduced protein cross-linking due to copper accumulation, allowing the liver to function normally (Zischka et al. 2011). Further, unlike other treatments, MB-OB3b successfully treated rodents with advanced liver failure, indicating that MB may obviate the need for liver transplants in some patients (Lichmannegger et al. 2016). Lastly, data from rodent studies indicate that MB need only be provided at intervals, i.e., not daily, suggesting that WD can be managed via short cycles of MB treatment followed by extensive drug holidays (Lichmannegger et al. 2016; Mullert et al. 2018). This can be quite beneficial to ensure patient compliance in following appropriate drug therapies.

Not only does copper accumulate in the liver of WD patients, it is common to also observe fat accumulation, often leading to the misdiagnosis of WD as non-alcoholic fatty liver disease (NAFLD) (Stattermayer et al. 2015). Perhaps not surprisingly, due to excessive consumption of fat and sugar, NAFLD is on the rise, particularly in Western nations (Vernon et al. 2011). Given that it is not uncommon for WD patients to also have poor diets, we hypothesized that high calorie diets could aggravate liver damage. Indeed we discovered that such diets increased the onset of liver damage in rodent models of WD, and that MB could also ameliorate this damage (Einer et al. 2019).

5.2 *Potential Applications for Treatment of Alzheimer's Disease and Some Cancers*

Clearly, MB has significant promise for the treatment of WD. It may also be useful for treating other metal-associated pathologies. One significant potential application is for the treatment of late onset Alzheimer's Disease (AD). A key element of late-onset AD is the presence of excess misfolded proteins resulting in neuronal-damaging plaques (Ballard et al. 2011; Davies et al. 2014; Gaggelli et al. 2006; Gamez and Caballero 2015). The causative agent(s) of late-onset AD is(are) not defined, but copper appears to play a key role (Ballard et al. 2011; Bossy-Wetzel et al. 2004; Brewer 2014; Bush 2008; Cherny et al. 2001; Gaggelli et al. 2006; Gamez and Caballero 2015; Huang et al. 2004; Miller et al. 2006; Squitti 2012; Squitti et al. 2014). It may be that MB can treat AD via sequestration of copper, limiting plaque development. Such an application, however, must be considered speculative at this time, and much more study required to determine what if any therapeutic value MB may have on AD patients.

One application, however, that may be promising is the use of MB to inhibit oncogenesis. That is, it was recently reported that copper is required for oncogenic

BRAF kinase-driven tumorigenesis. If copper uptake is disrupted through the addition of tetrathiolmolybdate, tumorigenesis is inhibited (Brady et al. 2014). As such, inhibiting kinase activity with the potent copper chelator MB could be used in the treatment of these cancers.

6 Conclusion and Potential Avenues for Future Work

MBs are novel RiPPs containing heterocyclic rings and thioamide groups that combined not only serve to enable MBs to bind copper with extraordinarily high affinity, but also can reduce other metal such as gold and mercury. Indeed, MBs appear to have multiple roles in methanotrophy, and they also have intriguing medical applications for the treatment of diseases associated with copper stress, e.g., Wilson Disease and BRAF-driven tumorigenesis. Although much has been learned in the past 10 years about the genetics, biosynthesis and diversity of MBs, there is still much to be discovered, including:

- How diverse are MBs? To date, only a handful of forms have been extensively characterized, although a large number of diverse microbes appear to have *mbn*-like gene clusters. Under what conditions are these genes expressed? What is the function(s) of the putative MB-like compounds that some non-methanotrophs appear to make? Do all these compounds serve to bind copper, and if so, for what purpose (i.e., is it for copper uptake or copper detoxification)? What if any other metals do these MBs bind? Can these MBs also be considered to be moonlighting proteins?
- Much more information is needed to determine the complete biosynthetic pathway of MB and the role of different genes in the conversion of the precursor polypeptide to an active copper chelating agent. Much has been learned about the role of several genes (e.g., *mbnN*, *mbnB*, *mbnC*, and *mbnT*), but the function(s) of others is/are still to be experimentally determined (e.g., *mbnF*, *mbnS*, *mbnP*, and *mbnH*, amongst other genes).
- Although MB production is clearly evident in laboratory cultures, expression of *mbn* genes and production of MB in situ where methanotrophic activity has been observed has yet to be done. How prevalent is MB in natural systems? How does MB production in situ affect copper availability to the broader microbial community? For example, what is the resulting effect on net greenhouse gas emissions?
- What is the role of MB in the observed methylmercury degradation by methanotrophs that can express MB? How widespread is methanotrophic-mediated methylmercury degradation in situ?
- It is apparent that MB synthesis and *mbn* gene expression are regulated by the availability of copper, but more work is needed to determine how copper controls MB production, and the regulatory elements involved.
- How can MB be modified to enhance its use in specific application(s), e.g., as a therapeutic agent? Along these lines, how can MB production be increased for

such applications, i.e., through heterologous expression of required *mbn* genes for MB biosynthesis and secretion in a more amenable host or determination of selective growth/reactor conditions for enhanced production from methanotrophs?

References

Ala A, Walker AP, Ashkan A, Dooley JS, Schilsky M (2007) Wilson's disease. Lancet 369:397–408

Anttila J, Petri Heinonen P, Nenonen T, Pino A, Iwaï H, Kauppi E, Soliymani R, Baumann M, Saksi J, Suni N, Haltia T (2011) Is coproporphyrin III a copper-acquisition compound in *Paracoccus denitrificans*? Biochim Biophys Acta 1807:311–318

Arnison PG, Bibb MJ, Bierbaum G, Bowers AA, Bugni TS, Bulaj G, Camarero JA, Campopiano DJ, Challis GL, Clardy J, Cotter PD, Craik DJ, Dawson M, Dittmann E, Donadio S, Dorrestein PC, Entian K-D, Fischbach MA, Gravelli JS, Goransson U, Gruber CW, Haft DH, Hemscheidt TK, Hertweck C, Hill C, Horswill AR, Jaspars M, Kelly WL, Klinman JP, Kiuipers OP, Link AJ, Liu W, Marahiel MA, Mitchell DA, Moll GN, Moore B, Muller R, Nair SK, Nes IF, Norris GE, Olivera BM, Onaka H, Patchett ML, Piel J, Reaney MJT, Rubuffat S, Ross RP, Sthl H-G, Schmidt EW, Selsted ME, Severinov K, Shen B, Sivonen K, Smith L, Stein T, Sussmuth RD, Tagg JR, Tang G-L, Truman AW, Vederas JC, Walsh CT, Walton JD, Wenz SC, Willey JM, van der Donk WA (2013) Ribosomally synthesized and post-translationally modified peptide natural products: overview and recommendations for universal nomenclature. Nat Prod Rep 30:108–160

Balasubramanian R, Smith S, Rawat S, Yatsunyk LA, Stemmier TL, Rosenzweig AC (2010) Oxidation of methane by a biological dicopper centre. Nature 465:115–120

Ballard C, Gauthier S, Cobett A, Brayne C, Aarsland D, Jones E (2011) Alzheimer's disease. Lancet 377:1019–1031

Banerjee R, Meier KK, Munck C, Lipscomb JD (2013) Intermediate P∗ from soluble methane monooxygenase contains a diferrous cluster. Biochemistry 52(25):4331–4342

Banerjee R, Proshlyakov Y, Lipscomb JD, Proshlyakov DA (2015) Structure of the key species in the enzymatic oxidation of methane to methanol. Nature 518:431–435

Baral BS, Bandow NL, Vorobev A, Freemeier BC, Bergman BH, Herdendorf T, Fuentes N, Ellias L, Turpin E, Semrau JD, DiSpirito AA (2014) Mercury binding by methanobactin from *Methylocystis* strain SB2. J Inorg Biochem 141:161–169

Barkay T, Wagner-Döbler I (2005) Microbial transformations of mercury: potentials, challenges, and achievements in controlling mercury toxicity in the environment. Adv Appl Microbiol 57:1–52

Basle A, El Ghazouani A, Lee J, Dennison C (2018) Insight into metal removal from peptides that sequester copper for methane oxidation. Chemistry 24(18):4515–4518

Basu P, Katterle B, Andersson KK, Dalton H (2003) The membrane-associated form of methane mono-oxygenase from *Methylococcus capsulatus* (Bath) is a copper/iron protein. Biochem J 369:417–427

Behling LA, Hartsel SC, Lewis DE, DiSpirito AA, Choi DW, Masterson LR, Veglia G, Gallagher WH (2008) NMR, mass spectrometry and chemical evidence reveal a different chemical structure for methanobactin that contains oxazolone rings. J Am Chem Soc 130(38):12604–12605

Bishop GR, Davidson VL (1998) Electron transfer from the aminosemiquinone reaction intermediate of methylamine dehydrogenase to amicyanin. Biochemistry 37(31):11026–11032

Bishop GR, Brooks HB, Davidson VL (1996) Evidence for a tryptophan tryptophylquione aminosemiquione intermediate in the physologic reaction between methylamine dehydrogenase and amicyanin. Biochemistry 35:8948–8954

Bossy-Wetzel E, Schwarzenbacher R, Lipton SA (2004) Molecular pathways to neurodegeneration. Nat Med 10:S2–S9. https://doi.org/10.1038/nm1067

Brady DC, Crowe MS, Turski ML, Hobbs GA, Yao X, Chaikuad A, Knapp S, Xio K, Cambell SL, Thiele DJ, Counter CM (2014) Copper is required for oncogenic BRAF signalling and tumorigenesis. Nature 509:492–496

Brantner CA, Buchholz LA, McSwain CL, Newcomb LL, Remsen CC, Collins MLP (1997) Intracytoplasmic membrane formation in Methylomicrobium album BG8 is stimulated by copper in the growth medium. Can J Microbiol 43(7):672–676

Brewer GJ (2014) Alzheimer's disease causation by copper toxicity and treatment with zinc. Front Aging Neurosci 6:92

Brown K, Tegoni M, Prudêncio M, Pereira AS, Besson S, Moura JJ, Moura I, Cambillau C (2000) A novel type of catalytic copper cluster in nitrous oxide reductase. Nat Struct Mol Biol 7 (3):191–195

Bush AI (2008) Drug development based on metals hypothesis of Alzheimer's disease. J Alzheimers Dis 15(2):223–240

Cao L, Calkarau O, Rosenzweig AC, Ryde U (2018) Quantum refinement does not support dinuclear copper sites in crystal structures of particular methane monooxygenase. Angew Chem 130:168–172

Chan SI, Wang VCC, Lai JCH, Yu SSF, Chen PPY, Chen KHC, Chen CL, Chan MK (2007) Redox potentiometry studies of particulate methane monooxygenase: support for a trinuclear copper cluster active site. Angew Chem Int Ed 46(12):1992–1994

Chang SL, Wallar BJ, Lipscomb JD, Mayo KH (1999) Solution structure of component B from methane monooxygenase derived through heteronuclear NMR and molecular modeling. Biochemistry 38:5799–5812

Chang J, Gu W, Park D, Semrau JD, DiSpirito AA, Yoon S (2018) Methanobactin from *Methylosinus trichosporium* OB3b inhibits N_2O reduction in denitrifiers. ISME J 12:2086–2089

Chaturvedi KS, Hung CS, Crowley JR, Stapleton AE, Henderson JP (2012) The siderophore yersiniabactin binding copper to protect pathogens during infections. Nat Chem Biol 8 (8):731–736

Cherny R, Atwood CS, Xilinas ME, Gray DN, Jones WDM, McLean CA, Barnham KJ, Volitakis I, Fraser FW, Kim Y-S, Huang X, Goldstein LE, Moir RD, Lim JT, Beyreuther K, Zheng H, Tanzi RE, Masters CL, Bush A (2001) Treatment with copper-zinc chelator markedly and rapidly inhibits B-amyloid accumulation in Alzheimer's disease transgenic mice. Neuron 30:665–676

Choi DW, Kunz RC, Boyd ES, Semrau JD, Antholine WE, Han JI, Zahn JA, Boyd JM, de la Mora AM, DiSpirito AA (2003) The membrane-associated methane monooxygenase (pMMO) and pMMO-NADH:quinone oxidoreductase complex from *Methylococcus capsulatus* Bath. J Bacteriol 185(19):5755–5764

Choi DW, Antholine WE, Do YS, Semrau JD, Kisting CJ, Kunz RC, Campbell D, Rao V, Hartsel SC, DiSpirito AA (2005) Effect of methanobactin on the activity and electron paramagnetic resonance spectra of the membrane-associated methane monooxygenase in *Methylococcus capsulatus* Bath. Microbiology 151(Pt 10):3417–3426

Choi DW, Do YS, Zea CJ, McEllistrem MT, Lee SW, Semrau JD, Pohl NL, Kisting CJ, Scardino LL, Hartsel SC, Boyd ES, Geesey GG, Riedel TP, Shafe PH, Kranski KA, Tritsch JR, Antholine WE, DiSpirito AA (2006) Spectral and thermodynamic properties of Ag(I), Au(III), Cd(II), Co (II), Fe(III), Hg(II), Mn(II), Ni(II), Pb(II), U(IV), and Zn(II) binding by methanobactin from *Methylosinus trichosporium* OB3b. J Inorg Biochem 100(12):2150–2161

Choi DW, Semrau JD, Antholine WE, Hartsel SC, Anderson RC, Carey JN, Dreis AM, Kenseth EM, Renstrom JM, Scardino LL, Van Gorden GS, Volkert AA, Wingad AD, Yanzer PJ, McEllistrem MT, de la Mora AM, DiSpirito AA (2008) Oxidase, superoxide dismutase, and hydrogen peroxide reductase activities of methanobactin from types I and II methanotrophs. J Inorg Biochem 102(8):1571–1580

Choi DW, Bandow NL, McEllistrem MT, Semrau JD, Antholine WE, Hartsel SC, Gallagher W, Zea CJ, Pohl NL, Zahn JA, DiSpirito AA (2010) Spectral and thermodynamic properties of methanobactin from γ-proteobacterial methane oxidizing bacteria: a case for copper competition on a molecular level. J Inorg Biochem 104(12):1240–1247

Colby J, Dalton H (1976) Some properties of a soluble methane mono-oxygenase from *Methylococcus capsulatus* strain Bath. Biochem J 157:495–497

Colby J, Dalton H (1978) Resolution of the methane monooxygenase of *Methylococcus capsulatus* (Bath) into three components. Purification and properties of component C, a flavoprotein. Biochem J 171:461–468

Colby J, Dalton H (1979) Characterization of the second prosthetic group of the flavoprotein NADH-acceptor reductase (component C) of the methane monooxygenase from *Methylococcus capsulatus* (Bath). Biochem J 177:903–908

Coleman NV, Le NB, Ly MA, Ogawa HE, McCarl V, Wilson NL, Holmes AJ (2012) Hydrocarbon monooxygenase in Mycobacterium: recombinant expression of a member of the ammonia monooxygenase superfamly. ISME J 6:171–182

Connor EE, Mwamuka J, Gole A, Murphy CJ, Wyatt DD (2005) Gold nanoparticles are taken up by human cells but do not cause acute cytotoxicity. Small 1(3):325–327

Culpepper MA, Cutsail GEI, Gunderson WA, Hoffman BM, Rosenzweig AC (2014) Identification of the valence and coordination environment of the particulate methane monooxygenase copper centers by advanced EPR characterization. J Am Chem Soc 136(33):11767–11775

Dalton H, Prior SD, Leak DJ, Stanley SH (1984) Regulation and control of methane monooxygenase. In: Crawford RL, Hanson RS (eds) Microbial growth in C1 compounds. American Society for Microbiology, Washington, DC

Dalyl A, Padmanaban M (2014) Wilson's disease: etiology, diagnosis, and treatment. Dis Mon 60:450–459

Davidson VL (2001) Pyrroloquinoline quinone (PQQ) from methanol dehydrogenase and tryptophan trypophyloquione (TTQ) from methylamine dehydrogenase. Adv Protein Chem 58:95–140

Davidson VL (2005) Structure and mechanism of tryptophyloquione enzymes. Bioorg Chem 33:159–170

Davies KM, Bohic S, Carmona A, Ortega R, Cottam V, Hare DJ, Finberg JPM, Reyes S, Halliday GM, Mercer JFB, Double KL (2014) Copper pathology in vulnerable brain regions in Parkinson's disease. Neurobiol Aging 35:858–866

de Bie P, Muller P, Wijmenga C, Klomp LWJ (2007) Molecular pathogenesis of Wilson and Menkes disease: correlation of mutations with molecular defects and disease phenotypes. J Med Genet 44:678–688

Dedysh SN, Knief C, Dunfield PF (2005) *Methylocella* species are facultatively methanotrophic. J Bacteriol 187:4665–4670

DeWitt JG, Bentsen JG, Rosenzweig AC, Hedman B, Green J, Pilkington S, Papaefthymiou GC, Dalton H, Hodgson KO, Lippard SJ (1991) X-ray absorption, Mossbauer, and EPR studies of the dinuclear iron center in the hydroxylase component of methane monooxygenase. Am J Chem Soc 113:9219–9235

DiSpirito AA, Zahn JA, Graham DW, Kim HJ, Larive CK, Derrick TS, Cox CD, Taylor A (1998) Copper-binding compounds from *Methylosinus trichosporium* OB3b. J Bacteriol 180 (14):3606–3613

DiSpirito AA, Zahn JA, Graham DW, Kim HJ, Alterman M, Larive C (2004) Methanobactin: a copper binding compound having antibiotic and antioxidant activity from methanotropic bacteria. US Patent No. 7,199,099. United States Patent and Trademark Office, Washington, DC

DiSpirito AA, Choi D-W, Semrau JD, Keeney D (2011) Use of methanobactin. In: United States Patent Office (ed) The regents of the University of Michigan and Iowa State University Research Foundation

DiSpirito AA, Choi D-W, Semrau JD, Keeney D (2014) Use of methanobactin. In: United States Patent Office (ed) Iowa State University Research Foundation and Regents of the University of Michigan

DiSpirito AA, Semrau JD, Murrell JC, Gallagher WH, Dennison C, Vuilleumier S (2016) Methanobactin and the link between copper and bacterial methane oxidation. Microbiol Mol Biol Rev 80(2):387–409

Dunfield PF, Dedysh SN (2014) Methylocella: a gourmand among methanotrophs. Trends Microbiol 22(7):368–369

Dunfield PF, Belova SE, Vorob'ev AV, Cornish SL, Dedysh SN (2010) *Methylocapsa aurea* sp. nov., a facultative methanotroph possessing a particulate methane monooxygenase, and emended description of the genus *Methylocapsa*. Int J Syst Evol Microbiol 60:2659–2664

Dusek P, Roos PM, Litwin T, Schneider SA, Flaten TP, Aaseth J (2014) The neurotoxicity of iron, copper and manganese in Parkinson's and Wilson's diseases. J Trace Elem Med Biol 31:193–203

Einer C, Leitzinger C, Lichtmannegger J, Eberhagen C, Rieder T, Borchard S, Wimmer R, Denk G, Popper B, Neff F, Polishchuk EV, Polishchuk RS, Hauck SM, von Toerne C, Karst U, Mullert J-C, Baral BS, DiSpirito AA, Kremer AE, Semrau JD, Weiss KH, Hohenester S, Zischka H (2019) A high caloric diet augments mitochondrial disfunction and triggers severe liver damage in Wilson disease rats. J Hepatol 7(3):571–596

El Ghazouani A, Basle A, Firbank SJ, Knapp CW, Gray J, Graham DW, Dennison C (2011) Copper-binding properties and structures of methanobactins from *Methylosinus trichosporium* OB3b. Inorg Chem 50(4):1378–1391

El Ghazouani A, Baslé A, Gray J, Graham DW, Firbank SJ, Dennison C (2012) Variations in methanobactin structure influences copper utilization by methane-oxidizing bacteria. Proc Natl Acad Sci USA 109(22):8400–8404

Elango N, Radhakrishnan R, Froland WA, Wallar BJ, Earhart CA, Lipscomb JD, Ohlendorf DH (1997) Crystal structure of the hydroxylase component of methane monooxygenase from *Methylosinus trichosporium* OB3b. Protein Sci 6:556–568

Fentona R, Chemizmu KD (2009) Fenton reaction-controversy concerning the chemistry. Ecol Chem Engin 16(3):347–358

Fitch MW, Graham DW, Arnold RG, Agarwal SK, Phelps P, Speitel GE, Georgiou G (1993) Phenotypic characterization of copper-resistant mutants of *Methylosinus trichosporium* Ob3b. Appl Environ Microbiol 59(9):2771–2776

Fontecave M, Menage S, Duboc-Toia C (1998) Functional models of non-heme diiron enzymes. Coord Chem Rev 178–180:1555–1572

Fox BG, Lipscomb JD (1988) Purification of a high specific activity methane monooxygenase hydroxylase component from a type II methanotroph. Biochem Biophys Res Commun 154:165–170

Fox BG, Surerus KK, Münck E, Lipscomb JD (1988) Evidence for a mu-oxo-bridged binuclear iron cluster in the hydroxylase component of methane monooxygenase – Mössbauer and electron-paramagnetic-resonance studies. J Biol Chem 263(22):10553–10556

Fox BG, Froland WA, Dege JE, Lipscomb JD (1989) Methane monooxygenase from *Methylosinus trichosporium* Ob3b – purification and properties of a 3-component system with high specific activity from a type-II methanotroph. J Biol Chem 264(17):10023–10033

Fox BG, Liu Y-N, Dege JE, Lipscomb JD (1991) Complex formation between the protein components of the methane monooxygenase from *Methylosinus trichosporium* OB3b. J Biol Chem 266:540–550

Fox BG, Hendrich MP, Surerus KK, Andersson KK, Froland WA, Lipscomb JD, Münck E (1993) Mössbauer, EPR, and ENDOR studies of the hydroxylase and reductase components of methane monooxygenase from *Methylosinus trichosporium* Ob3b. J Am Chem Soc 115(9):3688–3701

Fridovick I (1978) The biology of oxygen radicals. Science 201(4359):875–880

Gaggelli E, Kozlowski H, Valensin D, Valensin G (2006) Copper homeostasis and neurodegenerative disorders (Alzheimer's, Prion, and Parkinson's diseases and amyotrophic lateral sclerosisa). Chem Rev 106:1995–2044

Gamez P, Caballero AB (2015) Copper in Alzheimer's disease: implications in amyloid aggregation and neurotoxicity. AIP Adv 5:092503

Ghosh P, Han G, De M, Kim CK, Rotello VM (2008) Gold nanoparticles in delivery applications. Adv Drug Deliv Rev 60:1307–1315

Gilch S, Vogel M, Lorenz MW, Meyer O, Schmidt I (2009) Interaction of the mechanism-based inactivator acetylene with ammonia monooxygenase of *Nitrosomonas europaea*. Microbiology 155:279–284

Gitlin JD (2003) Wilson disease. Gastroenterology 125(6):1868–1877

Green J, Dalton H (1985) Protein B of soluble methane monooxygenase from *Methylococcus capsulatus* (Bath). A novel regulatory protein of enzyme activity. J Biol Chem 260 (29):15795–15801

Gu W, Farhan U-HM, Baral BS, Turpin EA, Bandow NL, DiSpirito AA, Lichtmannegger J, Kremmer E, Zischka H, Semaru JD (2016) A TonB dependent transporter is responsible for methanobactin uptake by *Methylosinus trichosporium* OB3b. Appl Environ Microbiol 82:1917–1923

Gu W, Baral BS, DiSpirito AA, Semrau JD (2017) An aminotransferase is responsible for the deamination of the N-terminal leucine and required for formation of oxazolone ring A in Methanobactin of *Methylosinus trichosporium* OB3b. Appl Environ Microbiol 82:e01619-01616

Guo S, Wang E (2007) Synthesis and electrochemical applications of gold nanoparticles. Anal Chim Acta 598:181–192

Hakemian AS, Rosenzweig AC (2007) The biochemistry of methane oxidation. Annu Rev Biochem 76:223–241

Hakemian AS, Kondapalli KC, Telser J, Hoffman BM, Stemmler TL, Rosenzweig AC (2008) The metal centers of particulate methane monooxygenase from *Methylosinus trichosporium* OB3b. Biochemistry 47(26):6793–6801

Huang X, Atwood CS, Moir RD, Hartshorn MA, Tanzi RE, Bush AL (2004) Trace metal contamnation initiates the apparent auto-aggregation, amyloidosis, and oligomerization of Alzheimer's Aβ peptides. J Biol Inorg Chem 9:954–960

Hyman MR, Arp DJ (1990) The small-scale production o [U-^{14}C]acetylene rom Ba^{14}CO$_3$: application to labeling of ammonia monooxygenase in autotrophic nitrifying acteria. Anal Biochem 190:348–353

Hyman MR, Arp DJ (1992) $^{14}C_2H_2$- and $^{14}CO_2$-labeling studies of the *de novo* synthesis of polypeptides by *Nitrosomonas europaea* during recovery from acetylene and light inactivation of ammonia monooxygenase. J Biol Chem 267:1534–1545

Hynes RK, Knowles R (1978) Inhibition by acetylene of ammonia oxidation in *Nitrosomonas europaea*. FEMS Microbiol Lett 4:319–321

Jeffery CJ (1999) Moonlighting proteins. Trends Biochem Sci 24(1):8–11

Jeffery CJ (2003) Moonlighting proteins: old proteins learning new tricks. Trends Genet 19 (8):415–417

Jeffery CJ (2009) Moonlighting proteins – an update. Mol Biosyst 5(4):345–350

Johnson CL (2006) Methanobactin: a potential novel biopreservative used against the foodborne pathogen *Listeria monocytogenes*. Iowa State University

Kalidass B, Ul-Haque MF, Baral BS, DiSpirito AA, Semrau JD (2015) Competition between metals for binding to methanobactin enables expression of soluble methane monooxygenase in the presence of copper. Appl Environ Microbiol 81(3):1024–1031

Kehrer JP (2000) The Harber-Weiss reaction mechanisms of toxicity. Toxicology 149(1):43–50

Kenney GE, Rosenzweig AC (2012) Chemistry and biology of the copper chelator methanobactin. ACS Chem Biol 7:260–268

Kenney GE, Rosenzweig AC (2013) Genome mining for methanobactins. BMC Biol 11:17

Kenney GE, Rosenzweig AC (2018) Chalkophores. Annu Rev Biochem 87:645–676

Kenney GE, Goering AW, Ross MO, Dehart CJ, Thomas PM, Hoffman BM, Kelleher NL, Rosenzweig AC (2016) Characterization of methanobactin from Methylosinus sp. LW4. J Am Chem Soc 138(35):11124–11127

Kenney GE, Dassama LMK, Pandelia ME, Gizzi AS, Martinie RJ, Gao P, DeHart CJ, Schachner LF, Skinner OS, Ro SY, Zhu X, Sadek M, Thomas PM, Almo SC, Bollinger JM Jr, Krebs C, Kelleher NL, Rosenzweig AC (2018) The biosynthesis of methanobactin. Science 359 (6382):1411–1416

Kim HJ, Graham DW, DiSpirito AA, Alterman MA, Galeva N, Larive CK, Asunskis D, Sherwood PMA (2004) Methanobactin, a copper-acquisition compound from methane-oxidizing bacteria. Science 305(5690):1612–1615

Koh E-I, Robinson A, Bandara N, Rogers B, Henderson JP (2017) Copper import in *Escherichia coli* by the yersiniabactin metallophore system. Nat Chem Biol 13:1016–1023

Krentz BD, Mulheron HJ, Semrau JD, DiSpirito AA, Bandow NL, Haft DH, Vuilleumier S, Murrell JC, McEllistrem MT, Hartsel SC, Gallagher WH (2010) A comparison of methanobactins from *Methylosinus trichosporium* OB3b and *Methylocystis* strain SB2 predicts methanobactins are synthesized from diverse peptide precursors modified to create a common core for binding and reducing copper ions. Biochemistry 49(47):10117–10130

Lawton TJ, Rosenzweig AC (2017) Biocatalysis for methane conversion: big progress on breaking a small substrate. Curr Opin Chem Biol 35:142–149

Leak DJ, Dalton H (1986) Growth yields of methanotrophs 2. A theoretical analysis. Appl Microbiol Biotechnol 23(6):477–481

Lee SK, Fox BG, Froland WA, Lipscomb JD, Münck E (1993a) A transient intermediate of the methane monooxygenase catalytic cycle containing a $Fe^{IV}Fe^{IV}$ cluster. J Am Chem Soc 115:6450–6451

Lee SK, Nesheim JC, Lipscomb JD (1993b) Transient intermediates of the methane monooxygenase catalytic cycle. J Biol Chem 268:21569–21577

Lichmannegger J, Leitinger C, Winner R, Schmitt S, Schulz S, Kabiri Y, Eberhagen C, Rieder T, Janik D, Neff F, Aichler M, DiSpirito AA, Bandow NL, Baral BS, Flatler A, Kremmer E, Denk G, Hohenester S, Eckardt-Schupp F, Dencher N, Adamski J, Merle U, Gotthardt DN, Kroemer G, Weiss KH, Zischka H (2016) Methanobactin: a new effective treatment strategy against acute liver failure in a Wilson disease rat model. J Clin Invest 126:2721–2735

Lieberman RL, Rosenzweig AC (2005) Crystal structure of a membrane-bound metalloenzyme that catalyses the biological oxidation of methane. Nature 434(7030):177–182

Lieberman RL, Shrestha DB, Doan PE, Hoffman BM, Stemmler TL, Rosenzweig AC (2003) Purified particulate methane monooxygenase from *Methylococcus capsulatus* (Bath) is a dimer with both mononuclear copper and a copper-containing cluster. Proc Natl Acad Sci USA 100(7):3820–3825

Lieberman RL, Kondapalli KC, Shrestha DB, Hakemian AS, Smith SM, Telser J, Kuzelka J, Gupta R, Borovik AS, Lippard SJ, Hoffman BM, Rosenzweig AC, Stemmler TL (2006) Characterization of the particulate methane monooxygenase metal centers in multiple redox states by X-ray absorption spectroscopy. Inorg Chem 45(20):8372–8381

Lipscomb JD (1994) Biochemistry of the soluble methane monooxygenase. Annu Rev Microbiol 48:371–399

Lu X, Gu W, Zhao L, Fagan UHM, DiSpirito AA, Semrau JD, Gu B (2017) Methylmercury uptake and degradation by methanotrophs. Sci Adv 3:e1700041

Maeda M, Zhao B, Ozaki Y, Yoneyama T (2003) Nitrate leaching in an Andisol treated with different types of fertilizers. Environ Pollut 121(3):477–487

Mandal S, Phadtare S, Sastry M (2005) Interfacing biology with nanoparticles. Curr Appl Phys 5:118–127

Martinho M, Choi DW, Dispirito AA, Antholine WE, Semrau JD, Münck E (2007) Mössbauer studies of the membrane-associated methane monooxygenase from *Methylococcus capsulatus* Bath: evidence for a Diiron center. J Am Chem Soc 129(51):15783–15785

Miller LM, Wang Q, Telivala TP, Smith RJ, Lanzirotti A, Miklossy J (2006) Synchrotron-based infrared and X-ray imaging shows focalized accumulation of Cu and Zn co-localized within β-amyloid deposites in Alzheimer's disease. J Struct Biol 155:30–36

Moffett JW, Tuit CB, Ward BB (2012) Chelator-induced inhibition of copper metalloenzymes in denitrifying bacteria. Limnol Oceanogr 57(1):272–280

Mullert J-C, Lichtmannegger J, Zischka H, Sperling M, Karst U (2018) High spatial resolution of LA-ICP-MS demonstrates massive liver copper depletion in Wilson disease rats upon methanobactin treatmenty. J Trace Elem Med Biol 49:119–127

Murrell JC, McDonald IR, Gilbert B (2000) Regulation of expression of methane monooxygenases by copper ions. Trends Microbiol 8(5):221–225

Myronova N, Kitmitto A, Collins RF, Miyaji A, Dalton H (2006) Three-dimensional structure determination of a protein supercomplex that oxidizes methane to formaldehyde in *Methylococcus capsulatus* (Bath). Biochemistry 45(39):11905–11914

Neilands JB (1995) Siderophores – structure and function of microbial iron transport compounds. J Biol Chem 270(45):26723–26726

Nguyen HHT, Elliott SJ, Yip JHK, Chan SI (1998) The particulate methane monooxygenase from *Methylococcus capsulatus* (Bath) is a novel copper-containing three-subunit enzyme – isolation and characterization. J Biol Chem 273(14):7957–7966

Nieboer E, Richardson DHS (1980) The replacement of the nondescript term 'heavy metals' by a biologically and chemically significant classification of metal ions. Environ Pollut B 1(1):3–26

Pelludat C, Rakin A, Jacobi CA, Schubert S, Heesemann J (1998) The Yersiniabactin biosynthetic gene cluster of *Yersinia enterocolitia*: organization and siderophore-dependent regulation. J Bacteriol 180(3):538–546

Phelps PA, Agarwal SK, Speitel GE, Georgiou G (1992) *Methylosinus trichosporium* OB3b mutants having constitutive epression of soluble methane monooxygenase in the presence of high-levels of copper. Appl Environ Microbiol 58(11):3701–3708

Polte J, Ahner TT, Delissen F, Sokolov S, Emmerling F, Thunemann AF, Kraehnert R (2010) Mecahnism of gold nanoparticle formation in the classical citrate synthesis method derived from coupled in situ XANES and SAXS evaluation. J Am Chem Scc 132:1296–1301

Prior SD, Dalton H (1985) Acetylene as a suicide substrate and active site probe for methane monooxygenase from *Methylococcus capsulatus* (Bath). FEMS Microbiol Lett 29:105–109

Reim A, Luke C, Krause S, Pratscher J, Frenzel P (2012) One millimetre makes the difference: high-resolution analysis of methane-oxidizing bacteria and their specific activity at the oxic-anoxic interface in a flooded paddy soil. ISME J 6(11):2128–2139

Roberts EA (2011) Wilson's disease. Medicine 39(10):602–604

Rosenzweig AC, Frederick CA, Lippard SJ (1992) Crystalization and reliminary X-ray analysis of the methane monooxygenase hydroxylase protein from *Methylococcus capsulatus* (Bath). J Mol Biol 227:583–585

Rosenzweig AC, Frederick CA, Lippard SJ, Nordlund P (1993) Crystal stucture of a bacterial non-heme iron hydroxylase protein from *Methylococcus capsulatus* (Bath). Nature 366:357–543

Saha K, Agasti S, Kim C, Li X, Rotello V (2012) Gold nanoparticles in chemical and biological sensing. Chem Rev 112:2739–2779

Schilsky ML (2001) Treatment of Wilson's disease: what the the relative roles of penicillamine, trientine, and zinc supplementation? Curr Gastroenterol Rep 3:54–59

Schilsky ML (2014) Liver transplantation for Wilson's disease. Ann NY Acad Sci 1315:45–49

Semrau JD, Chistoserdov A, Lebron J, Costello AM, Davagnino J, Kenna E, Holmes AJ, Finch R, Murrell JC, Lidstrom ME (1995) Particulate methane monooxygenase genes in methanotrophs. J Bacteriol 177:3071–3079

Semrau JD, DiSpirito AA, Murrell JC (2008) Life in the extreme: thermoacidophilic methanotrophy. Trends Microbiol 16(5):190–193

Semrau JD, DiSpirito AA, Yoon S (2010) Methanotrophs and copper. FEMS Microbiol Rev 34 (4):496–531

Semrau JD, diSpirito AA, Vuilleumier S (2011) Facultative methanotrophy: false leads, true results, and suggestions for future research. FEMS Microbiol Lett 323(1):1–12

Semrau JD, Jagadevan S, Dispirito AA, Khalifa A, Scanlan J, Bergman BH, Freemeier BC, Baral BS, Bandow NL, Vorobev A, Haft DH, Vuilleumier S, Murrell JC (2013) Methanobactin and MmoD work in concert to act as the 'copper-switch' in methanotrophs. Environ Microbiol 15 (11):3077–3086

Semrau JD, DiSpirito AA, Gu W, Yoon S (2018) Metals and methanotrophy. Appl Environ Microbiol 84(6)

Smith DDS, Dalton H (1989) Solubilization of the methane monooxygenase from *Methylococcus capsulatus* Bath. Eur J Biochem 182:667–671

Squitti R (2012) Copper disfunction in Alzheimer's disease: from meta-analysis of biochemical studies to new insights into genetics. J Trace Elem Med Biol 26:93–96

Squitti R, Siotto M, Polimanti R (2014) Low-copper diet as a preventive strategy for Alzheimer's disease. Neurobiol Aging 35:540–550

Stanley SH, Prior SD, Leak DJ, Dalton H (1983) Copper stress underlies the fundamental change in intracellular location of methane mono-oxygenase in methane-oxidizing oganisms – studies in batch and continuous cultures. Biotechnol Lett 5(7):487–492

Stattermayer AF, Traussnigg S, Dienes HP, Aigner E, Stauber R, Lackner K, Hofer H, Stift J, Wrba F, Stadlmayr A, Datz C, Strasser M, Maieron A, Trauner M, Ferenci P (2015) Hepatic steatosis in Wilson disease—role of copper and PNPLA3 mutations. J Hepatol 63(1):156–163

Stoecker K, Bendinger B, Schöning B, Nielsen PH, Nielsen JL, Baranyi C, Toenshoff ER, Daims H, Wagner M (2006) Cohn's Crenothrix is a filamentous methane oxidizer with an unusual methane monooxygenase. Proc Natl Acad Sci USA 103(7):2363–2367

Strong PJ, Xie S, Clarke WP (2015) Methane as a resource: can the methanotrophs add value? Environ Sci Technol 49:4001–4018

Strong PJ, Kalyuzhnaya M, Silverman J, Clarke WP (2016) A methanotroph-based biorefinery: potential scenarios for generating multiple products from a single fermentation. Bioresour Technol 215:314–323

Strong PJ, Karthikeyan OP, Zhu J, Clarke W, Wu W (2017) Agro-environmental sustainability. In: Singh JS, Seneviratne G (eds) Managing environmental pollution, vol 2. Springer, Cham, pp 19–40

Sullivan MJ, Gates AJ, Appia-Ayme C, Rowley G, Richardson DJ (2013) Copper control of bacterial nitrous oxide emission and its impact on vitamin B12-dependent metabolism. Proc Natl Acad Sci 110(49):19926–19931

Summer KH, Lichtmannegger J, Bandow N, Choi DW, DiSpirito AA, Michalke B (2011) The biogenic methanobactin is an effective chelator for copper in a rat model for Wilson disease. J Trace Elem Med Biol 25(1):36–41

Takeguchi M, Miyakawa K, Okura I (1998) Purification and properties of particulate methane monooxygenase from *Methylosinus trichosporium* OB3b. J Mol Catal Chem 132(2–3):145–153

Tonge GM, Harrison DEF, Knowles CJ, Higgins IL (1975) Properties and partial purification of the methane-oxidizing enzyme systems from *Methylosinus trichosporium* OB3b. FEBS Lett 58:293–299

Torre A, Metivier A, Chu F, Laurens LML, Beck DAC, Pienkos PT, Lidstrom ME, Kalyuzhnaya MG (2015) Genome-scale metabolic reconstructions and theoretical investigation of methane conversion in Methylomicrobium buryatense strain 5G(B1). Microb Cell Fact 14(1):188

Tumanova LV, Tukhvatullin IA, Burbaev DSH, Gvozdev RI, Andersson KK (2008) The binuclear iron site of membrane-bound methane hydroxylase from *Methylococcus capsulatus* (strain M). Russ J Bioinorgan Chem 34:177–185

Twining BS, Mylon SE, Benoit G (2007) Potential role of copper availability in nitrous oxide accumulation in a temperate lake. Limnol Oceanogr 52(4):1354–1366

Vernon G, Baranova A, Younossi ZM (2011) Systematic review: the epidemiology and natural history of non-alcoholic fatty liver disease and non-alcoholic steatohepatitis in adults. Aliment Pharmacol Ther 34(3):274–285

Vorobev A, Jagadevan S, Baral BS, Dispirito AA, Freemeier BC, Bergman BH, Bandow NL, Semrau JD (2013) Detoxification of mercury by methanobactin from *Methylosinus trichosporium* OB3b. Appl Environ Microbiol 79(19):5918–5926

Wallar BJ, Lipscomb JD (1996) Dioxygen activation by enzymes containing binuclear non-heme iron clusters. Chem Rev 96(7):2625–2657

Walshe JM (2007) Cause of death in Wilson's disease. Mov Disord 22(15):2216–2220

Walters KJ, Gassner GT, Lippard SJ, Wagner G (1999) Structure of the soluble methane monooxygenase regulatory protein B. Proc Natl Acad Sci USA 96:7877–7882

Weiss KH, Stremmel W (2012) Evolving perspectives in Wilson disease: diagnosis, treatment and monitoring. Curr Gastroenterol Rep 14(1):1–7

Woodland MP, Dalton H (1979) Purification and characterization of component A of the methane monooxygenase from *Methylococcus capsulatus* (Bath). J Biol Chem 259:53–60

Zahn JA, DiSpirito AA (1996) Membrane-associated methane monooxygenase from *Methylococcus capsulatus* (Bath). J Bacteriol 178(4):1018–1029

Zhang JZ, Wallar BJ, Popescu CV, Renner DB, Thomas DD, Lipscomb JD (2006) Methane monooxygenase hydroxylase and B component interactions. Biochemistry 45:2913–2926

Zhou J, Ralston J, Sedev R, Beattie DA (2009) Functionalized gold nanoparticles: synthesis, structure and colloid stability. J Colloid Interface Sci 331:251–262

Zischka H, Lichtmannegger J, Schmitt S, Jagemann N, Schulz S, Wartini D, Jennen L, Rust C, Larochette N, Galluzzi L, Chajes V, Bandow N, Gilles VS, DiSpirito AA, Esposito I, Goettlicher M, Summer KH, Kroemer G (2011) Liver mitochondrial membrane crosslinking and destruction in a rat model of Wilson disease. J Clin Invest 121(4):1508–1518

Environmental Applications of Methanotrophs

Adrian Ho, Miye Kwon, Marcus A. Horn, and Sukhwan Yoon

Contents

Abstract Methanotrophs are microorganisms that are able to utilize methane as the electron donor and carbon source. For long, methanotrophs have been widely studied for their application in environmental biotechnology, due mainly to the exclusive ownership of the unique enzymes that mediate oxidation of methane to methanol, namely the particulate methane monooxygenases (pMMO) and soluble methane monooxygenases (sMMO). Utilizing these methane monooxygenases, methanotrophs are capable of co-oxidizing a broad range of organic pollutants including chlorinated ethenes. Thus, methanotrophs have long been studied and utilized as biocatalysts for in situ bioremediation of soil and aquatic environments contaminated with these xenobiotic compounds. Due to the growing concerns in anthropogenically induced climate change and global warming, methanotrophs have increasingly gained attention also for greenhouse gas mitigation purposes. Active methane removal using methanotrophic biofilters of diverse configurations have proven to be effective for treatments of gases with relatively high methane concentrations, e.g., landfill gases and animal husbandry tank exhausts. Furthermore, improving the atmospheric methane sink capability of agricultural soils has been one of the foremost foci of climate-smart soil research. This

A. Ho · M. A. Horn
Institute for Microbiology, Leibniz Universität Hannover, Hannover, Germany
e-mail: adrian.ho@fmb.uni-hannover.de; horn@ifmb.uni-hannover.de

M. Kwon · S. Yoon (✉)
Department of Civil and Environmental Engineering, KAIST, Daejeon, South Korea
e-mail: miyekwon@kaist.ac.kr; syoon80@kaist.ac.kr

E. Y. Lee (ed.), *Methanotrophs*, Microbiology Monographs 32,
https://doi.org/10.1007/978-3-030-23261-0_8

chapter provides an extensive overview of scientific and engineering breakthroughs geared towards practical applications of methanotroph biotechnology in managing impending environmental problems.

1 Introduction

Methanotrophs are groups of organisms that are capable of utilizing methane as the energy and carbon source (Hanson and Hanson 1996; Semrau et al. 2010). Traditionally, only two subgroups belonging to the *Proteobacteria* phylum (aerobic alphaproteobacterial and gammaproteobacterial methanotrophs) had been recognized as methanotrophs; however, the term 'methanotroph' is now used in a broader context, encompassing the aerobic verrucomicrobial methanotrophs, anaerobic archaeal methanotrophs, and denitrifying methanotrophs of the NC10 phylum, besides the conventional proteobacterial methanotrophs (Raghoebarsing et al. 2006; Knittel and Boetius 2009; Op den Camp et al. 2009). These organisms, except for the archaeal anaerobic methanotrophs, possess and utilize particulate (pMMO) and/or soluble methane monooxygenases (sMMO), which are the only identified enzymes capable of catalyzing CH_4 oxidation to CH_3OH (Semrau et al. 2010). Most, if not all, previous studies on environmental applications of methanotrophs have focused on aerobic proteobacterial methanotrophs possessing these MMOs. Therefore, the literature review in this chapter will be limited mostly to these conventional aerobic methanotrophs, despite the recently invigorated scientific interests in verrucomicrobial methanotrophs and anaerobic methanotrophy.

The methane monooxygenases are key to environmental applications of methanotrophs (Jiang et al. 2010; Wendlandt et al. 2010). Both pMMO and sMMO have the identical physiological function of catalyzing CH_4 turnover to CH_3OH in methanotrophs; however, these two groups of MMOs have distinct evolutionary history and do not share any structural similarity (Elango et al. 1997; Leahy et al. 2003; Lieberman and Rosenzweig 2005; Khadka et al. 2018). While pMMO are membrane-integrated cuproenzymes that share the evolutionary lineage with ammonia monooxygenases (AMO), sMMO are structurally and evolutionarily affiliated to soluble di-iron monooxygenases, which include diverse alkene/aromatic monooxygenases (Leahy et al. 2003; Khadka et al. 2018). Ecological studies have suggested that pMMO is the prevalent form of methane monooxygenases in most environments inhabited by proteobacterial methanotrophs. The *pmoA* gene, encoding for the pMMO, has been recovered in a much larger quantity than the *mmoX* gene which encodes for the sMMO, in most soil and aquatic environments examined to date (Rahman et al. 2011). Until recently, no RNA-targeted analysis had been successful in detecting *mmoX* transcript in environmental samples, suggesting that the *mmoX* is rarely transcribed by methanotrophs in natural environmental settings (Chen et al. 2007; Liebner and Svenning 2013; Kumaresan et al. 2018). All isolated proteobacterial methanotrophs, except for the strains belonging to the *Methylocella* and *Methyloferula* genera, possess pMMO. Contrastingly, the genes encoding sMMO are only found in selected subgroups of methanotrophs across diverse phylogenetic groups within the *Proteobacteria* (Semrau et al. 2010; Kenney et al. 2016). These different lines of evidence point towards pMMO being more relevant to overall methanotrophic activity in natural environments, although in vitro incubation in Cu-depleted medium elevates sMMO expression and activity in methanotrophs possessing both MMOs, i.e. *Methylosinus trichosporium* strain OB3b and *Methylococcus* strain Bath (Murrell et al. 2000; Kenney et al. 2016).

Attempts to artificially generate methane-oxidizing bacteria using synthetic biology approaches have been unsuccessful, and the methanotrophs of the phyla *Proteobacteria*, *Verrucomicrobia*, and NC10 remain as the only groups of organisms capable of wielding either MMO. Ex vivo activities of isolated MMO enzymes, both sMMO and pMMO, have been examined, but without success in reproducing in vivo activity levels or maintaining their reactivity over extended period of time (Fox et al. 1989; Zahn and DiSpirito 1996). Thus, to utilize MMOs for environmental applications, whole-cell activities of the sMMO- or pMMO-expressing methanotrophs appears as the most feasible option. Such application of methanotrophs capitalize on largely two catalytic capabilities of the MMOs: CH_4 oxidation to CH_3OH, and involuntary cometabolic mono-oxidation of various mono- and multi-carbon organics. During the decades spanning 1980s through 2000s, when the public interest in environmental bioremediation was at its apex, cometabolic degradation of organic contaminants was the foremost research focus of methanotroph biotechnology (Semrau 2011). More recently, however, with the growing concern in global warming and climate change, the focus has shifted towards biological mitigation of CH_4 emissions from its major sources, e.g., landfills, animal feeding operations, and rice paddies. Besides, recent research aims to develop strategies to augment the methane sink function of the indigenous high-affinity methanotrophs in agricultural soils. In this chapter, the history of methanotroph biotechnology is summarized and the future research directions geared towards practical applications of methanotrophs are posited.

2 Cometabolic Organic Pollutant Degradation

Methanotrophs are capable of degrading various organic compounds utilizing methane monooxygenases. Particularly, their capability to oxidize various halogenated organic compounds has been widely studied. The most commonly practiced approach for bioremediation of chlorinated organic compounds has been the reductive dechlorination approach, whereby dechlorination serve as the electron acceptor reaction for selected groups of anaerobic microorganisms, e.g., *Dehalobacter* spp. and *Dehalococcoides* spp. (Holliger et al. 1998; Löffler et al. 2013). The mechanism of methanotrophic cometabolic degradation is entirely distinct from that of reductive dehalogenation, in that dehalogenation occurs via an oxidative pathway with O_2 as the electron acceptor (Keck et al. 1989; Alvarez-Cohen and Speitel 2001). Unlike in reductive dechlorination, in which halogenated organic compounds are utilized as growth substrates, these organic contaminants cannot serve as growth substrates for methanotrophs and can be co-oxidized only in the presence of CH_4. Thus, supply of CH_4 (the electron donor and carbon source) and O_2 (electron acceptor) need to be ensured to maintain activity of methanotrophs as they cometabolically oxidize halogenated organic compounds.

Methanotrophic co-oxidation of halogenated hydrocarbons is catalyzed by MMOs (Semrau et al. 2010). In general, the copper-free sMMOs have been observed with higher reactivity towards non-CH_4 substrates than the copper-dependent pMMOs (Lee et al. 2006; Yoon and Semrau 2008). Thus, initial research efforts for halogenated hydrocarbon degradation were focused on utilization of sMMO-expressing methanotrophs (Oldenhuis et al. 1991; van Hylckama Vlieg et al. 1996). Although the high metabolic rate of sMMO towards chlorinated hydrocarbons is certainly attractive, the sMMO-utilizing approach has limitations in its practical applications. The expression of sMMO is tightly regulated by copper and the near-complete absence of copper necessitated for sMMO expression is rare in terrestrial environments (Murrell et al. 2000). In fact, *mmoX* transcripts are rarely detected in molecular analyses targeting RNA in CH_4-oxidizing soils (Chen et al. 2007, 2008; Lee et al. 2009). Thus, the effectiveness of sMMO-catalyzed halogenated hydrocarbon oxidation is questionable in in situ bioremediation. As cometabolic substrates are often competitive inhibitors to methane oxidation and cometabolic oxidation often generates toxic intermediates, the high affinity and reactivity of sMMO towards chlorinated hydrocarbon compounds may act against establishment of methanotroph population when the contaminants are present at high concentrations (Alvarez-Cohen and McCarty 1991; Lee et al. 2006). Therefore, several researches have investigated pMMO-mediated halogenated hydrocarbon as an alternative, despite of the low reactivity of pMMO towards non-CH_4 compounds (Anderson and McCarty 1997; Lontoh and Semrau 1998; Han et al. 1999).

The xenobiotic halogenated hydrocarbons targeted by methanotrophic cometabolic degradation and the kinetic parameters experimentally determined with laboratory strains of methanotrophs expressing sMMO or pMMO are summarized in Table 1. Majority of these biodegradation experiments were performed with *M. trichosporium* OB3b (*Alphaproteobacteria*) and *M. capsulatus* Bath (*Gammaproteobacteria*) at the sMMO-expressing condition, i.e., in absence of copper. Degradation of trichloroethylene (TCE), *cis*- and *trans*-dichloroethylene (*c*- and *t*-DCE), and vinyl chloride (VC), among many other one- or two-carbon halogenated compounds have also been confirmed with pMMO-expressing methanotrophs, including those lacking *mmoX* gene (Han et al. 1999).

Biodegradation of the chlorinated ethenes using the methanotrophic cometabolism attracted particular interests during the decades spanning the 1980s to 2000s. Although methanotrophic cometabolic degradation had a major drawback in its field application, in that it demands continuous supply of CH_4 and O_2 to the target contaminated sites, cometabolic degradation has certain advantages over reductive dechlorination in removal of chlorinated ethenes. The most toxic form of the chlorinated ethenes is VC. Even though *Dehalococcoides* spp. capable of

Table 1 Apparent whole-cell Michaelis-Menten rate constants determined for cometabolic degradation of halogenated hydrocarbons by methanotrophs expressing either particulate or soluble methane monooxygenases

Compound[a]	Methanotroph[b]	MMO[c]	V_{max} [nmol min^{-1}·(mg protein)$^{-1}$][d]	K_m (μM)	References
TCE	BG8	pMMO	4.3	59	Han et al. (1999)
	OB3b	pMMO	214	15	Lontoh et al. (1999)
	OB3b	sMMO	580[e]	145	Oldenhuis et al. (1991)
	Mixed	pMMO	1.3	44	Forrester et al. (2005)
VC	BG8	pMMO	7	30	Han et al. (1999)
	OB3b	pMMO	42	26	Lee et al. (2006)
	OB3b	sMMO	2100	160	Lee et al. (2006)
	Mixed	pMMO	9.9	47	Forrester et al. (2005)
c-DCE	BG8	pMMO	0.12	0.8	Han et al. (1999)
	Mixed	pMMO	0.52	3.1	Forrester et al. (2005)
	OB3b	sMMO	364[e]	30	Oldenhuis et al. (1991)
t-DCE	BG8	pMMO	43	60	Han et al. (1999)
	OB3b	sMMO	662[e]	148	Oldenhuis et al. (1991)
	OB3b	pMMO	61	42	Lee et al. (2006)
1,1-DCE	BG8	pMMO	0.23	2.5	Han et al. (1999)
	OB3b	sMMO	12[e]	5	Oldenhuis et al. (1991)
DCM	BG8	pMMO	33	73	Han et al. (1999)
	OB3b	pMMO	235	32	Lontoh et al. (1999)
	OB3b	sMMO	66[e]	4	Oldenhuis et al. (1991)

(continued)

Table 1 (continued)

Compound[a]	Methanotroph[b]	MMO[c]	V_{max} [nmol min^{-1}·(mg protein)$^{-1}$][d]	K_m (μM)	References
CF	OB3b	sMMO	1096[e]	34	Oldenhuis et al. (1991)
DBM	BG8	pMMO	45	171	Han et al. (1999)
1,1,1-TCA	OB3b	sMMO	48	214	Oldenhuis et al. (1991)

[a]TCE, trichloroethylene; VC, vinyl chloride; *c*-DCE, *cis*-dichloroethylene; *t*-DCE, *trans*-dichloroethylene; 1,1-DCE, 1,1-dichloroethylene; DCM, dichloromethane; BF, bromoform; CF, chloroform; DBM, dibromomethane; 1,1,1-TCA, 1,1,1-trichloroethane

[b]BG8, *Methylomicrobium album* strain BG8; OB3b, *Methylosinus trichosporium* strain OB3b; Mixed, Uncharacterized mixed methanotrophic culture

[c]pMMO, particulate methane monooxygenase; sMMO, soluble methane monooxygenase

[d]Methanotrophs were cultured at 30 °C with addition of 20 mM formate

[e]Converted from the reported units of mL min^{-1} (mg cells)$^{-1}$, assuming that 50% of the total cell dry weight is protein

reducing VC (C_2H_3Cl) to innocuous ethene (C_2H_4) have been discovered and microbial consortia capable of complete reduction of chlorinated ethenes have been fully commercialized, VC remains to be the most recalcitrant intermediate in the reductive dechlorination pathway (He et al. 2003a, b; Ritalahti et al. 2005; Popat and Deshusses 2009). Thus, VC accumulation is unavoidable in bioremediation practices employing reductive dechlorination. In methanotrophic cometabolism, regardless of whether sMMO or pMMO is involved, oxidization of TCE and *c*- and *t*-DCE does not leave stable toxic intermediates and VC is the most rapidly degraded compound of the chlorinated ethenes (van Hylckama Vlieg and Janssen 2001; Lee et al. 2006).

Methanotrophs are also less vulnerable to adverse changes to environmental conditions than VC-respiring *Dehalococcoides* spp., which are easily inactivated by O_2 exposure or modest changes to pH or salinity (Amos et al. 2008; Islam et al. 2016; Matturro et al. 2016; Yang et al. 2017a; Ho et al. 2018). Methanotrophic degradation of chlorinated ethenes have been examined only at the neutral-pH conditions; however, diverse groups of acidophilic and halophilic methanotrophs harboring pMMO and/or sMMO have been isolated, suggesting that acidophilic or halophilic biological degradation of chlorinated ethenes, including VC, may be feasible (Kip et al. 2011; Semrau 2011). No VC-to-ethene reduction activity has been observed in acidic VC-to-ethene reduction or saline environments to date (Kittelmann and Friedrich 2008; Yang et al. 2017b). Reductive dehalogenation of chlorinated ethenes at pH 5.5 had resulted in permanent stoichiometric accumulation of VC in *D. mccartyi* cultures, suggesting complete inactivation of VC reductase and potential VC accumulation from PCE or TCE in contaminated acidic soil (Yang et al. 2017a). Biostimulation or, if necessary, bioaugmentation of acidophilic or halophilic methanotrophs may be a plausible alternative for removal of residual VC from such environments recalcitrant to the reductive dechlorination approach.

Several different field application methods were tested as the means to stimulate methanotrophic cometabolism in situ for removal of chlorinated ethenes. In the pilot scale experiment performed at Moffett Naval Air Station using biostimulation approach, the methanotrophic activity in the artificially contaminated aquifer was stimulated in situ by alternating pulse injections of air and CH_4 at the injection wells (Fig. 1a). The concentrations of TCE, *t*-DCE, and VC at the extraction wells were significantly lower with the air and CH_4 injections than without, indicating the occurrence of cometabolic biodegradation at the site. The bioremediation of the PCE/TCE-contaminated Savannah River Site employed horizontal injection wells installed below the water table for direct injection of CH_4/air mixed gas and horizontal vacuum extraction wells positioned in the vadose zone to capture stripped chlorinated ethenes for incineration (Fig. 1b) (Brockman et al. 1995; Hazen et al. 1993). Addition of CH_4/air mixed gas and gaseous nitrogen (nitrous oxide) and phosphorous (triethyl phosphate) sources resulted in active TCE/PCE removal in situ. Similar pilot scale studies using similar biostimulation approaches proved to be effective in removing chlorinated ethenes from contaminated soils and groundwater (Eguchi et al. 2001; Takeuchi et al. 2005).

In an attempt to avoid the drawbacks of such biostimulation strategy (e.g., inconsistent contacts between contaminant and stimulated bacterial population, spatially uneven biostimulation, and competitive inhibition of MMO's methane oxidation activity by chlorinated ethenes), an unconventional bioaugmentation approach was also tested (Taylor et al. 1993; Duba et al. 1996). Dense pre-incubated cultures of *M. trichosporium* OB3b were applied to porous media to establish in situ microbial filters and TCE-contaminated plumes were passed through these microbial filters. In both laboratory-scale experiments and the pilot-scale study at a TCE-contaminated site at the Chico Municipal Airport, TCE removal activities of the biofilters were sustained for >20 days, albeit with the removal efficiency gradually decreasing over time.

Whether sMMO- or pMMO-expressing methanotrophs are involved in the observed in situ chlorinated ethene degradation still remains controversial, as no comparative quantification of *pmoA* and *mmoX* transcripts or expressed proteins in situ has been reported to date. The presence of *mmoX* transcripts in the Savannah test site was confirmed with gene probes; however, due to the lack of information with regards to the quantitative comparison between *pmoA* and *mmoX* transcripts, the relative contribution of sMMO activity to TCE degradation remained unknown (Hazen et al. 2009).

3 Potential Environmental Applications of Methanobactin

The majority of methanotrophs in the environment utilize pMMO to carry out the first step of CH_4 oxidation and thus require sufficient supply of Cu^{2+} ions. Some alphaproteobacterial methanotrophs, previously termed type II methanotrophs, are capable of synthesizing and utilizing a Cu^{2+}-chelator termed methanobactin (Mbn)

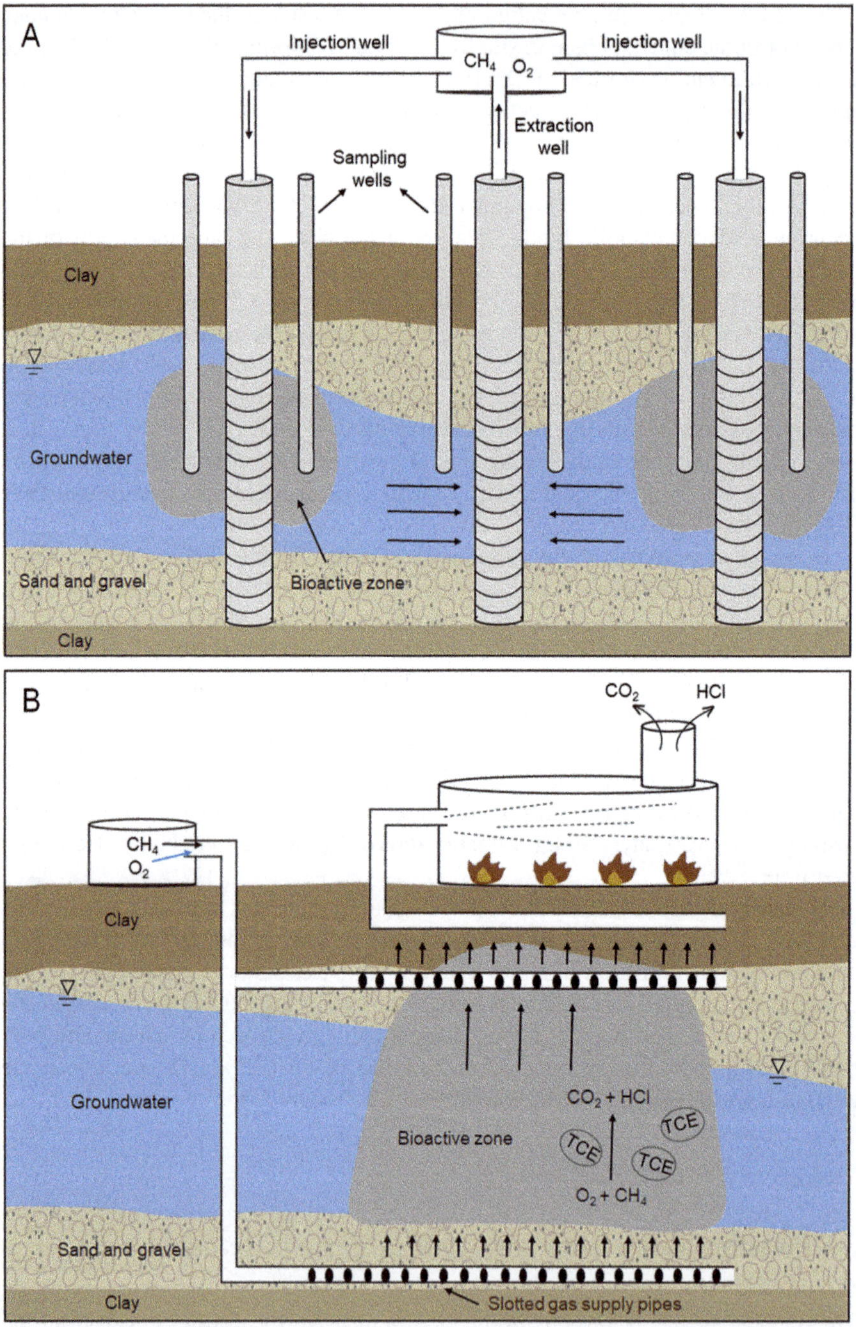

Fig. 1 Schematic depictions of the pilot-scale field tests of bioremediation of TCE-contaminated groundwater and soil via in situ CH_4 biostimulation of indigenous methanotrophs. (**a**) The field test at Moffet Naval Air Base, where pulse injections of air- and CH_4-saturated groundwater at the injection wells supported TCE removal from contaminated groundwater. (**b**) The field test at Savannah River Site, where mixed gas of CH_4 and air was supplied into contaminated aquifer

(Kim et al. 2004; DiSpirito et al. 2016). Recent studies of Mbn and Mbn-synthesizing methanotrophs have suggested several potential environmental applications of this novel copper-binding compound. Although Mbn binds Cu ions with high specificity, Mbn also forms complexes with other trace metals, most notably Hg^{2+} (Baral et al. 2014). Abstraction of Hg^{2+} by Mbn mitigated the mercury toxicity to methanotrophs of diverse phylogenetic groups, including those not able to synthesize Mbn (Vorobev et al. 2013); however, whether Mbn-bound Hg is less bioavailable and toxic to other micro- or macro-organisms still awaits to be examined. More recently, the Mbn-producing *M. trichosporium* strain OB3b was examined for the capability to uptake and demethylate methyl-mercury (CH_3Hg^+) (Lu et al. 2017). Methanobactin was found to have a crucial role in the demethylation reaction, as the *mbnA*$^-$ deletion mutant defective in the capability to Mbn lacked the capability. These experimental evidences suggest the possibility of utilizing Mbn or Mbn-producing methanotrophs for detoxifying mercury in heavy-metal contaminated sites.

Quite a few of the most consequential reactions in the biological N-cycling are catalyzed by cuproenzymes, i.e., the enzymes that require Cu for their activities. Ammonia monooxygenases (AMO), NirK-type NO_2^- reductases, and NosZ-type N_2O reductases are all crucial enzymes in the N-cycling that require Cu for their activities (Hulse et al. 1989; Brown et al. 2000; Khadka et al. 2018). Whether Mbn-bound Cu^{2+} can be taken up by non-methanotrophic organisms has not yet been closely examined. If the ability to uptake and utilize Mbn-bound Cu is not a widespread trait among the organisms wielding aforementioned cuproenzymes, Cu sequestration by Mbn may significantly alter the nitrogen flux in the environment. Indeed, in a recent experiment investigating the effect of Mbn on denitrification, the activities of NosZ and NirK were evidently hindered by the presence of Mbn (Chang et al. 2018). It is certainly plausible that AMO activities may also be affected by the presence of Mbn or Mbn-synthesizing methanotrophs. Such alteration of the nitrogen cycling reactions may have utilities in environmental engineering and agricultural practices. Methanobactin-induced inhibition of N_2O reduction can be utilized for recovery of N_2O from wastewater treatment plants with biological nitrogen removal (BNR) systems. Theoretically, denitrification can be stopped at N_2O by adding Mbn-enriched additives to anoxic tanks of activated sludge reactors. N_2O has been studied as the replacement for hydrazine as rocket fuels and has newfound utility in manufacturing of organic light emitting diode (OLED) displays (Yamazaki et al. 2014; Zakirov et al. 2001). Recovery of such valuable resource from otherwise wasted nitrogenous contaminants is crucial in developing more sustainable sewage treatment systems. The impacts of Mbn on ammonia oxidizing bacteria and ammonia oxidizing archaea have not yet been examined, but two possible effects can be

Fig. 1 (continued) through horizontal injection wells. The rising gas containing stripped TCE was extracted with another set of overlying horizontal injection wells and incinerated before release. The figures (**a**) and (**b**) were adapted from the graphical presentations in Semprini and McCarty (1991) and Hazen et al. (1993), respectively

expected. Methanobactin may suppress NH_3 oxidation by sequestering Cu ions away from ammonia monooxygenase-utilizing ammonia oxidizing bacteria and archaea, as was observed with NosZ-utilizing denitrifiers. Another possible outcome is protection of these sensitive organisms from copper and heavy metal toxicity (Park and Ely 2008). Either way, Mbn may prove useful, as a nitrification inhibitor for enhancement of nitrogen fertilizer efficiency or NH_3 oxidation facilitator in wastewater treatment plants debilitated by heavy metal influx.

4 Methane Emission Mitigation

The main growth substrate of methanotrophs, CH_4, is a potent greenhouse gas (GHG) with ~25 times higher global warming potential than CO_2 over 100-year period and the International Panel on Climate Change (IPCC) estimated the contribution of CH_4 to the net global GHG to be approximately 16% (Ciais et al. 2013). The largest anthropogenic CH_4 sources include landfills, livestock farming, rice farming, and mining industry (Ciais et al. 2013). Utilization of methanotrophic metabolism has long been considered as a viable strategy for mitigation of CH_4 emissions from these sources (summarized in Table 2).

Table 2 Lab- and field-scale experimentations of various CH_4 emission mitigation strategies utilizing CH_4-oxidizing capability of methanotrophs

Targeted source	Strategy	Scale	Materials	Methane removal rate (g h^{-1} m^{-3})	References
Landfill	Biofilter (active)	Pilot-scale	Loamy sand, clay, gravel	80	Gebert and Gröngröft (2006)
Landfill	Biofilter (passive)	Lab-scale	Compost	63	Streese and Stegmann (2003)
Landfill	Biofilter (passive)	Lab-scale	Inorganic material	13.3–29.2	Nikiema et al. (2005)
Manure storage	Biofilter (passive)	Pilot-scale	Perlite, garden compost	5.5–22	Melse and van der Werf (2005)
Enteric fermentation	Bio-integrated building material	Lab-scale	Aerated concrete	<0.01	Ganendra et al. (2014)
Enteric fermentation	Biofilter (passive)	Pilot-scale	Aerated concrete	2.8	Ganendra et al. (2015)
Dairy farm	Biofilter (passive)	Pilot-scale	Landfill cover soil, perlite	16	Pratt et al. (2012)
Anaerobic digester	Biofilter (passive)	Lab-scale	Activated carbon	348	Wu et al. (2017)

Landfill gases (LFG) with high CH_4 contents (e.g., >30%) are often collected and utilized for energy production and those with lower CH_4 contents are mostly flared before being released to the atmosphere. Neither is applicable, however, for the landfills gases with CH_4 contents below the flammable threshold (the lower flammability limit for CH_4 is 5.0% in air at 25 °C). Both passive biofiltration and forced-ventilation biofiltration have been devised and tested for CH_4 removal from the low-CH_4 LFG. The best example of the former is the pilot-scale biofilter tested for treatment of passively emitted gas from a low-organic land disposal site in Hamburg, Germany (Gebert et al. 2003). The biofilter was built into the landfill cover soils and filled with layers of gravels, sand and clay pellets and operated without artificial watering or nutrient supply (Fig. 2a). In spite of the reduced performance during the wintry months, the biofilter removed 62% of CH_4 emitted during the 1-year span of the test period (Gebert and Gröngröft 2006). The fatty acid profile-based analysis of the microbial composition indicated that the CH_4-oxidizing population in the biofilter was enriched with alphaproteobacterial methanotrophs of the *Methylosinus* and *Methylocystis* genera (Gebert et al. 2004).

Several researchers have examined CH_4-removal capacities of lab-scale and/or pilot-scale biofiltration experiments using forced ventilation with higher volumetric processing rates (Streese and Stegmann 2003; Nikiema et al. 2005). The lab-scale CH_4 biofilter developed by Nikiema et al. (2005) was constructed to the height and the internal diameter of 135 cm (subdivided into three sections of 45 cm) and 15 cm, respectively, and was operated as a biotrickling biofilter wetted with nutrient solution fed at the rate of 1.5 L day^{-1} (Fig. 2b). For treating 4.2 L min^{-1} gas influx containing 7000–7500 ppmv CH_4, CH_4-to-CO_2 conversion rates of up to 41% and elimination capacities up to 29 g m^{-3} h^{-1} were achieved with the biofilter packed with inorganic filter beds. The biofiltration system developed by Streese and Stegmann (2003) used a mixture of squeezed wood fibers, peat, and compost as the filter material and instead of wetting the filter beds with trickled nutrient medium, LFG was humidified in a separate antechamber with scrubbers before being distributed into the parallel biofilters. The pilot-scale biofiltration system with the total filter volume of 4 m^3 treated a mixed stream carrying 0.2–2.5% CH_4 with 25 m^3 h^{-1} total flow rate at an elimination capacity of 20–40 g m^{-3} h^{-1}.

Methane emissions from livestock farming have become increasingly problematic, due to the recent growth in global demands for meat and dairy products (Godfray et al. 2010). Enteric fermentation in cattle rumens is estimated to be responsible for approximately 25% of the current total anthropogenic CH_4 emissions, and additional CH_4 emission results from manure storage and processing (Ciais et al. 2013; Hou et al. 2015). Utilization of methanotrophs has been, for long, considered as a viable strategy for reducing CH_4 emissions from livestock farming (Melse and van der Werf 2005; Yoon et al. 2009; Ganendra et al. 2015). Melse and van der Werf (2005) examined the feasibility of CH_4 removal from off-gas from a liquid manure storage tank with a pilot-scale biofilter packed with a mixture of perlite and compost. Presumably due to the low temperature (average of 12.0 °C during the period of study) and the low CH_4 concentrations (<5 g m^{-3}), the observed elimination capacities were lower than the biofilters developed for LFG treatments. Nevertheless,

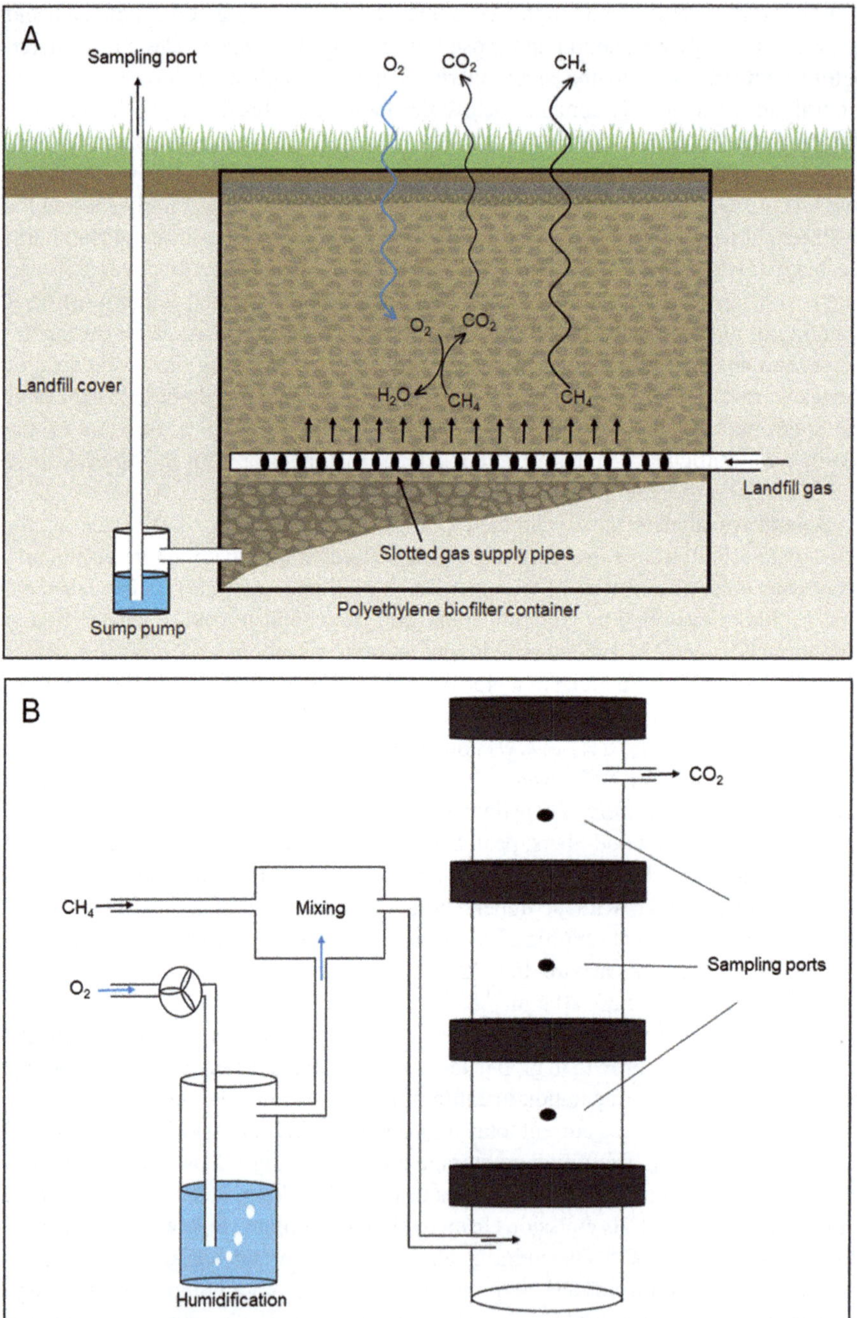

Fig. 2 Schematics of (**a**) a passive biofilter and (**b**) a forced-ventilation biofilter designed and experimented for removal of CH_4 from landfill gas. The figures (**a**) and (**b**) were adapted from Gebert et al. (2003) and Nikiema et al. (2005), respectively

substantial removal of CH_4 was possible using this approach at an estimated cost of \$26 per ton CO_2eq. Incorporation of immobilized methanotrophs into the porous building materials has also been examined as an alternative strategy (Ganendra et al. 2014). Of the porous materials tested as the potential support for methanotrophic biofilms, aerated concrete was determined as the most suitable material, as the highest rates of CH_4 oxidation was observed at both high (*ca.* 20% v/v) and low (*ca.* 50 ppmv) concentrations of CH_4 with methanotrophs immobilized in this medium (Ganendra et al. 2014). A biofilter using this same material as the filter bed was also tested in ruminant respiration chambers; however, the device failed to achieve sufficiently high removal efficiency (Ganendra et al. 2015).

Similar methanotroph-utilizing approaches have been devised and tested for removal of CH_4 from other environmental hotspots, which include coal mines and oil sands tailings ponds (Limbri et al. 2014; La et al. 2018). Methane biofiltration may also provide a solution for the CH_4 leakage problem in shale gas mining, although no reliable literature has been published regarding such possibility (Alvarez et al. 2012). As such, microbial conversion of CH_4 to CO_2 remains to be one of the most realistic approaches for GHG emission reduction in the foreseeable future. Provided that microbial CH_4 emission reduction can be coupled to resource recovery, further cuts in the net costs will be possible. Such possibilities abound; methanotrophs (*M. capsulatus* Bath) have long been used for production of commercial single-cell protein and methanobactin has commercial potential as a cost-efficient cure for Wilson's disease (Lichtmannegger et al. 2016). Single cell protein and/or methanobactin can be collected from methanotrophic biomass grown in the CH_4 removal devices and processed for commercial production, relieving, at least in part, financial burden for the prospective operators.

The major dilemma in utilizing methanotrophic activity for reducing CH_4 emissions is the oft low concentration of emitted CH_4 (Scheutz et al. 2009; Yoon et al. 2009; Ganendra et al. 2015). The reported observed half-saturation constants ($K_{m,app}$) for CH_4 oxidation by conventional laboratory strains of methanotrophs, range from 0.11 to 23 μM (equivalent to 77–16,100 ppmv, assuming equilibrium at 25 °C) (Knief and Dunfield 2005; Baani and Liesack 2008). Most, if not all, sources of CH_4, e.g., rice paddy soils, animal husbandry storage tank, and animal barn housings, emit CH_4 at concentrations in the order of several hundred ppmv at the maximum, as CH_4 diffuses rapidly after emission (Yoon et al. 2009). These concentrations are rarely high enough to ensure sustained CH_4 oxidation by the conventional methanotrophs (Knief and Dunfield 2005). A potential solution to this dilemma is to establish and maintain high-affinity CH_4-oxidizing population in the bioreactors (Yoon et al. 2009). The existence of methanotrophs capable of oxidizing CH_4 at concentrations as low as the atmospheric concentration (1.8 ppmv) has been suggested since the turn of the century (Bull et al. 2000). The analyses of *pmoA* genes of upland soils with atmospheric CH_4 sink capability identified the USCα (upland soil cluster α) clade of alphaproteobacterial methanotrophs with supposedly high affinity to CH_4 (Kolb et al. 2005). The genomic potential of the USCα methanotroph has recently been characterized using a targeted cell sorting approach coupled to metagenomic analysis (Pratscher et al. 2018). Evidences of high-affinity CH_4 oxidation were also found in reports of experiments performed with pure cultures of alphaproteobacterial methanotrophs. Upon starvation, *Methylocystis* sp. LR1 exhibited half saturation constants as low as ~300 nM (equivalent to a gaseous concentration of ~210 ppmv at 25 °C) and *Methylocystis* strain SC2, possessing two dissimilar *pmoCAB* operons, was observed to oxidize CH_4 with a $K_{m,app}$ value of 0.11 μM with the expression and activity of higher-affinity pMMO2 (Dunfield and Conrad 2000; Baani and Liesack 2008). The wildtype *Methylocystis* strain SC2 was able to sustain its viability on CH_4 concentration as low as the atmospheric concentration, suggesting its potential utility in CH_4 emission mitigation. These experimental evidences should not be ignored despite the recently casted doubts on the existence of methanotrophs specifically suited for consumption of atmospheric CH_4 (Cai et al. 2016). There is no doubt that better understanding of physiology of such high-affinity methanotrophs and the ability to wield these organisms in engineered systems are crucial for successful development of practical biological CH_4 emission control strategies.

5 Climate-Smart Soils and Methanotrophs

The increasing demand for food, feed, and fibres to meet the global human population growth necessitates a sustainable circular economy to combat climate change. Meeting the global demands for these commodities entails conversion of native to arable lands and intensification of agriculture. Land-use related GHG emissions, including conversion to agricultural usage, account for approximately 25% of the global anthropogenic GHG emissions (Ciais et al. 2013; Tubiello et al. 2015). Climate-smart organic-based fertilization is one of the emerging GHG emission reduction strategies to offset the land-use related GHG emissions. In this approach, bio-based residues from various waste streams (e.g., livestock manure, plant waste materials, and anaerobic digester effluents) are recycled as soil additives to minimize GHG (mostly CH_4 and N_2O) emissions in agricultural soils, while sustaining or improving soil fertility, quality, and carbon storage capacity (Cayuela et al. 2010; Paustian et al. 2016). Recent research results suggest that optimization of the choice and combination of the residue materials may enable further reduction (Cayuela et al. 2010; Ho et al. 2017; Brenzinger et al. 2018).

The incorporation of bio-based residues in agricultural soils alters the abundance and composition of indigenous microbial populations and interactions among soil microorganisms (Hartmann et al. 2014; Sengupta and Dick 2015). Such shift in the microbial community composition and abundances may affect the methanotrophic population and the overall net methane emissions (Ho et al. 2015; Malghani et al. 2016). Upland agricultural soils are generally regarded as a relatively weak sink for atmospheric methane, when compared to native upland soils (Maxfield et al. 2008; Levine et al. 2011; Ho et al. 2015; Tate 2015; Malghani et al. 2016). However, the potential for methane oxidation in agricultural soils can be stimulated by the addition of specific organic residues (e.g., compost), thereby augmenting the methane sink function (Ho et al. 2015, 2019). Although the stimulatory effect was transient (20–30 days), the elevated methane uptake rate was comparable to those of native upland soils and could offset up to 16% of total GHG emissions, bringing into question whether repeated compost amendment in agricultural soils may or may not be able to sustain higher methane uptake rates for prolonged periods (Ho et al. 2015). Nevertheless, crop (common wheat) yield remained unchanged, suggesting that compost amendment was neither particularly beneficial nor detrimental to crop growth (Ho et al. 2017). Although slow mineralization rate of composted organics allow long-term deposition of organic carbon in the soil, compost is often deprived of nutrients and thus, is less beneficial in terms of plant growth enhancement (Ryals et al. 2015). A recent research suggested that the trade-off between higher crop yields and lower CH_4 emission may be reached considering the combination of compost and nutrient-rich bio-based residues, e.g., sewage sludge (Brenzinger et al. 2018). Organic amendments may also be combined with specific beneficial microorganisms (e.g., methanotrophs) to further improve the soil CH_4 uptake, as well as for promoting plant growth and suppressing plant pathogens (Berg 2009; Singh and Strong 2016; Carrión et al. 2018). Therefore, combinations of bio-based residues and microbial inocula at the optimum ratios to improve fertilization efficiency while curbing GHG emissions would be a potentially viable strategy to be implemented in climate-smart agriculture.

6 Harnessing the Methanotroph Interactome for Environmental Applications

Methanotrophs are typically employed as pure cultures for biotechnological applications, fueling the development of novel isolation strategies (Svenning et al. 2003; Kim et al. 2018; Kwon et al. 2018). In the environment, however, microorganisms rarely live in seclusion and methanotrophs are not exceptions. Soil and aquatic microorganisms co-exist and interact in the environment to modulate Earth's nutrient cycles. The response of methanotrophs to physico-chemical environmental parameters are widely documented; however, the interactions of methanotrophs with the biotic components of the Earth's environments are known to a lesser extent (Semrau et al. 2010; Ho et al. 2013). The investigations of methanotroph interactomes (i.e., mathanotrophs and accompanying non-methanotrophic community members) for environmental applications have, in fact, just recently gained scientific attention (Ho et al. 2016).

Several examples of direct and indirect modes of interactions constituting methanotroph interactomes can be found in recent literature (Fig. 3). A recent study suggested that methanotrophs may benefit from interaction with non-methanotrophic heterotrophs (e.g., *Rhizobium* spp.) that can provide the methanotrophs with the essential micronutrient vitamin B_{12} (Fig. 3a) (Iguchi et al. 2011). Co-culturing of *Methylovulum miyakonense* strain HT12 with the vitamin B_{12}-producing *Rhizobium* spp. stimulated CH_4 uptake and cell growth. In another recent work, co-habitation with methylotroph was found to influence the regulation of methanol dehydrogenase

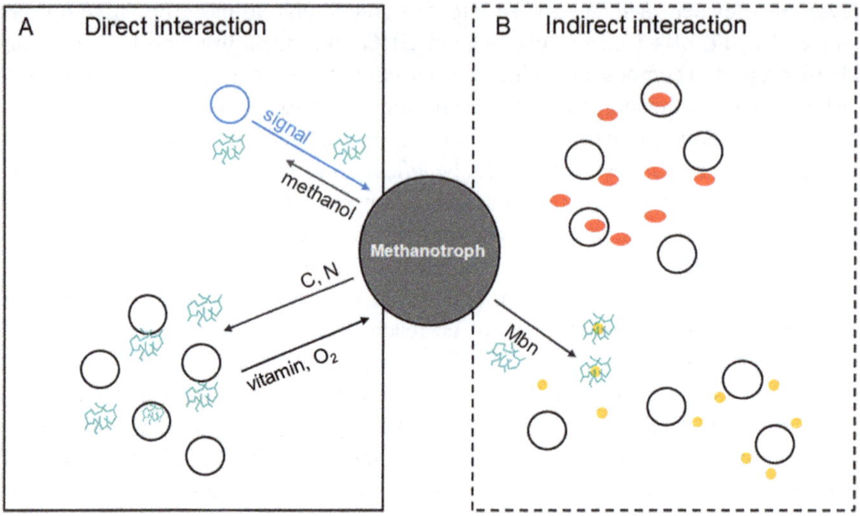

Fig. 3 Schematic representation of direct (**a**) and indirect (**b**) interactions within methanotroph interactomes. The empty circles of different colors denote non-methanotrophs. Indirect interactions represented in the panel (**a**), the blue circle and arrow depict chemical signaling of methylotrophs inducing methanol release from methanotrophs (Krause et al. 2017), whereas the black circle and arrow depict the flux of essential nutrients, e.g. vitamin B_{12}, and oxygen by other non-methanotrophs to the methanotrophs. In return, the methanotrophs provide assimilable carbon and nitrogen sources to the neighboring non-methanotrophs. In indirect interactions represented in the panel (**b**), release of methanobactin (Mbn, in green color) by methanotroph is depicted, which may impose a monopoly over bioavailable copper (solid yellow circles), effectively suppressing other microbial processes catalyzed by copper-dependent enzymes (Chang et al. 2018). Other non-methanotrophs may competitively uptake necessary micronutrients (solid red circles), which may, likewise, bar utilization by methanotrophs

in a gammaproteobacterial methanotroph, *Methylobacter tundripalodum* strain 31/32, inducing methanol release from the methanotroph (Krause et al. 2017). As a result, the methylotrophs in the co-culture thrived with methanotrophs despite the absence of non-CH_4 organic carbon substances in the medium. Conversely, as already mentioned above, *M. trichosporium* strain OB3b, and possibly most alphaproteobacterial methanotrophs with capability to synthesize methanobactin, may indirectly suppress other soil microbial processes such as N_2O reduction via secretion of methanobactin, imposing a 'copper monopoly' which depletes bioavailable copper needed for other cuproenzymes (Fig. 3b) (Chang et al. 2018).

Taking such interactome approach enables stable methanotrophic growth under conditions previously known to be unfit for methanotroph cultivation. Aerobic methanotrophs and strictly anaerobic methanogens conventionally thrive under seemingly contrasting conditions; however, a recent study have shown tight spatial organization of these microorganisms in a membrane bioreactor provided with acetate as the only substrate and maintained under oxygen-limiting condition. The methanogens provided CH_4 to the methanotrophs and the methanotrophs relieved O_2 tension in the microenvironments in return, to allow for survival of methanogens ('t Zandt et al.

2018). Driven by the incentives to generate higher value bio-products from biogas, van der Ha et al. (2012) successfully established stable enriched co-culture of methanotrophs with eukaryotic microalgae on biogas containing CH_4 and CO_2 but not O_2. Interestingly, such interactions between methanotrophs and non-methanotrophs may even occur in the presence of a physical barrier between the organisms with a cocktail of volatile organic compounds as the mediator of the interactome (Veraart et al. 2018). These studies exemplify diverse modes of biotic interactions within the methanotroph interactome at multi-trophic levels to modulate methanotrophic activity and cell growth.

Working with mixed microbial cultures, capitalizing on such interactions, often enhances growths and activities of the microorganisms within, and/or eliminates the necessity to operate under sterile conditions, lowering production costs (Ho et al. 2014; Strong et al. 2016). Methanotrophs, interacting with heterotrophs also enable development of environmental applications that are not feasible with pure methanotroph cultures. The association of aerobic methanotrophs and denitrifiers have been capitalized for developing nitrogen removal processes with CH_4 as the sole external carbon source (Eisentraeger et al. 2001; Modin et al. 2007; Zhu et al. 2016). A mixed community of methanotrophs and heterotrophs growing on organic substrates 'leaked' from the methanotrophs was able to collectively degrade a broad spectrum of micropollutants, a feat that could not be achieved with pure cultures of methanotrophs (Benner et al. 2015). Although still at its infancy, these studies provide an impetus for harnessing the methanotroph interactome for environmental applications. In future studies, further investigation of the key drivers regulating the methanotroph interactome is warranted for optimizing the intricate balances between the co-existing microorganisms to improve the effectiveness, stability, and reproducibility of the artificial communities engineered for specific environmental applications, as well as for devising novel environmental processes (Hays et al. 2015).

7 Conclusion and Future Perspective

Methanotrophs have now been studied for over 100 years since their first discovery, but continue to offer surprises. The new scientific discoveries regarding methanotroph physiology and ecology in the past few decades, e.g., copper regulation of MMOs, synthesis and utilization of methanobactin, high-affinity methane oxidation, and the discovery of the NC-10 denitrifying methanotrophs capable of generating O_2 internally from dismutation of NO, were among the most exciting discoveries in the field of microbiology and have had substantial implications in the development of novel environmental biotechnologies. As reviewed in this chapter, utilization of MMO-utilizing methanotrophs has already been vigorously studied for their practical engineering applications. Nevertheless, the new scientific discoveries are continuing to expand the horizon.

The recent discovery of the denitrifying anaerobic methanotrophs opened up new opportunities for improving sustainability of wastewater treatment processes (Shi et al. 2013; van Kessel et al. 2018). Combining anaerobic ammonia oxidation (anammox) and denitrifying anaerobic methane oxidation (DAMO) by both archaea and *Methylomirabilis*-like bacteria in a laboratory-scale membrane biofilm reactor enabled simultaneous removal of NO_3^- and NH_4^+ without aeration (Luesken et al. 2011; Shi et al. 2013). The DAMO reactions may also be utilized for simultaneous oxidation of residual CH_4 and nitrogenous nutrients in wastewater treatment plant effluents (van Kessel et al. 2018). Likewise, the discoveries of the extremely acidophilic methanotrophs of the *Verrucomicrobia* phylum may lead to development of environmental cleanup technologies for extreme acidic environments utilizing the cometabolic activity of the pMMO expressed by these organisms (Semrau 2011). The collaborative efforts across disciplines are necessary, for bridging these hypothetical or nascent technologies to actual field applications. Better scientific understanding of the novel physiologies, capitalizing on the recent advances in molecular techniques, sequencing technology, and cell-imaging techniques, would be crucial for effectively wielding these microorganisms of great potential utility. Designing of innovative bioreactor configurations or field application methods for establishing and sustaining stable populations of these unconventional culture-resistant organisms and harnessing them with minimum energy consumption and greenhouse gas emissions would require rigorous efforts from the engineering side. As such, the relentless collaborative efforts of environmental microbiologists, microbial ecologists, and environmental and chemical engineers will be critical for success of future research on environmental applications of methanotrophs.

Acknowledgement This work was supported by the C1 Gas Refinery Program through the National Research Foundation of Korea (NRF), which is funded by the Ministry of Science, ICT and Future Planning (grant number: NRF-2015-M3D3A1A01064881) and the Deutsche Forschungsgemeinschaft (grant number: HO6234/1-1) and the Leibniz Universität Hannover.

References

Alvarez RA, Pacala SW, Winebrake JJ, Chameides WL, Hamburg SP (2012) Greater focus needed on methane leakage from natural gas infrastructure. Proc Nat Acad Sci 109:6435–6440

Alvarez-Cohen L, McCarty PL (1991) Product toxicity and cometabolic competitive inhibition modeling of chloroform and trichloroethylene transformation by methanotrophic resting cells. Appl Environ Microbiol 57:1031–1037

Alvarez-Cohen L, Speitel GE (2001) Kinetics of aerobic cometabolism of chlorinated solvents. Biodegradation 12:105–126

Amos BK, Ritalahti KM, Cruz-Garcia C, Padilla-Crespo E, Loffler FE (2008) Oxygen effect on *Dehalococcoides* viability and biomarker quantification. Environ Sci Technol 42:5718–5726

Anderson JE, McCarty PL (1997) Transformation yields of chlorinated ethenes by a methanotrophic mixed culture expressing particulate methane monooxygenase. Appl Environ Microbiol 63:687–693

Baani M, Liesack W (2008) Two isozymes of particulate methane monooxygenase with different methane oxidation kinetics are found in *Methylocystis* sp. strain SC2. Proc Nat Acad Sci 105:10203–10208

Baral BS, Bandow NL, Vorobev A, Freemeier BC, Bergman BH, Herdendorf TJ, Fuentes N, Ellias L, Turpin E, Semrau JD, DiSpirito AA (2014) Mercury binding by methanobactin from *Methylocystis* strain SB2. J Inorg Biochem 141:161–169

Benner J, De Smet D, Ho A, Kerckhof F-M, Vanhaecke L, Heylen K, Boon N (2015) Exploring methane-oxidizing communities for the co-metabolic degradation of organic micropollutants. Appl Microbiol Biotechnol 99:3609–3618

Berg G (2009) Plant–microbe interactions promoting plant growth and health: perspectives for controlled use of microorganisms in agriculture. Appl Microbiol Biotechnol 84:11–18

Brenzinger K, Drost SM, Korthals G, Bodelier PLE (2018) Organic residue amendments to modulate greenhouse gas aes from agricultural soils. Front Microbiol 9:3035

Brockman FJ, Payne W, Workman DJ, Soong A, Manley S, Hazen TC (1995) Effect of gaseous nitrogen and phosphorus injection on *in situ* bioremediation of a trichloroethylene-contaminated site. J Hazard Mater 41:287–298

Brown K, Tegoni M, Prudêncio M, Pereira AS, Besson S, Moura JJ, Moura I, Cambillau C (2000) A novel type of catalytic copper cluster in nitrous oxide reductase. Nat Struct Mol Biol 7:191–195

Bull ID, Parekh NR, Hall GH, Ineson P, Evershed RP (2000) Detection and classification of atmospheric methane oxidizing bacteria in soil. Nature 405:175–178

Cai Y, Zheng Y, Bodelier PLE, Conrad R, Jia Z (2016) Conventional methanotrophs are responsible for atmospheric methane oxidation in paddy soils. Nat Commun 7:11728

Carrión VJ, Cordovez V, Tyc O, Etalo DW, de Bruijn I, de Jager VCL, Medema MH, Eberl L, Raaijmakers JM (2018) Involvement of *Burkholderiaceae* and sulfurous volatiles in disease-suppressive soils. ISME J 12:2307–2321

Cayuela ML, Oenema O, Kuikman PJ, Bakker RR, van Groenigen JW (2010) Bioenergy by-products as soil amendments? Implications for carbon sequestration and greenhouse gas emissions. Glob Change Biol Bioenergy 2:201–213

Chang J, Gu W, Park D, Semrau JD, DiSpirito AA, Yoon S (2018) Methanobactin from *Methylosinus trichosporium* OB3b inhibits N_2O reduction in denitrifiers. ISME J 12:2086–2089

Chen Y, Dumont MG, Cébron A, Murrell JC (2007) Identification of active methanotrophs in a landfill cover soil through detection of expression of 16S rRNA and functional genes. Environ Microbiol 9:2855–2869

Chen Y, Dumont MG, McNamara NP, Chamberlain PM, Bodrossy L, Stralis-Pavese N, Murrell JC (2008) Diversity of the active methanotrophic community in acidic peatlands as assessed by mRNA and SIP-PLFA analyses. Environ Microbiol 10:446–459

Ciais P, Sabine C, Bala G, Bopp L, Brovkin V, Canadell J, Chhabra A, DeFries R, Galloway J, Heimann M, Jone C, Quéré CL, Myneni RB, Piao S, Thornton P (2013) Carbon and other biogeochemical cycles. In: Stocker TF, Qin D, Plattner G-K, Tignor M, Allen SK, Boschung J et al (eds) Climate change 2013: the physical science basis contribution of working group I to the fifth assessment report of the intergovernmental panel on climate change. Cambridge University Press, Cambridge, pp 465–570

DiSpirito AA, Semrau JD, Murrell JC, Gallagher WH, Dennison C, Vuilleumier S (2016) Methanobactin and the link between copper and bacterial methane oxidation. Microbiol Mol Biol Rev 80:387–409

Duba AG, Jackson KJ, Jovanovich MC, Knapp RB, Taylor RT (1996) TCE remediation using *in situ*, resting-state bioaugmentation. Environ Sci Technol 30:1982–1989

Dunfield PF, Conrad R (2000) Starvation alters the apparent half-saturation constant for methane in the type II methanotroph *Methylocystis* strain LR1. Appl Environ Microbiol 66:4136–4138

Eguchi M, Kitagawa M, Suzuki Y, Nakamuara M, Kawai T, Okamura K, Sasaki S, Miyake Y (2001) A field evaluation of *in situ* biodegradation of trichloroethylene through methane injection. Water Res 35:2145–2152

Eisentraeger A, Klag P, Vansbotter B, Heymann E, Dott W (2001) Denitrification of groundwater with methane as sole hydrogen donor. Water Res 35:2261–2267

Elango NA, Radhakrishnan R, Froland WA, Wallar BJ, Earhart CA, Lipscomb JD, Ohlendorf DH (1997) Crystal structure of the hydroxylase component of methane monooxygenase from *Methylosinus trichosporium* OB3b. Protein Sci 6:556–568

Forrester SB, Han J-I, Dybas MJ, Semrau JD, Lastoskie CM (2005) Characterization of a mixed methanotrophic culture capable of chloroethylene degradation. Environ Eng Sci 22:178–186

Fox BG, Froland WA, Dege JE, Lipscomb JD (1989) Methane monooxygenase from *Methylosinus trichosporium* OB3b. Purification and properties of a three-component system with high specific activity from a type II methanotroph. J Biol Chem 264:10023–10033

Ganendra G, De Muynck W, Ho A, Hoefman S, De Vos P, Boeckx P, Boon N (2014) Atmospheric methane removal by methane-oxidizing bacteria immobilized on porous building materials. Appl Microbiol Biotechnol 98:3791–3800

Ganendra G, Mercado-Garcia D, Hernandez-Sanabria E, Peiren N, De Campeneere S, Ho A, Boon N (2015) Biofiltration of methane from ruminants gas effluent using autoclaved aerated concrete as the carrier material. Chem Eng J 277:318–323

Gebert J, Gröngröft A (2006) Performance of a passively vented field-scale biofilter for the microbial oxidation of landfill methane. Waste Manage 26:399–407

Gebert J, Gröengröeft A, Miehlich G (2003) Kinetics of microbial landfill methane oxidation in biofilters. Waste Manage 23:609–619

Gebert J, Gröngröft A, Schloter M, Gattinger A (2004) Community structure in a methanotroph biofilter as revealed by phospholipid fatty acid analysis. FEMS Microbiol Lett 240:61–68

Godfray HCJ, Beddington JR, Crute IR, Haddad L, Lawrence D, Muir JF, Pretty J, Robinson S, Thomas SM, Toulmin C (2010) Food security: the challenge of feeding 9 billion people. Science 327:812–818

Han J-I, Lontoh S, Semrau JD (1999) Degradation of chlorinated and brominated hydrocarbons by *Methylomicrobium album* BG8. Arch Microbiol 172:393–400

Hanson RS, Hanson TE (1996) Methanotrophic bacteria. Microbiol Rev 60:439–471

Hartmann M, Frey B, Mayer J, Mäder P, Widmer F (2014) Distinct soil microbial diversity under long-term organic and conventional farming. ISME J 9:1177–1194

Hays SG, Patrick WG, Ziesack M, Oxman N, Silver PA (2015) Better together: engineering and application of microbial symbioses. Curr Opin Biotechnol 36:40–49

Hazen TC, Looney BB, Enzien M, Franck MM, Fliemans CB, Eddy CA (1993) *In-situ* bioremediation via horizontal wells. I&EC Special Symposium, American Chemical Society, Atlanta, GA

Hazen TC, Chakraborty R, Fleming JM, Gregory IR, Bowman JP, Jimenez L, Zhang D, Pfiffner SM, Brockman FJ, Sayler GS (2009) Use of gene probes to assess the impact and effectiveness of aerobic *in situ* bioremediation of TCE. Arch Microbiol 191:221–232

He J, Ritalahti KM, Aiello MR, Löffler FE (2003a) Complete detoxification of vinyl chloride by an anaerobic enrichment culture and identification of the reductively dechlorinating population as a *Dehalococcoides* species. Appl Environ Microbiol 69:996–1003

He J, Ritalahti KM, Yang K-L, Koenigsberg SS, Löffler FE (2003b) Detoxification of vinyl chloride to ethene coupled to growth of an anaerobic bacterium. Nature 424:62–65

Ho A, Kerckhof F-M, Luke C, Reim A, Krause S, Boon N, Bodelier PLE (2013) Conceptualizing functional traits and ecological characteristics of methane-oxidizing bacteria as life strategies. Environ Microbiol Rep 5:335–345

Ho A, de Roy K, Thas O, De Neve J, Hoefman S, Vandamme P, Heylen K, Boon N (2014) The more, the merrier: heterotroph richness stimulates methanotrophic activity. ISME J 8:1945–1948

Ho A, Reim A, Kim SY, Meima-Franke M, Termorshuizen A, de Boer W, van der Putten WH, Bodelier PL (2015) Unexpected stimulation of soil methane uptake as emergent property of agricultural soils following bio-based residue application. Glob Change Biol 21:3864–3879

Ho A, Angel R, Veraart AJ, Daebeler A, Jia Z, Kim SY, Kerckhof F-M, Boon N, Bodelier PLE (2016) Biotic interactions in microbial communities as modulators of biogeochemical processes: methanotrophy as a model system. Front Microbiol 7:1285

Ho A, Ijaz UZ, Janssens TKS, Ruijs R, Kim SY, de Boer W, Termorshuizen A, van der Putten WH, Bodelier PLE (2017) Effects of bio-based residue amendments on greenhouse gas emission from agricultural soil are stronger than effects of soil type with different microbial community composition. Glob Change Biol Bioenergy 9:1707–1720

Ho A, Mo Y, Lee HJ, Sauheitl L, Jia Z, Horn MA (2018) Effect of salt stress on aerobic methane oxidation and associated methanotrophs; a microcosm study of a natural community from a non-saline environment. Soil Biol Biochem 125:210–214

Ho A, Lee HJ, Reumer M, Meima-Franke M, Raaijmakers C, Zweers H, de Boer W, can der Putten WH, Bodelier PLE (2019) Unexpected role of canonical aerobic methanotrophs in upland agricultural soils. Soil Biol Biochem 131:1–8

Holliger C, Wohlfarth G, Diekert G (1998) Reductive dechlorination in the energy metabolism of anaerobic bacteria. FEMS Microbiol Rev 22:383–398

Hou Y, Velthof GL, Oenema O (2015) Mitigation of ammonia, nitrous oxide and methane emissions from manure management chains: a meta-analysis and integrated assessment. Global Change Biol 21:1293–1312

Hulse CL, Tiedje JM, Averill BA (1989) Evidence for a copper-nitrosyl intermediate in denitrification by the copper-containing nitrite reductase of *Achromobacter cycloclastes*. J Am Chem Soc 111:2322–2323

Iguchi H, Yurimoto H, Sakai Y (2011) Stimulation of methanotrophic growth in cocultures by cobalamin excreted by rhizobia. Appl Environ Microbiol 77:8509–8515

in ’t Zandt MH, van den Bosch TJM, Rijkers R, van Kessel MAHJ, MSM J, Welte CU (2018) Co-cultivation of the strictly anaerobic methanogen *Methanosarcina barkeri* with aerobic methanotrophs in an oxygen-limited membrane bioreactor. Appl Microbiol Biotechnol 102:5685–5694

Islam T, Torsvik V, Larsen Ø, Bodrossy L, Øvreås L, Birkeland N-K (2016) Acid-tolerant moderately thermophilic methanotrophs of the class *Gammaproteobacteria* isolated from tropical topsoil with methane seeps. Front Microbiol 7:851

Jiang H, Chen Y, Jiang P, Zhang C, Smith TJ, Murrell JC, Xing X-H (2010) Methanotrophs: multifunctional bacteria with promising applications in environmental bioengineering. Biochem Eng J 49:277–288

Keck J, Sims RC, Coover M, Park K, Symons B (1989) Evidence for cooxidation of polynuclear aromatic hydrocarbons in soil. Water Res 23:1467–1476

Kenney GE, Sadek M, Rosenzweig AC (2016) Copper-responsive gene expression in the methanotroph *Methylosinus trichosporium* OB3b. Metallomics 8:931–940

Khadka R, Clothier L, Wang L, Lim CK, Klotz MG, Dunfield PF (2018) Evolutionary history of copper membrane monooxygenases. Front Microbiol 9:2493

Kim HJ, Graham DW, DiSpirito AA, Alterman MA, Galeva N, Larive CK, Asunskis D, Sherwood PMA (2004) Methanobactin, a copper-acquisition compound from methane-oxidizing bacteria. Science 305:1612–1615

Kim J, Kim DD, Yoon S (2018) Rapid isolation of fast-growing methanotrophs from environmental samples using continuous cultivation with gradually increased dilution rates. Appl Microbiol Biotechnol 102:5707–5715

Kip N, Ouyang W, van Winden J, Raghoebarsing A, van Niftrik L, Pol A, Pan Y, Bodrossy L, van Donselaar EG, Reichart GJ, Jetten MS, Damsté JS, Op den Camp HJ (2011) Detection, isolation, and characterization of acidophilic methanotrophs from sphagnum mosses. Appl Environ Microbiol 77:5643–5654

Kittelmann S, Friedrich MW (2008) Novel uncultured *Chloroflexi* dechlorinate perchloroethene to trans-dichloroethene in tidal flat sediments. Environ Microbiol 10:1557–1570

Knief C, Dunfield PF (2005) Response and adaptation of different methanotrophic bacteria to low methane mixing ratios. Environ Microbiol 7:1307–1317

Knittel K, Boetius A (2009) Anaerobic oxidation of methane: progress with an unknown process. Annu Rev Microbiol 63:311–334

Kolb S, Knief C, Dunfield PF, Conrad R (2005) Abundance and activity of uncultured methanotrophic bacteria involved in the consumption of atmospheric methane in two forest soils. Environ Microbiol 7:1150–1161

Krause SMB, Johnson T, Samadhi Karunaratne Y, Fu Y, Beck DAC, Chistoserdova L, Lidstrom ME (2017) Lanthanide-dependent cross-feeding of methane-derived carbon is linked by microbial community interactions. Proc Nat Acad Sci 114:358–363

Kumaresan D, Stephenson J, Doxey AC, Bandukwala H, Brooks E, Hillebrand-Voiculescu A, Whiteley AS, Murrell JC (2018) Aerobic proteobacterial methylotrophs in Movile Cave: genomic and metagenomic analyses. Microbiome 6:1

Kwon M, Ho A, Yoon S (2018) Novel approaches and reasons to isolate methanotrophic bacteria with biotechnological potentials: recent achievements and perspectives. Appl Microbiol Biotechnol 103:1–8

La H, Hettiaratchi JPA, Achari G, Dunfield PF (2018) Biofiltration of methane. Bioresour Technol 268:759–772

Leahy JG, Batchelor PJ, Morcomb SM (2003) Evolution of the soluble diiron monooxygenases. FEMS Microbiol Rev 27:449–479

Lee S-W, Keeney DR, Lim D-H, Dispirito AA, Semrau JD (2006) Mixed pollutant degradation by *Methylosinus trichosporium* OB3b expressing either soluble or particulate methane monooxygenase: can the tortoise beat the hare? Appl Environ Microbiol 72:7503–7509

Lee S-W, Im J, DiSpirito AA, Bodrossy L, Barcelona MJ, Semrau JD (2009) Effect of nutrient and selective inhibitor amendments on methane oxidation, nitrous oxide production, and key gene presence and expression in landfill cover soils: characterization of the role of methanotrophs, nitrifiers, and denitrifiers. Appl Microbiol Biotechnol 85:389–403

Levine UY, Teal TK, Robertson GP, Schmidt TM (2011) Agriculture's impact on microbial diversity and associated fluxes of carbon dioxide and methane. ISME J 5:1683–1691

Lichtmannegger J, Leitzinger C, Wimmer R, Schmitt S, Schulz S, Kabiri Y, Eberhagen C, Rieder T, Janik D, Neff F, Straub BK, Schirmacher P, DiSpirito AA, Bandow N, Baral BS, Flately A, Kremmer E, Denk G, Reiter FP, Hohenester S, Eckardt-Schupp F, Dencher NA, Adamski J, Sauer V, Niemietz C, Schmidt HH, Merle U, Gotthardt DN, Kroemer G, Weiss KH, Zischka H (2016) Methanobactin reverses acute liver failure in a rat model of Wilson disease. J Clin Invest 126:2721–2735

Lieberman RL, Rosenzweig AC (2005) Crystal structure of a membrane-bound metalloenzyme that catalyses the biological oxidation of methane. Nature 434:177

Liebner S, Svenning MM (2013) Environmental transcription of *mmoX* by methane-oxidizing *Proteobacteria* in a subarctic palsa peatland. Appl Environ Microbiol 79:701–706

Limbri H, Gunawan C, Thomas T, Smith A, Scott J, Rosche B (2014) Coal-packed methane biofilter for mitigation of greenhouse gas emissions from coal mine ventilation air. PLoS One 9:e94641

Löffler FE, Ritalahti KM, Zinder SH (2013) *Dehalococcoides* and reductive dechlorination of chlorinated solvents. In: Stroo HF, Leeson A, Ward CH (eds) Bioaugmentation for groundwater remediation. Springer, New York, pp 39–88

Lontoh S, Semrau JD (1998) Methane and trichloroethylene degradation by *Methylosinus trichosporium* OB3b expressing particulate methane monooxygenase. Appl Environ Microbiol 64:1106–1114

Lontoh S, DiSpirito AA, Semrau JD (1999) Dichloromethane and trichloroethylene inhibition of methane oxidation by the membrane-associated methane monooxygenase of *Methylosinus trichosporium* OB3b. Arch Microbiol 171:301–308

Lu X, Gu W, Zhao L, Farhan Ul Haque M, DiSpirito AA, Semrau JD, Gu B (2017) Methylmercury uptake and degradation by methanotrophs. Sci Adv 3:e1700041

Luesken FA, Sánchez J, van Alen TA, Sanabria J, Op den Camp HJM, Jetten MSM, Kartal B (2011) Simultaneous nitrite-dependent anaerobic methane and ammonium oxidation processes. Appl Environ Microbiol 19:6802–6807

Malghani S, Reim A, von Fischer J, Conrad R, Kuebler K, Trumbore SE (2016) Soil methanotroph abundance and community composition are not influenced by substrate availability in laboratory incubations. Soil Biol Biochem 101:184–194

Matturro B, Presta E, Rossetti S (2016) Reductive dechlorination of tetrachloroethene in marine sediments: biodiversity and dehalorespiring capabilities of the indigenous microbes. Sci Total Environ 545–546:445–452

Maxfield PJ, Hornibrook ERC, Evershed RP (2008) Acute impact of agriculture on high-affinity methanotrophic bacterial populations. Environ Microbiol 10:1917–1924

Melse RW, van der Werf AW (2005) Biofiltration for mitigation of methane emission from animal husbandry. Environ Sci Technol 39:5460–5468

Modin O, Fukushi K, Yamamoto K (2007) Denitrification with methane as external carbon source. Water Res 41:2726–2738

Murrell JC, McDonald IR, Gilbert B (2000) Regulation of expression of methane monooxygenases by copper ions. Trend Microbiol 8:221–225

Nikiema J, Bibeau L, Lavoie J, Brzezinski R, Vigneux J, Heitz M (2005) Biofiltration of methane: an experimental study. Chem Eng J 113:111–117

Oldenhuis R, Oedzes JY, van der Waarde JJ, Janssen DB (1991) Kinetics of chlorinated hydrocarbon degradation by *Methylosinus trichosporium* OB3b and toxicity of trichloroethylene. Appl Environ Microbiol 57:7–14

Op den Camp HJM, Islam T, Stott MB, Harhangi HR, Hynes A, Schouten S, Jetten MS, Birkeland NK, Pol A, Dunfield PF (2009) Environmental, genomic and taxonomic perspectives on methanotrophic *Verrucomicrobia*. Environ Microbiol Rep 1:293–306

Park S, Ely RL (2008) Candidate stress genes of *Nitrosomonas europaea* for monitoring inhibition of nitrification by heavy metals. Appl Environ Microbiol 74:5475–5482

Paustian K, Lehmann J, Ogle S, Reay D, Robertson GP, Smith P (2016) Climate-smart soils. Nature 532:49–57

Popat SC, Deshusses MA (2009) Reductive dehalogenation of trichloroethene vapors in an anaerobic biotrickling filter. Environ Sci Technol 43:7856–7861

Pratscher J, Vollmers J, Wiegand S, Dumont MG, Kaster A-K (2018) Unravelling the identity, metabolic potential and global biogeography of the atmospheric methane-oxidizing upland soil cluster α. Environ Microbiol 20:1016–1029

Pratt C, Walcroft AS, Tate KR, Ross DJ, Roy R, Reid MH, Veiga PW (2012) Biofiltration of methane emissions from a dairy farm effluent pond. Agric Ecosyst Environ 152:33–39

Raghoebarsing AA, Pol A, van de Pas-Schoonen KT, Smolders AJP, Ettwig KF, Rijpstra WIC, Schouten S, Sinninghe Damsté JS, Op den Camp HJM, Jetten MSM, Strous M (2006) A microbial consortium couples anaerobic methane oxidation to denitrification. Nature 440:918

Rahman MT, Crombie A, Chen Y, Stralis-Pavese N, Bodrossy L, Meir P, McNamara NP, Murrell JC (2011) Environmental distribution and abundance of the facultative methanotroph *Methylocella*. ISME J 5:1061–1066

Ritalahti KM, Löffler FE, Rasch EE, Koenigsberg SS (2005) Bioaugmentation for chlorinated ethene detoxification: bioaugmentation and molecular diagnostics in the bioremediation of chlorinated ethene-contaminated sites. Ind Biotechnol 1:114–118

Ryals R, Hartman MD, Parton WJ, DeLonge MS, Silver WL (2015) Long-term climate change mitigation potential with organic matter management on grasslands. Ecol Appl 25:531–545

Scheutz C, Kjeldsen P, Bogner JE, De Visscher A, Gebert J, Hilger HA, Huber-Humer M, Spokas K (2009) Microbial methane oxidation processes and technologies for mitigation of landfill gas emissions. Waste Manage Res 27:409–455

Semprini L, McCarty PL (1991) Comparison between model simulations and field results for in-situ biorestoration of chlorinated aliphatics: Part 1. Biostimulation of methanotrophic bacteria. Groundwater 29:365–374

Semrau J (2011) Bioremediation via methanotrophy: overview of recent findings and suggestions for future research. Front Microbiol 2:209

Semrau J, DiSpirito A, Yoon S (2010) Methanotrophs and copper. FEMS Microbiol Rev 34:496–531

Sengupta A, Dick WA (2015) Bacterial community diversity in soil under two tillage practices as determined by pyrosequencing. Microbial Ecol 70:853–859

Shi Y, Hu S, Lou J, Lu P, Keller J, Yuan Z (2013) Nitrogen removal from wastewater by coupling anammox and methane-dependent denitrification in a membrane biofilm reactor. Environ Sci Technol 47:11577–11583

Singh JS, Strong PJ (2016) Biologically derived fertilizer: a multifaceted bio-tool in methane mitigation. Ecotoxicol Environ Saf 124:267–276

Streese J, Stegmann R (2003) Microbial oxidation of methane from old landfills in biofilters. Waste Manage 23:573–580

Strong P, Laycock B, Mahamud S, Jensen P, Lant P, Tyson G, Pratt S (2016) The opportunity for high-performance biomaterials from methane. Microorganisms 4:11

Svenning MM, Wartiainen I, Hestnes AG, Binnerup SJ (2003) Isolation of methane oxidising bacteria from soil by use of a soil substrate membrane system. FEMS Microbiol Ecol 44:347–354

Takeuchi M, Nanba K, Iwamoto H, Nirei H, Kusuda T, Kazaoka O, Owaki M, Furuya K (2005) *In situ* bioremediation of a *cis*-dichloroethylene-contaminated aquifer utilizing methane-rich groundwater from an uncontaminated aquifer. Water Res 39:2438–2444

Tate KR (2015) Soil methane oxidation and land-use change – from process to mitigation. Soil Biol Biochem 80:260–272

Taylor RT, Hanna ML, Shah NN, Shonnard DR, Duba AG, Durham WB, Jackson KJ, Knapp RB, Wijesinghe AM, Knezovich JP, Jovanovich MC (1993) *In situ* bioremediation of trichloroethylene-contaminated water by a resting-cell methanotrophic microbial filter. Hydrol Sci J 38:323–342

Tubiello FN, Salvatore M, Ferrara AF, House J, Federici S, Rossi S, Biancalani R, Condor Golec RD, Jacobs H, Flammini A, Prosperi P, Gardenas-Galindo P, Schmidhuber J, Sanz Sanchez MJ, Srivastava N, Smith P (2015) The contribution of agriculture, forestry and other land use activities to global warming, 1990–2012. Glob Chang Biol 21:2655–2660

van der Ha D, Nachtergaele L, Kerckhof F-M, Rameiyanti D, Bossier P, Verstraete W, Boon N (2012) Conversion of biogas to bioproducts by algae and methane oxidizing bacteria. Environ Sci Technol 46:13425–13431

van Hylckama Vlieg JET, de Koning W, Janssen DB (1996) Transformation kinetics of chlroinated ethenes by *Methylosinus trichosporium* OB3b and detection of unstable epoxides by on-line gas chromatography. Appl Environ Microbiol 62:3304–3312

van Hylckama Vlieg JET, Janssen DB (2001) Formation and detoxification of reactive intermediates in the metabolism of chlorinated ethenes. J Biotechnol 85:81–102

van Kessel MAHJ, Stultiens K, Slegers MFW, Guerrero Cruz S, Jetten MSM, Kartal B, Op den Camp HJM (2018) Current perspectives on the application of N-damo and anammox in wastewater treatment. Curr Opin Biotechnol 50:222–227

Veraart AJ, Garbeva P, van Beersum F, Ho A, Hordijk CA, Meima-Franke M, Zweers AJ, Bodelier PLE (2018) Living apart together – bacterial volatiles influence methanotrophic growth and activity. ISME J 12:1163–1166

Vorobev A, Jagadevan S, Baral BS, DiSpirito AA, Freemeier BC, Bergman BH, Bandow NL, Semrau JD (2013) Detoxification of mercury by methanobactin from *Methylosinus trichosporium* OB3b. Appl Environ Microbiol 79:5918–5926

Wendlandt K-D, Stottmeister U, Helm J, Soltmann B, Jechorek M, Beck M (2010) The potential of methane-oxidizing bacteria for applications in environmental biotechnology. Eng Life Sci 10:87–102

Wu YM, Yang J, Fan XL, Fu SF, Sun MT, Guo RB (2017) Elimination of methane in exhaust gas from biogas upgrading process by immobilized methane-oxidizing bacteria. Bioresour Technol 231:124–128

Yamazaki S, Koyama J, Miyake H (2014) Inventors; Semiconductor Energy Laboratory Co., Ltd., assignee. Semiconductor device having an oxide semiconductor with a crystalline region and manufacturing method thereof. United States patent US 8,633,480 B2

Yang Y, Cápiro NL, Yan J, Marcet TF, Pennell KD, Löffler FE (2017a) Resilience and recovery of *Dehalococcoides mccartyi* following low pH exposure. FEMS Microbiol Ecol 93:fix130

Yang Y, Higgins SA, Yan J, Simsir B, Chourey K, Iyer R, Hettich RL, Baldwin B, Oqles DM, Löffler FE (2017b) Grape pomace compost harbors organohalide-respiring Dehalogenimonas species with novel reductive dehalogenase genes. ISME J 11:2767–2780

Yoon S, Semrau JD (2008) Measurement and modeling of multiple substrate oxidation by methanotrophs at 20 °C. FEMS Microbiol Lett 287:156–162

Yoon S, Carey J, Semrau J (2009) Feasibility of atmospheric methane removal using methanotrophic biotrickling filters. Appl Microbiol Biotechnol 83:949–956

Zahn JA, DiSpirito AA (1996) Membrane-associated methane monooxygenase from *Methylococcus capsulatus* (Bath). J Bacteriol 178:1018–1029

Zakirov V, Sweeting M, Lawrence T, Sellers J (2001) Nitrous oxide as a rocket propellant. Acta Astronaut 48:353–362

Zhu J, Wang Q, Yuan M, Tan G-YA, Sun F, Wang C, Wu W, Lee PH (2016) Microbiology and potential applications of aerobic methane oxidation coupled to denitrification (AME-D) process: a review. Water Res 90:203–215

Index

E. Y. Lee (ed.), *Methanotrophs*, Microbiology Monographs 32,
https://doi.org/10.1007/978-3-030-23261-0

B

E

I

Q

R

Lightning Source UK Ltd.
Milton Keynes UK
UKHW020611110919
349584UK00001B/18/P

9 783030 232603